Springer Handbook of Auditory Research

Volume 64

More information about this series at http://www.springer.com/series/2506

The ASA Press

The ASA Press imprint represents a collaboration between the Acoustical Society of America and Springer dedicated to encouraging the publication of important new books in acoustics. Published titles are intended to reflect the full range of research in acoustics. ASA Press books can include all types of books published by Springer and may appear in any appropriate Springer book series.

ASA Press

Karina S. Cramer • Allison B. Coffin
Richard R. Fay • Arthur N. Popper
Editors

Auditory Development and Plasticity

In Honor of Edwin W Rubel

ASA Press

Springer

Editors
Karina S. Cramer
Department of Neurobiology & Behavior
University of California, Irvine
Irvine, CA, USA

Allison B. Coffin
VCAPP Department
Washington State University
Vancouver, WA, USA

Richard R. Fay
Loyola University Chicago
Chicago, IL, USA

Arthur N. Popper
Department of Biology
University of Maryland
College Park, MD, USA

ISSN 0947-2657 ISSN 2197-1897 (electronic)
Springer Handbook of Auditory Research
ISBN 978-3-319-21529-7 ISBN 978-3-319-21530-3 (eBook)
DOI 10.1007/978-3-319-21530-3

Library of Congress Control Number: 2017953007

Printed on acid-free paper

This Springer imprint is published by Springer Nature
The registered company is Springer International Publishing AG
The registered company address is: Gewerbestrasse 11, 6330 Cham, Switzerland

This volume is dedicated to Edwin W Rubel, an extraordinary scholar, mentor, colleague, and friend.[1]

[1] Photo courtesy of Rudolf Rubsamen.

Acoustical Society of America

The purpose of the Acoustical Society of America (www.acousticalsociety.org) is to generate, disseminate, and promote the knowledge of acoustics. The Acoustical Society of America (ASA) is recognized as the world's premier international scientific society in acoustics, and counts among its more than 7,000 members, professionals in the fields of bioacoustics, engineering, architecture, speech, music, oceanography, signal processing, sound and vibration, and noise control.

Since its first meeting in 1929, the ASA has enjoyed a healthy growth in membership and in stature. The present membership of approximately 7,000 includes leaders in acoustics in the United States of America and around the world. The ASA has attracted members from various fields related to sound including engineering, physics, oceanography, life sciences, noise and noise control, architectural acoustics; psychological and physiological acoustics; applied acoustics; music and musical instruments; speech communication; ultrasonics, radiation, and scattering; mechanical vibrations and shock; underwater sound; aeroacoustics; macrosonics; acoustical signal processing; bioacoustics; and many more topics.

To assure adequate attention to these separate fields and to new ones that may develop, the Society establishes technical committees and technical groups charged with keeping abreast of developments and needs of the membership in their specialized fields. This diversity and the opportunity it provides for interchange of knowledge and points of view has become one of the strengths of the Society.

The ASA's publishing program has historically included the *The Journal of the Acoustical Society of America, JASA-Express Letters, Proceedings of Meetings on Acoustics*, the magazine *Acoustics Today*, and various books authored by its members across the many topical areas of acoustics. In addition, ASA members are involved in the development of acoustical standards concerned with terminology, measurement procedures, and criteria for determining the effects of noise and vibration.

Series Preface

The following preface is the one that we published in Volume 1 of the Springer Handbook of Auditory Research back in 1992. As anyone reading the original preface, or the many users of the series, will note, we have far exceeded our original expectation of eight volumes. Indeed, with books published to date and those in the pipeline, we are now set for over 70 volumes in SHAR, and we are still open to new and exciting ideas for additional books.

We are very proud that there seems to be a consensus, at least among our friends and colleagues, that SHAR has become an important and influential part of the auditory literature. While we have worked hard to develop and maintain the quality and value of SHAR, the real value of the books is very much because of the numerous authors who have given their time to write outstanding chapters and to our many coeditors who have provided the intellectual leadership to the individual volumes. We have worked with a remarkable and wonderful group of people, many of whom have become great personal friends of both of us. We also continue to work with a spectacular group of editors at Springer. Indeed, several of our past editors have moved on in the publishing world to become senior executives. To our delight, this includes the current president of Springer US, Dr. William Curtis.

But the truth is that the series would and could not be possible without the support of our families, and we want to take this opportunity to dedicate all of the SHAR books, past and future, to them. Our wives, Catherine Fay and Helen Popper, and our children, Michelle Popper Levit, Melissa Popper Levinsohn, Christian Fay, and Amanda Fay Sierra, have been immensely patient as we developed and worked on this series. We thank them and state, without doubt, that this series could not have happened without them. We also dedicate the future of SHAR to our next generation of (potential) auditory researchers—our grandchildren—Ethan and Sophie Levinsohn; Emma Levit; and Nathaniel, Evan, and Stella Fay.

Preface 1992

The Springer Handbook of Auditory Research presents a series of comprehensive and synthetic reviews of the fundamental topics in modern auditory research. The volumes are aimed at all individuals with interests in hearing research including advanced graduate students, postdoctoral researchers, and clinical investigators. The volumes are intended to introduce new investigators to important aspects of hearing science and to help established investigators to better understand the fundamental theories and data in fields of hearing that they may not normally follow closely.

Each volume presents a particular topic comprehensively, and each serves as a synthetic overview and guide to the literature. As such, the chapters present neither exhaustive data reviews nor original research that has not yet appeared in peer-reviewed journals. The volumes focus on topics that have developed a solid data and conceptual foundation rather than on those for which a literature is only beginning to develop. New research areas will be covered on a timely basis in the series as they begin to mature.

Each volume in the series consists of a few substantial chapters on a particular topic. In some cases, the topics will be ones of traditional interest for which there is a substantial body of data and theory, such as auditory neuroanatomy (Vol. 1) and neurophysiology (Vol. 2). Other volumes in the series deal with topics that have begun to mature more recently, such as development, plasticity, and computational models of neural processing. In many cases, the series editors are joined by a coeditor having special expertise in the topic of the volume.

Richard R. Fay, Chicago, IL, USA
Arthur N. Popper, College Park, MD, USA

SHAR logo by Mark B. Weinberg, Potomac, Maryland, used with permission.

Volume Preface

Auditory function depends on highly ordered structures and extremely precise connectivity. Understanding the ontogeny of these pathways represents a unique interdisciplinary challenge in auditory research. This volume presents some of the most recent advances in the field. The volume is also presented as a celebration of the career of Dr. Edwin W Rubel, a scholar, mentor, and friend who has contributed a vast range of new discoveries in this area. Indeed, Ed's broad contributions to auditory system development have substantially impacted the field and spawned new generations of investigators pursuing new questions in these areas. This volume spans a wide range of topics within auditory development, from cellular specification of the auditory periphery to physiological development of the central auditory system and maturation of auditory perception. These chapters, from colleagues whose work has been touched by Ed's contributions, pay homage to the breadth of this area and the advances that have been made possible.

In Chap. 2, Matthew Kelley and Jennifer Stone review sensory hair cell development and examine the innate plasticity (or lack thereof) that characterizes hair cell regeneration in vertebrates. Next, in Chap. 3, Hillary McGraw, Catherine Drerup, Teresa Nicolson, and Alex Nechiporuk review the development of the lateral line, a sensory system present in aquatic vertebrates in which lateral line organs (neuromasts) contain clusters of hair cells and supporting cells, analogous to the inner ear.

Auditory brainstem development is then considered in Chap. 4, in which Jason Sanchez and Yong Lu review the role of the excitatory neurotransmitter glutamate in the auditory brainstem, drawing on the vast literature from both birds and mammals. Next, in Chap. 5, Michael Burger describes the role of inhibition in the auditory brainstem, relying primarily on the elegant avian model pioneered by Ed Rubel. Then, in Chap. 6, Leonard Kaczmarek examines functional maturation and plasticity of the auditory brainstem through the lens of potassium current changes.

The last two chapters focus on perceptual development of auditory processing and vocal communication in species that learn their vocalization. In Chap. 7, Lynne Werner reviews the maturation of the human outer and middle ear and development of perceptual features of sound, relying largely on behavioral studies in human infants and children. Then, in Chap. 8, Sarah Woolley addresses vocal learning in songbirds, including the critical role of auditory experience, highlighting some important parallels with language learning in humans.

As most other SHAR volumes, this volume closely complements earlier ones in the series. Most notably, this volume represents the advances in the field of auditory development since *Development of the Auditory System* (Vol. 9, 1998, edited by Rubel, Popper, and Fay) and *Plasticity of the Auditory System* (Vol. 23, 2004, edited by Parks, Rubel, Fay, and Popper). Other related volumes include *Development of the Inner Ear* (Vol. 26, 2005, edited by Kelley, Wu, Popper, and Fay), *Hair Cell Regeneration, Repair, and Protection* (Vol. 33, 2008, edited by Salvi, Popper, and Fay), and *Human Auditory Development* (Vol. 42, 2012, edited by Werner, Fay, and Popper).

Karina S. Cramer, Irvine, CA, USA
Allison B. Coffin, Vancouver, WA, USA
Richard R. Fay, Chicago, IL, USA
Arthur N. Popper, College Park, MD, USA

Contents

Contributors

R. Michael Burger Department of Biological Sciences, Lehigh University, Bethlehem, PA, USA

Allison B. Coffin Department of Integrative Physiology and Neuroscience, Washington State University Vancouver, Vancouver, WA, USA

Karina S. Cramer Department of Neurobiology and Behavior, University of California, Irvine, Irvine, CA, USA

Catherine M. Drerup Division of Developmental Biology, Eunice Kennedy Shriver National Institute of Child Health and Human Development, National Institutes of Health, Bethesda, MD, USA

Leonard K. Kaczmarek Departments of Pharmacology and Cellular and Molecular Physiology, Yale University School of Medicine, New Haven, CT, USA

Matthew W. Kelley Laboratory of Cochlear Development, National Institute on Deafness and Other Communication Disorders, National Institutes of Health, Bethesda, MD, USA

Yong Lu Department of Anatomy and Neurobiology, Northeast Ohio Medical University, Rootstown, OH, USA

Hillary F. McGraw Division of Cell Biology and Biophysics, University of Missouri Kansas City, Kansas City, MO, USA

Alex V. Nechiporuk Department of Cell, Developmental and Cancer Biology, Oregon Health & Science University, Portland, OR, USA

Teresa Nicolson Vollum Institute, Oregon Health & Science University, Portland, OR, USA

Jason Tait Sanchez Roxelyn and Richard Pepper Department of Communication Sciences and Disorders, The Hugh Knowles Hearing Research Center, and Department of Neurobiology and the Interdepartmental Neuroscience Program, Northwestern University, Evanston, IL, USA

Jennifer S. Stone Department of Otolaryngology/Head and Neck Surgery and Virginia Merrill Bloedel Hearing Research Center, University of Washington School of Medicine, Seattle, WA, USA

Lynne A. Werner Department of Speech and Hearing Sciences, University of Washington, Seattle, WA, USA

Sarah M.N. Woolley Department of Psychology, Jerome L. Greene Science Center, New York, NY, USA

Chapter 1
Auditory System Development: A Tribute to Edwin W Rubel

Karina S. Cramer and Allison B. Coffin

Abstract Auditory function depends on highly ordered structures and extremely precise connectivity. Understanding the ontogeny of these pathways represents a unique interdisciplinary challenge in auditory research. The field of auditory system development has expanded significantly over the past decades, when the Springer Handbook of Auditory Research (SHAR) last dedicated a volume to this topic (Rubel et al., Development of the auditory system. New York: Springer-Verlag, 1998). Progress has been driven by new approaches and by interdisciplinary research at the molecular, cellular, systems, and behavioral levels. This volume presents some of the most recent advances. It is written in celebration of the career of Dr. Edwin W Rubel, who has contributed a vast range of new discoveries in this area. Ed has not only pioneered this research, but he has also served as a dedicated colleague and mentor committed to cultivating future investigators. He has trained nearly 100 graduate students and postdoctoral fellows, and his research group and colleagues have established an exemplary culture of collaborative support. The authors of the chapters in this book present a tribute to Ed and to his positive influence on their research and their scientific careers.

Keywords Auditory brainstem • Dendrite • Frequency-place map • Hair cells • Nucleus laminaris • Nucleus magnocellularis • Regeneration

K.S. Cramer (✉)
Department of Neurobiology and Behavior, University of California, Irvine, 2205 McGaugh Hall, Irvine, CA 92697, USA
e-mail: cramerk@uci.edu

A.B. Coffin
Department of Integrative Physiology and Neuroscience, Washington State University Vancouver, 14204 NE Salmon Creek Ave., Vancouver, WA 98686, USA
e-mail: allison.coffin@wsu.edu

K.S. Cramer et al. (eds.), *Auditory Development and Plasticity*, Springer Handbook of Auditory Research 64, DOI 10.1007/978-3-319-21530-3_1

1.1 Contributions to Science

Throughout his career, Edwin W Rubel has defined key scientific questions, often arising in response to his own experimental observations. These insights stem from his ability to examine questions across levels and his eagerness to embrace new technological approaches. Significantly, in the field of auditory research, Ed has been enormously successful at bridging basic science and translational research. His contributions are briefly summarized, with selected representative examples cited.

1.1.1 Early Career

During his graduate studies, Ed focused not on the auditory system but on somatosensory cortex. He recorded electrophysiological responses from single neurons in somatosensory cortex of newborn and adult cats (Rubel 1971). The results demonstrate that at the time of birth the somatotopic organization is adultlike in precision and responsivity, suggesting that this aspect of sensory coding develops very early throughout the neuroaxis. This study, one of the first of its kind in developing animals, set the stage for Ed's groundbreaking studies of developing auditory areas.

1.1.2 Auditory Brainstem

The Rubel laboratory has dedicated much of its efforts to the study of the auditory brainstem, the focus of a highly successful, uninterrupted research program since the 1970s. The early work (Rubel 1978), which began at Yale University (Fig. 1.1), carefully and quantitatively described the anatomy, organization, and ontogeny of the chick nucleus magnocellularis (NM) and n. laminaris (NL), the first and second central nuclei along the ascending auditory pathway. These studies carefully documented the tonotopic organization of the nuclei and their connectivity (Parks and Rubel 1975; Rubel and Parks 1975) as well as their development (Rubel et al. 1976; Smith and Rubel 1979). This work subsequently led to the identification of the neural circuitry underlying computation of interaural time differences in chicks (Young and Rubel 1983; Overholt et al. 1992). More recently, the Rubel laboratory has identified specific mechanisms that lead to precisely timed action potentials needed for coincidence detection in NL (Seidl et al. 2014; Cramer and Rubel 2016), and, as with many of their other studies, they demonstrated corresponding mechanisms in the mammalian auditory brainstem (Seidl and Rubel 2016).

Alongside these seminal studies of auditory brainstem function and development, the Rubel laboratory has contributed a major body of research on the cellular and molecular influences of afferent innervation and activity on auditory areas of the brainstem. Experimental removal of peripheral auditory input causes cell loss in NM in chicks and in the cochlear nucleus in mammals; however, this cell loss occurs only when cochlear ablation or silencing of the nerve is performed early in

Fig. 1.1 Ed Rubel and his family during the Yale years. Back row, left, Ed's mother, Ruth; right, Ed Rubel. Front row, Ed's wife and children, Wendy Rubel, Lisa Rubel, and Trevor Rubel. (Photograph taken by Ed's father, Nathan Rubel)

development (Born and Rubel 1985; Hashisaki and Rubel 1989). In subsequent studies of the molecular basis for this critical period (Harris et al. 2008), Ed and his colleagues have shown that distinct sets of genes are regulated by deafferentation during the critical period versus later in postnatal development when the critical period is closed. More recently, they have developed an inducible mutant mouse in which the diphtheria toxin receptor is expressed under control of the *Pou4f3* promoter, which is activated in hair cells (Golub et al. 2012; Tong et al. 2015). Thus, injection of diphtheria toxin can be used to eliminate hair cells at any age. Studies in these mice showed that hair cell loss accounts for deafferentation-induced cell loss in the cochlear nucleus during the critical period, and that spiral ganglion neurons are similarly susceptible to loss following deafferentation. These studies have had a major impact on our understanding not only of developmental factors that are required for neuronal survival, but also of the nature of critical periods in general.

1.1.3 Dendritic Regulation

Dendritic morphology plays a particularly important role in the computation of interaural time differences by auditory brainstem nuclei. Ed and his colleagues have contributed greatly to the understanding of this function. They found that NL neurons display a highly ordered tonotopic gradient of dendritic arbor size (Smith and Rubel 1979). Ed and his group carried out a beautiful series of experiments on the structure, development, and cellular plasticity of the dendrites in the chick NL and its mammalian counterpart, the medial superior olive (MSO), demonstrating that deafferentation of dendrites rapidly leads to their atrophy (Benes et al. 1977). They later showed that dendrites are sensitive to the strength of synaptic input (Sorensen and Rubel 2006; Wang and Rubel 2012) and that the response to deafferentation relies on dendritic regulation of the neurotrophin receptor TrkB (Schecterson et al. 2012). Recent work from Ed and his colleagues has demonstrated expression of the fragile X mental retardation protein (FMRP) in branches of NL dendrites. Remarkably, this expression is conserved across species in NL and in MSO, suggesting a role for FMRP in regulating dendritic morphology dynamics in the auditory brainstem (Wang et al. 2014).

1.1.4 Cochlea Frequency-Place Map

It was previously thought that the frequency-place representation in the cochlea remained constant throughout the life of an organism. In the 1980s, Ed and colleagues showed that this "map" shifts in predictable ways in birds and mammals during development (Lippe and Rubel 1983; Sanes et al. 1989). They then went on to show that this shift took place through changes in the passive membrane properties of the cochlear partition (Norton et al. 1991; Mills and Rubel 1998). Together, these studies showed that maturation of frequency representation occurs over a protracted period of functional development.

1.1.5 Hair Cell Regeneration

A keen understanding of auditory system structure, function, and development has permitted Ed to carry out studies with important clinical implications. A major discovery was that of inner ear hair cell regeneration in birds, made by Ed's group and by Doug Cotanche at the University of Pennsylvania (Cotanche 1987; Cruz et al. 1987), with subsequent papers describing the process in more detail (Corwin and Cotanche 1988; Ryals and Rubel 1988). They discovered that hair cell regeneration in birds restores hearing and balance after ototoxic drug administration. They went on to develop new paradigms for analyses of the mechanisms underlying inner ear hair cell regeneration in birds and for understanding the lack of regeneration in mammals. This contribution ushered in a new field of hearing and balance research with high translational potential. Many laboratories and an international consortium are currently working to determine the mechanisms underlying this process in birds and to develop approaches to promote regeneration of hair cells in the mature mammalian inner ear.

1.1.6 Prevention of Hearing Loss

In a distinct translational approach, the Rubel laboratory has worked toward prevention of hair cell loss. Since the early 1980s they sporadically conducted research to understand more fully how inner ear hair cells die as a result of ototoxic drug exposure. This line of investigation took on more importance in 2001, when Ed teamed up with David Raible and began a research program using the larval zebrafish (*Danio rerio*) lateral line system. This system allowed them to understand more fully the pathways, using genetics and dynamic imaging and to screen for genes that modulate the sensitivity of hair cells to the damaging effects of ototoxic drugs. Soon, Julian Simon joined the team and Dr. Felipe Santos, a resident in the Department of Otolaryngology, recognized that this platform could be used to screen large numbers of drugs and drug-like small molecules for potential to protect hair cells from the ototoxic effects of aminoglycoside antibiotics and chemotherapeutic therapies (Owens et al. 2008). New candidate drugs have been discovered, enhanced through medicinal chemistry and compound evaluations, and proven effective in mammals. This program has now become fully translational, with the discovery and patenting of novel compounds that are safe and protect hearing resulting from inner ear hair cell loss, and the establishment of a small biotech company, Oricula Therapeutics, to advance this drug to therapeutic usage. Efforts are underway to take the new drug entities for US Food and Drug Administration (FDA) approval and clinical use.

1.1.7 Scientific Community

While Ed has made far-reaching and groundbreaking contributions to auditory system function and development, he is most proud of the community he has built and fostered throughout his career. Rubel laboratory alumni form a highly collaborative scientific family, even among individuals who never overlapped in their time in the laboratory. This camaraderie has grown from gatherings at Rubel laboratory retreats, annual dinners at the Association for Research in Otolaryngology midwinter meetings, and other conferences. The retreats were held in a rustic camp at Pack Forest near Mt. Ranier, where current and past Rubel laboratory members gave scientific presentations and discussed their ongoing research. A group photograph from the 2002 retreat is shown in Fig. 1.2. The immersive nature of these retreats allowed for extensive interactions through talks as well as leisure activities (Fig. 1.3). The success of this close community is a reflection of Ed, who has been a consistent catalyst in fostering communication and collaboration among his trainees and colleagues. Ed has thus brought together several generations of scientists who interact at multiple levels. He is shown in Fig. 1.4 with his graduate mentor, Jack Johnson; his postdoctoral fellow, Dan Sanes; and his former graduate student, Melissa Caras, who is currently a postdoctoral fellow with Dan Sanes. Ed and many of his trainees

Fig. 1.2 The Rubel laboratory during the 2002 Rubelab retreat at Pack Forest near Mt. Ranier. Front row: Mae del Puerto, Natalie Hardie, Laurie Johnson, Dale Cunningham, Melissa Lofgren, Michael Burger, Karina Cramer, Stephanie Smith, Jennifer Stone, Brenda Ryals, Julie Harris, Elizabeth Oesterle, Tiffany Kao, Keiko Hirose, Sydia Ty, Yong Lu. Back row: Alan Cheng, Lisa Cunningham, Olivia Bermingham-McDonogh, Staci Sorensen, Brandon Warren, Oswald Steward, Tom Parks, Remy Pujol, Glen MacDonald, Josh Gittelman

Fig. 1.3 Leisure activities at the Rubelab retreat. Left to right, Alan Cheng, Tom Parks, Ed Rubel, and Yong Lu

Fig. 1.4 Four generations of scholars: Seattle, August 2015. Seated in front, Jack Johnson, Ed's graduate advisor at Michigan State University. Back, left to right: Ed Rubel, Melissa Caras, and Dan Sanes

and colleagues gathered at the University of Washington in August, 2015 to present and discuss recent progress at the conference, "Auditory Development: From Cochlea to Cognition," shown in Fig. 1.5.

At the University of Washington, Ed spearheaded and founded the Virginia Merrill Bloedel Hearing Research Center. Together with Prentise and Virginia Bloedel, who provided generous endowments for the Center, Ed formulated a plan for a Research Center aimed at facilitating fundamental and patient-directed research on hearing, balance, and communication disorders, and to train future investigators in these fields. The Center was designated to be university-based and independent of any department, in order to support in perpetuity research and research training throughout the institution. The Center rapidly became recognized as one of the leading hearing research and research training centers in the world. Ed had the pleasure and honor of being the founder and the first director, and has held the Virginia Merrill Bloedel Endowed Chair for Hearing Science from 1989 to the present.

1.2 Overview of Chapters

Ed Rubel's broad contributions to auditory system development have substantially impacted the field and spawned new generations of investigators pursuing new questions in these areas. This volume spans a wide range of topics within auditory

Fig. 1.5 Attendees at the Auditory Development Conference held in Seattle, Washington in August 2015. (Photograph courtesy of Rudolf Rubsamen)

development, from cellular specification of the auditory periphery to physiological development of the central auditory system and maturation of auditory perception. These chapters, from colleagues whose work has been touched by Ed's contributions, pay homage to the breadth of this area, and the advances that have been made possible.

1.2.1 Development of the Mechanosensory Periphery

In Chap. 2, Matthew Kelley and Jennifer Stone review sensory hair cell development and examine the innate plasticity (or lack thereof) that characterizes differential hair cell regeneration in vertebrates. Nonmammalian vertebrates robustly regenerate hair cells throughout life. In contrast, the mammalian vestibular system retains modest regenerative capacity, while the mature cochlea is not capable of regeneration. Understanding inner ear development therefore informs future regenerative efforts. The inner ear develops from a thickened ectodermal structure, the otic placode, which then invaginates to form the otic vesicle. Molecular markers such as *Sox2* characterize the proneurosensory region of the otic vesicle. Some of these early proneurosensory cells delaminate, proliferate, and differentiate as auditory and vestibular neurons, while the remaining prosensory cells differentiate as hair cells and supporting cells. The transcription factor *Atoh1* is a potent driver of hair cell fate within the prosensory domain, although additional factors, including Wnt and Notch signaling, act upstream or jointly with *Atoh1* to pattern the prosensory domain. Identification of downstream *Atoh1* targets is an area of current interest, as they may be important for stimulating hair cell regeneration in the mammalian cochlea.

Vestibular organs in nonmammalian vertebrates, and lateral line organs in aquatic vertebrates, undergo continuous turnover of hair cells. These organs have a basal level of supporting cell proliferation, with a subset of daughter cells assuming a hair cell fate. Conversely, the chicken auditory organ (basilar papilla) is quiescent under baseline conditions. However, all of these structures exhibit a robust response following hair cell damage, including supporting cell proliferation, and in the many species, direct conversion of supporting cells into hair cells. Phenotypic conversion likely drives the limited regenerative ability observed in mature mammalian vestibular organs, where damage results in little cell proliferation and incomplete hair cell replacement. Growth factor signaling, including fibroblast growth factor (FGF) and transforming growth factor (TGF), may regulate supporting cell proliferation, while hair cell differentiation during regeneration is likely driven by the same factors that regulate development, including Notch, Wnt, and *Atoh1*. However, the mature mammalian cochlea has lost the ability to respond to proregenerative signals such as TGF, suggesting that modulation of pro- and antiproliferative factors is necessary to achieve mammalian hair cell regeneration.

In Chap. 3, Hillary McGraw, Catherine Drerup, Teresa Nicolson, and Alex Nechiporuk review the development of the lateral line, a sensory system present in aquatic vertebrates in which lateral line organs (neuromasts) contain clusters of hair cells and supporting cells, analogous to the inner ear. Their chapter provides a counterpoint to Chap. 2 by Matthew Kelley and Jennifer Stone, highlighting some of the developmental similarities in vertebrate hair cell organs. Chapter 3 focuses on lateral line development in zebrafish, as most of our understanding is gleaned from this model organism. The lateral line develops from a series of specialized ectodermal placodes similar to the otic placode, and each placode is competent to give rise to hair cell–bearing neuromasts and associated sensory neurons. The posterior lateral line primordium, which gives rise to neuromasts on the trunk and tail, migrates down the trunk of the fish and deposits clusters of protoneuromasts from the trailing edge of the primordium. Cells within the leading edge of the primordium proliferate, providing a continuous source of cells for protoneuromast deposition. As in the inner ear, Wnt and FGF signaling play key roles in lateral line development, both in regulating proliferation and in neuromast patterning. Within deposited neuromasts, hair cell specification is dependent on Wnt, Notch, and FGF signaling, again demonstrating parallels with inner ear epithelia.

Studies of lateral line hair cells have offered insight into the development of ribbon synapses—specialized presynaptic structures containing clusters of glutamatergic vesicles found in all vertebrate hair cells. Using gain- and loss-of-function approaches, studies show that the ribbon protein Ribeye b is important for targeting of voltage-gated calcium channels (Cav1.3a) to synapses along the hair cell basolateral membrane. Conversely, Cav1.3a localization, and the associated calcium influx, modulates ribbon morphology and number. Hair cell ribbons synapse on afferent nerves of the lateral line ganglia. Lateral line axons can innervate hair cells from several adjacent neuromasts, but a single axon always synapses on hair cells with the same polarity. Lateral line axons, and the analogous spiral ganglion axons, carry information from the mechanosensory periphery to the brainstem.

1.2.2 Auditory Brainstem Development

In Chap. 4, Jason Sanchez and Yong Lu review the role of the excitatory neurotransmitter glutamate in the auditory brainstem, drawing on the vast literature from both birds and mammals. The auditory CNS expresses both α-amino-3-hydroxy-5-methyl-4-isoxazolepropionic acid (AMPA) and *N*-methyl-D-aspartate (NMDA) ionotropic glutamate receptors, as well as the relatively slow metabotropic glutamate receptors, and all play important roles in auditory processing. Developing synapses in the auditory brainstem are characterized by an early excitatory response driven by AMPA receptors and a later response via NMDA receptors, consistent with glutamatergic synapses in other brain regions. Several lines of evidence suggest a shift in AMPA receptor subunit composition during development, with receptors in mature synapses exhibiting high calcium permeability and strong inward rectification. NMDA receptors also exhibit a developmental switch in subunit composition, although overall NMDA receptors are downregulated at mature auditory synapses, coincident with hearing onset. Both ionotropic and metabotropic receptors likely contribute to modulation of calcium signaling in auditory nuclei such as the chicken NL and mammalian anteroventral cochlear nucleus (AVCN), where tightly controlled calcium signaling is critical for high-speed processing of sound stimuli. In addition, NMDA receptors may act in an activity-dependent manner to help control formation of binaural connections during auditory development.

Glutamatergic signaling can regulate synaptic plasticity in the auditory brainstem. Short-term postsynaptic depression can occur via presynaptic mechanisms, such as changes in glutamate release, and by postsynaptic mechanisms such as desensitization of AMPA receptors. Glutamate can also regulate long-term plasticity, which manifests as long-term potentiation or long-term depression; these processes may play a role in auditory learning in regions such as the dorsal cochlear nucleus (DCN) or the inferior colliculus (IC). In the IC, plasticity is important for adaptive adjustments of auditory and visual space maps that subserve stimulus localization. Glutamatergic signaling in the central auditory system is altered in congenitally deaf animals or following cochlear ablation, although the direction of the change differs between brain regions; in the IC, for example, glutamate signaling increases, whereas glutamate signaling decreases in the lateral superior olive (LSO). Glutamatergic signaling does not act in a vacuum; GABAergic signaling can modulate glutamate signaling to reduce synaptic depression, adjust neural gain, and regulate processing of interaural time differences (ITD) and interaural level differences (ILD) necessary for sound source localization. Therefore, the complex interplay of excitatory and inhibitory signaling is critical for auditory processing.

In Chap. 5, Michael Burger describes the role of inhibition in the auditory brainstem, relying primarily on the elegant avian model pioneered by Ed Rubel. In the chick auditory brainstem, input from the auditory nerve innervates the nucleus angularis (NA), which encodes several features of the acoustic stimulus, and the NM, which constitutes the beginning of the temporal processing pathway. Each NM neuron sends projections to both the contralateral and ipsilateral NL, in a close approximation of the delay line model first proposed by Jeffress (1948). Coordinated input from each NM to cells within NL is required for localization of the sound source. Inhibition is key for sound source localization by regulating temporal coding of ITD used to compute location in azimuth. Inhibitory input, primarily from the superior olivary nucleus (SON), sharpens the timing of excitatory inputs to the NL, which improves temporal processing by increasing the precision of phase locking to the temporal features of auditory stimuli.

In the chick, beginning at embryonic day 2 (E2), NM, NL, and SON all arise from rhombomere 5. Inhibitory SON fibers first synapse on NL neurons around E9 and inhibitory inputs appear morphologically mature at E17, coincident with maturation of excitatory inputs and hearing onset. Input from the SON to NL is primarily GABAergic and results in excitation of postsynaptic NL neurons early in development, with a switch to γ-aminobutyric acid (GABA)-mediated inhibition later, albeit one that is still depolarizing in nature and coupled to opening of low-threshold potassium channels. This mechanism of GABAergic inhibition in the mature chicken NL is different from the mechanism observed in the mammalian auditory brainstem, where GABA regulates postsynaptic hyperpolarization. In both cases, however, the end result is inhibitory, likely with similar physiological consequences for binaural processing.

In Chap. 6, Leonard Kaczmarek examines functional maturation and plasticity of the auditory brainstem through the lens of potassium current changes. The auditory system must be capable of firing at high rates with great temporal precision to encode high-frequency sounds faithfully and determine the location of a sound source. Voltage-gated potassium channels of the Kv3 family allow neurons to repolarize rapidly after an action potential, helping to drive the fast kinetics of auditory neurons. Kv3.1 and Kv3.3 are highly expressed in the auditory brainstem. Kv3.1 encodes two splice variants, with Kv3.1a expressed early in development and Kv3.1b increasing later, coincident with hearing onset. Stimulation can alter isoform expression and the phosphorylation state of the mature ion channel, which can in turn modify the response properties of the neuron.

Kv3.1 and Kv3.3 are highly expressed in glutamatergic bushy cells of the AVCN. These cells relay timing and intensity information to the LSO, MSO, and medial nucleus of the trapezoid body (MNTB). Neurons within the MNTB, which also express high levels of Kv3.1, convey inhibitory input to the LSO and MSO. Kv3.1b expression is tonotopically arrayed in the MNTB and the tonotopic arrangement is modulated by sound exposure, demonstrating stimulus-dependent plasticity. Kv3.1b channels in both the MNTB and the AVCN adapt the excitability of auditory brainstem neurons to features of the auditory environment, and stimulation modulates the spike timing precision within the MNTB. Stimulus-dependent plasticity is driven by several mechanisms. Phosphorylation of Kv3.1b markedly reduces current through this channel, and several studies suggest that Kv3.1b is basally phosphorylated, with stimulation leading to channel dephosphorylation and a corresponding increase in Kv3.1b current. Activity also regulates Kv3.1 expression at transcriptional and translational levels, allowing the neuron to precisely regulate Kv3.1 current. Chapter 6 concludes with unanswered questions in the field, including the contribution of other potassium channels to auditory processing and the role these channels play in auditory learning.

1.2.3 Development of Auditory Perception

The last two chapters in this volume focus on perceptual development of auditory processing and vocal communication in species that learn their vocalizations, humans (Chap. 7) and songbirds (Chap. 8). In Chap. 7, Lynne Werner reviews the maturation of the human outer and middle ear and development of perceptual features of sound, relying largely on behavioral studies in human infants and children. Behavioral studies in pregnant women, coupled with fetal imaging, show fetal motor responses to low frequencies at earlier developmental ages than responses to higher frequencies (above 1 kHz), although these results may be biased by frequency-dependent differences in acoustic conductance though the maternal tissue. After birth, conductive maturation of the outer and middle ear largely determines changes in absolute auditory sensitivity in human infants, although maturation of synapse morphology and transmission efficiency likely play additional roles.

Auditory brainstem response (ABR) and behavioral studies demonstrate that central auditory processing does not mature until months to years after birth. Frequency resolution appears mature by 3 months of age, while central conduction time may not mature until 3 years or later. Spatial hearing improves during early childhood, in correlation with changes in head size, and sound source localization in azimuth is relatively mature at 5 years of age. Perceptual features of auditory stimuli such as intensity discrimination, frequency discrimination, and temporal resolution all mature postnatally and with differing time courses. Young children have deficits in auditory stream formation and segregation and show larger deficits of masking noise. Studies of cochlear implant patients have informed our understanding of human auditory development. Individuals implanted at different times in their lives, and with differing auditory experience preimplant, form a set of natural experiments to understand how auditory experience, or the lack of it, influences the developing auditory system.

In Chap. 8, Sarah Woolley addresses vocal learning in songbirds, including the critical role of auditory experience, highlighting some important parallels to language learning in humans. Normal song learning in young male songbirds requires early exposure to conspecific adult song, followed by vocal practice and auditory feedback. Deafening a juvenile male or raising him in isolation, without access to tutor song, leads to development of abnormal song in the adult. Song maintenance in adult males shows reduced dependence on auditory feedback, although hearing helps maintain stereotyped song. In most songbirds, only the males sing, but the females' early experience with song helps to shape their perception of conspecific vocal communication signals. Even without early song experience, most songbirds show a preference for conspecific song over very different heterospecific song, demonstrating some innate preference for song features. However, birds cross-fostered to a different songbird species show preference for heterospecific song. Males of closely related species will even sing the heterospecific song of the foster male, demonstrating the importance of early song experience on song preference in adults.

Electrophysiology experiments have begun to determine how song exposure influences auditory development. Auditory midbrain neurons possess unique combinations of spectral and temporal tuning properties that determine how each neuron responds to specific features of song. In some species, auditory midbrain neurons are tuned for spectrotemporal features of conspecific vocalizations, demonstrating that midbrain neurons are capable of encoding the acoustic properties of song. Cross-fostering experiments, which expose birds to heterospecific song, result in weaker responses to song in midbrain neurons, suggesting that early song experience shapes auditory processing. Similarly, rearing birds without song experience alters the response properties of cortical neurons later in life, including reducing the selectivity of those neurons.

1.3 Conclusions

The work presented in this volume, written in honor of Ed Rubel, highlights advances in the field of auditory system development and indicates new research directions in this area. Through the foundational work of Ed and his students, postdocs, and colleagues, investigators have gained a greater understanding of molecular mechanisms of development, the role of experience, and the physiological changes that take place during maturation. While many questions remain, Ed's deep commitment to his colleagues, students, and collaborators has fostered a scientific community dedicated to auditory research.

Compliance with Ethics Requirements Karina Cramer declares that she has no conflict of interest. Allison Coffin declares that she has no conflict of interest.

References

Benes, F. M., Parks, T. N., & Rubel, E. W. (1977). Rapid dendritic atrophy following deafferentation: An EM morphometric analysis. *Brain Research*, 122(1), 1–13.

Born, D. E., & Rubel, E. W. (1985). Afferent influences on brain stem auditory nuclei of the chicken: Neuron number and size following cochlea removal. *The Journal of Comparative Neurology*, 231(4), 435–445.

Corwin, J. T., & Cotanche, D. A. (1988). Regeneration of sensory hair cells after acoustic trauma. *Science*, 240(4860), 1772–1774.

Cotanche, D. A. (1987). Regeneration of hair cell stereociliary bundles in the chick cochlea following severe acoustic trauma. *Hearing Research*, 30(2–3), 181–195.

Cramer, K. S., & Rubel, E. W. (2016). Glial cell contributions to auditory brainstem development. *Frontiers in Neural Circuits*, 10, 83.

Cruz, R. M., Lambert, P. R., & Rubel, E. W. (1987). Light microscopic evidence of hair cell regeneration after gentamicin toxicity in chick cochlea. *Archives of Otolaryngology –Head and Neck Surgery*, 113(10), 1058–1062.

Golub, J. S., Tong, L., Ngyuen, T. B., Hume, C. R., et al. (2012). Hair cell replacement in adult mouse utricles after targeted ablation of hair cells with diphtheria toxin. *The Journal of Neuroscience*, 32(43), 15093–15105.

Harris, J. A., Iguchi, F., Seidl, A. H., Lurie, D. I., & Rubel, E. W. (2008). Afferent deprivation elicits a transcriptional response associated with neuronal survival after a critical period in the mouse cochlear nucleus. *The Journal of Neuroscience*, 28(43), 10990–11002.

Hashisaki, G. T., & Rubel, E. W. (1989). Effects of unilateral cochlea removal on anteroventral cochlear nucleus neurons in developing gerbils. *The Journal of Comparative Neurology*, 283(4), 5–73.

Lippe, W., & Rubel, E. W. (1983). Development of the place principle: Tonotopic organization. *Science*, 219(4584), 514–516.

Mills, D. M., & Rubel, E. W. (1998). Development of the base of the cochlea: Place code shift in the gerbil. *Hearing Research*, 122(1–2), 82–96.

Norton, S. J., Bargones, J. Y., & Rubel, E. W. (1991). Development of otoacoustic emissions in gerbil: Evidence for micromechanical changes underlying development of the place code. *Hearing Research*, 51(1), 73–91.

Overholt, E. M., Rubel, E. W., & Hyson, R. L. (1992). A circuit for coding interaural time differences in the chick brainstem. *The Journal of Neuroscience*, 12(5), 1698–1708.

Owens, K. N., Santos, F., Roberts, B., Linbo, T., et al. (2008). Identification of genetic and chemical modulators of zebrafish mechanosensory hair cell death. *PLoS Genetics*, 4(2), e1000020.

Parks, T. N., & Rubel, E. W. (1975). Organization and development of brain stem auditory nuclei of the chicken: Organization of projections from n. magnocellularis to n. laminaris. *The Journal of Comparative Neurology*, 164(4), 435–448.

Rubel, E. W. (1971). A comparison of somatotopic organization in sensory neocortex of newborn kittens and adult cats. *The Journal of Comparative Neurology*, 143(4), 447–480.

Rubel, E. W. (1978). Ontogeny of structure and function in the vertebrates auditory system. In M. Jacobson (Ed.), *Handbook of sensory physiology* (pp. 135–237). New York: Springer.

Rubel, E. W., & Parks, T. N. (1975). Organization and development of brain stem auditory nuclei of the chicken: Tonotopic organization of n. magnocellularis and n. laminaris. *The Journal of Comparative Neurology*, 164(4), 411–433.

Rubel, E. W., Smith, D. J., & Miller, L. C. (1976). Organization and development of brain stem auditory nuclei of the chicken: Ontogeny of n. magnocellularis and n. laminaris. *The Journal of Comparative Neurology*, 166, 469–490.

Ryals, B. M., & Rubel, E. W. (1988). Hair cell regeneration after acoustic trauma in adult Coturnix quail. *Science*, 240(4860), 1774–1776.

Sanes, D. H., Merickel, M., & Rubel, E. W. (1989). Evidence for an alteration of the tonotopic map in the gerbil cochlea during development. *The Journal of Comparative Neurology*, 279(3), 436–444.

Schecterson, L. C., Sanchez, J. T., Rubel, E. W., & Bothwell, M. (2012). TrkB downregulation is required for dendrite retraction in developing neurons of chicken nucleus magnocellularis. *The Journal of Neuroscience*, 32(40), 14000–14009.

Seidl, A. H., & Rubel, E. W. (2016). Systematic and differential myelination of axon collaterals in the mammalian auditory brainstem. *Glia*, 64(4), 487–494.

Seidl, A. H., Rubel, E. W., & Barria, A. (2014). Differential conduction velocity regulation in ipsilateral and contralateral collaterals innervating brainstem coincidence detector neurons. *The Journal of Neuroscience*, 34(14), 4914–4919.

Smith, D. J., & Rubel, E. W. (1979). Organization and development of brain stem auditory nuclei of the chicken: Dendritic gradients in nucleus laminaris. *The Journal of Comparative Neurology*, 186(2), 213–239.

Sorensen, S. A., & Rubel, E. W. (2006). The level and integrity of synaptic input regulates dendrite structure. *The Journal of Neuroscience*, 26(5), 1539–1550.

Tong, L., Strong, M. K., Kaur, T., Juiz, J. M., et al. (2015). Selective deletion of cochlear hair cells causes rapid age-dependent changes in spiral ganglion and cochlear nucleus neurons. *The Journal of Neuroscience*, 35(20), 7878–7891.

Wang, Y., & Rubel, E. W. (2012). In vivo reversible regulation of dendritic patterning by afferent input in bipolar auditory neurons. *The Journal of Neuroscience*, 32(33), 11495–11504.

Wang, Y., Sakano, H., Beebe, K., Brown, M. R., et al. (2014). Intense and specialized dendritic localization of the fragile X mental retardation protein in binaural brainstem neurons: A comparative study in the alligator, chicken, gerbil, and human. *The Journal of Comparative Neurology*, 522(9), 2107–2128.

Young, S. R., & Rubel, E. W. (1983). Frequency-specific projections of individual neurons in chick brainstem auditory nuclei. *The Journal of Neuroscience*, 3(7), 1373–1378.

Chapter 2
Development and Regeneration of Sensory Hair Cells

Matthew W. Kelley and Jennifer S. Stone

Abstract Hair cells are sensory receptors for hearing and balance, and for detection of water movement in aquatic animals. In mammals, the vast majority of hair cells are formed during embryogenesis and early postnatal development, whereas in other vertebrates hair cells are formed throughout life. Destruction of hair cells is caused by genetic, environmental, or aging factors and results in sensorineural deficits that are irreversible in humans. Research in the 1980s demonstrated that nonmammalian vertebrates fully replace hair cells after damage and recover function, suggesting hair cell regeneration may someday be coaxed in humans as a treatment for some forms of hearing and balance deficits. To facilitate this possibility, subsequent studies explored the molecular and cellular bases of hair cell formation during development and after damage in mature animals. This chapter reviews the findings in each of these areas, describing similarities and differences across species, sensory organs, and age. For instance, while mature mammals have a limited innate ability to regenerate hair cells in the vestibular inner ear, no hair cells are replaced in the cochlea. Further, although the transcription factor (*Atoh1*) drives cells toward a hair cell fate during development in all types of animals and in nonmammals after damage, it has limited ability to promote hair cell regeneration in mature mammals. Finally, we discuss some of the hurdles that remain, as well as new technologies that may be used to move the field forward.

Keywords Development • Hair cell • Inner ear • Lateral line • Molecular regulation • Regeneration • Supporting cell

---

M.W. Kelley
Laboratory of Cochlear Development, National Institute on Deafness and Other Communication Disorders, National Institutes of Health, Bethesda, MD 20892, USA
e-mail: kelleymt@nidcd.nih.gov

J.S. Stone (✉)
Department of Otolaryngology/Head and Neck Surgery and Virginia Merrill Bloedel Hearing Research Center, University of Washington School of Medicine, Seattle, WA 98195, USA
e-mail: stoner@uw.edu

K.S. Cramer et al. (eds.), *Auditory Development and Plasticity*, Springer Handbook of Auditory Research 64, DOI 10.1007/978-3-319-21530-3_2

Abbreviations

Atoh1	Atonal homolog 1
E	Embryonic day
HMG	High mobility group
Lfng	Lunatic fringe
mTOR	Mechanistic target of rapamycin
P	Postnatal day
PI3K	Phosphoinoside-3 kinase

2.1 Introduction

Hair cells are sensory receptors located in specialized sensory epithelia in the inner ear of all vertebrates and along the body surfaces of aquatic vertebrates. They transduce biologically relevant stimuli into signals that are used to avoid predators, locate prey, and regulate motor activities, including vocalizations and locomotion. Hair cells in auditory organs respond to sound waves and enable hearing, while those in vestibular organs encode head movements and help to maintain posture, coordinate active motions, and stabilize gaze. In aquatic vertebrates, hair cells are also located along the body surface in lateral line sensory organs called neuromasts. These organs sense water movements, enabling aquatic animals to determine current direction and sense predators.

All hair cell epithelia are composed of three cellular elements: hair cells, nonsensory supporting cells that resemble glia in many respects, and peripheral fibers of the VIIIth cranial (cochleovestibular) nerve. The number of hair cells in each type of organ varies substantially within and across species. For example, some fish saccules contain hundreds of thousands of hair cells, while each zebrafish (*Danio rerio*) neuromast has around 10. In addition, the morphology and function of hair cells vary within and between organs. For instance, in mammals, cochlear hair cells are classified as “inner” or “outer” based on anatomical and physiological distinctions, while vestibular organs comprise type I and type II hair cells identified using similar criteria. In contrast, the criteria used to define supporting cells are limited, but they are typically elongated cells that extend from the lumen to the basal lamina. They provide structural and physiological support to hair cells and neurons and generate the overlying and underlying extracellular matrices of the sensory epithelium. Proportions of supporting cells to hair cells are approximately 1:1 in some organs (e.g., mouse utricle, Desai et al. 2005) but higher in other organs (e.g., chicken auditory epithelium, Goodyear and Richardson 1997). Peripheral fibers leading to and from the brain are distributed throughout most sensory epithelia and synapse on hair cells, although efferent fibers are limited with respect to the hair cell subtypes that they contact.

Section 2.2 of this chapter reviews hair cell formation during embryogenesis. Section 2.3 then discusses homeostasis and replacement of hair cells in mature animals. Although mechanisms of hair cell development are largely conserved across vertebrates, the ability to fully replace lost hair cells is confined to nonmammalian vertebrates. However, recent new discoveries indicate that there is partial replacement of hair cells in adult mammalian vestibular organs. Finally, factors that may inhibit hair cell regeneration in mammals, and potential methodologies to overcome those barriers, are discussed.

2.2 Hair Cell Development

2.2.1 Sensorineural Development in the Inner Ear

The inner ear is a remarkable structure. Initially derived from an ectodermal placode, individual cells proliferate to generate an appropriate number of cells, which then undergo complex lineage restriction and cell fate determination events to give rise to all of the cell types required to form the inner ear (Wu and Kelley 2012). Inner ear derivatives from the otocyst can be divided into three primary classes of cells: neuronal cells comprising the afferent neurons of cranial nerve VIII, sensory epithelial cells including mechanosensory hair cells and surrounding supporting cells, and nonsensory epithelial cells, which include all remaining otocyst-derived cells. It is important to note that many nonsensory structures such as the stria vascularis, endolymphatic duct, Reissner's membrane, and epithelial linings of the semicircular canals are essential for inner ear function. However, these structures are not addressed in this chapter.

At the time of closure, the otic vesicle appears largely homogeneous in terms of both cellular and gross morphologies. Consistent with these observations, experiments performed in amphibians in the early 1900s demonstrated that early rotation of the otocyst along any one of the main body axes leads to development of an essentially normal otocyst, suggesting cells in the early otocyst are largely equipotential. However, fairly soon after closure cellular identities along the three body axes become restricted. It has been shown that the dorsal–ventral axis is established through opposing gradients of sonic hedgehog (ventral) and Wnt (dorsal) (Riccomagno et al. 2002, 2005), while the anterior–posterior axis is defined through a gradient of retinoic acid (Bok et al. 2011). For further details see Wu and Kelley (2012).

A key outcome of axis specification is the formation of a proneurosensory cell population in the anteroventral region of the otocyst (Raft et al. 2007) (Fig. 2.1). Understanding of the molecular basis for the formation of this region is still fairly limited. However, the transcription factor *Tbx1*, which inhibits proneurosensory formation, is expressed outside of this region (Raft et al. 2004). Disruption of *Tbx1*

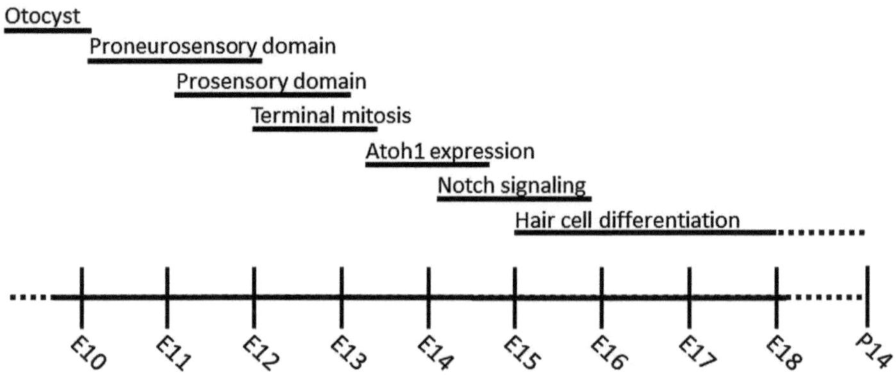

Fig. 2.1 Timeline of hair cell development in the mouse. The timing of specific events in the progression of cells from uncommitted neuroepithelial cells in the otocyst through several developmental restrictions en route to forming mature hair cells is illustrated

expression or function leads to increased neural and sensory structures at the expense of nonsensory regions of the inner ear.

Several markers of the proneurosensory region have been identified, including *Sox2* and *Lfng* (Morsli et al. 1998; Dabdoub et al. 2008). Otocyst-specific deletion of *Sox2* leads to loss of both neural and sensory structures, demonstrating the importance of this transcription factor in formation of the proneurosensory derivatives of the ear (Kiernan et al. 2005b; Dabdoub et al. 2008). In contrast, *Lfng* mutants are phenotypically normal, suggesting either a nonessential role in inner ear development or, possibly, functional compensation by one of two other *fringe* genes, *radical* and *manic* (Zhang et al. 2000).

Following specification of the proneurosensory domain, a subset of cells within this region becomes further determined toward a neural lineage (Ma et al. 2000; Appler and Goodrich 2011). These cells then undergo an epithelial-to-mesenchymal transition, delaminate through the basement membrane of the otocyst, and migrate a short distance medially toward the hindbrain. These statoacoustic neuroblasts divide extensively and then differentiate as auditory and vestibular neurons. Cells that remain within the proneurosensory region develop as prosensory cells (Brooker et al. 2006).

As development proceeds, the prosensory cells separate and redistribute to give rise to the sensory patches of the inner ear (Fig. 2.1). At present, three factors are known to be necessary and sufficient for formation of the prosensory domains. The first is *Sox2*, a high-mobility group (HMG) transcription factor known for its role in stem cell induction but that also plays a key role in the formation of neural progenitor cells (Zhang and Cui 2014; Takahashi and Yamanaka 2016). Two other factors that seem to play similar roles in specification of prosensory cells are the transcription factor Eya1 and its binding cofactor Six1 (Xu et al. 1999; Zheng et al. 2003). Mutations in either *Eya1* or *Six1* lead to branchial-oto-renal (BOR) syndrome, which cause defects in inner ear formation and hearing, and targeted deletion of either gene results in complete absence of sensory epithelia. *Eya1* is strongly expressed within the prosensory domain beginning at a relatively early stage in otocyst development, while *Six1* shows broader expression within the otocyst. More recently, transfection experiments have shown that Sox2, Eya1, and Six1 can bind to and activate specific regions within a highly conserved *atonal homolog 1* (*Atoh1*) enhancer (Ahmed et al. 2012a, b; Kempfle et al. 2016). These results establish key roles for each factor in the onset of expression of *Atoh1*, the most potent known inducer of hair cell fate (Sect. 2.2.2.1). However, it is not clear whether the effects of any or all of these factors are limited to regulation of *Atoh1*.

One study examined this issue by forcing ectopic expression of *Atoh1* in embryonic and early postnatal cochlear cells from *Sox2* mutants (*Sox2$^{Lcc/Lcc}$*, Puligilla and Kelley 2016). No *Atoh1* expression is observed in *Sox2* mutant ears. However, if the only role of Sox2 is to induce expression of *Atoh1*, then ectopic expression of *Atoh1* should be sufficient to induce limited hair cell formation. Interestingly, the results of these experiments indicated no hair cell formation following ectopic expression of *Atoh1* (Puligilla and Kelley 2016). This result suggests that Sox2 has critical roles in hair cell formation beyond its role in transcriptional activation of *Atoh1*. These roles might include activation of progenitor genes that change the state of the prosensory cells and/or epigenetic modifications that allow Atoh1 to modulate key target genes.

Another intriguing aspect of the role of Sox2 in prosensory development is the observation that Sox2 is downregulated in all auditory hair cells and in type I vestibular hair cells as they differentiate. This pattern of expression is similar to that of the central nervous system, in which *Sox2* is transiently expressed as progenitors transition from highly proliferative blast cells to terminally differentiated neurons (Bylund et al. 2003) and has led to the suggestion that Sox2 can also act as an inhibitor of hair cell differentiation (Dabdoub et al. 2008). However, although the role of Sox2 in promoting *Atoh1* expression has been examined extensively, its inhibitory role remains unclear (Liu et al. 2012b).

The remainder of the discussion of hair cell development refers primarily to the mammalian cochlea. As development proceeds, the number of prosensory cells continues to increase through ongoing proliferation. As discussed in the text that follows, the total number of prosensory cells is tightly regulated, as programmed cell death within the epithelium is minimal and mutations that lead to prolonged proliferation also result in overproduction of hair cells and supporting cells. However, the factors that regulate ongoing proliferation within the prosensory domain are poorly understood. Recent work has demonstrated key roles for fibroblast growth factor 9 (Fgf9) and Fgf20 in the overall level of proliferation in the prosensory domain (Huh et al. 2015). However, this function may be indirect, as Fgf9 and Fgf20 arising from the epithelium appear to signal to adjacent mesenchyme, which then reciprocally influences epithelial proliferation through an undetermined mechanism.

Following prosensory specification, the next key step in formation of the sensory epithelia is terminal mitosis. Within the developing cochlear duct, terminal mitosis of prosensory cells occurs in a highly stereotyped and conserved pattern, beginning in the apex and extending in a gradient toward the base (Ruben 1967; Lee et al.

2006). In mice, apical cells become postmitotic beginning on embryonic day 12.5 (E12.5), with the last basal cells leaving the cell cycle by E14.5 (Fig. 2.1). Recent studies examined the roles of several cell cycle regulatory factors in controlling terminal mitosis in both auditory and vestibular prosensory cells (Schimmang and Pirvola 2013). The cyclin-dependent kinase inhibitor Cdkn1b (formerly $p27^{kip1}$) plays a key role in the timing of cell cycle exit in the cochlear duct (Chen and Segil 1999). Cdkn1b protein and transcripts are expressed in a gradient that foreshadows the apical-to-basal gradient of terminal mitosis. Deletion of *Cdkn1b* leads to prolonged proliferation of prosensory cells, which in turn leads to formation of supernumerary hair cells and supporting cells and hearing loss. The formation of differentiated cell types in *Cdkn1b* mutants indicates that Cdkn1b is not essential for terminal mitosis. In fact, prosensory cells are thought to go through only one additional round of proliferation before cell cycle exit. Interestingly, in the absence of Cdkn1b, the gradient of terminal mitosis switches from apical-to-basal to basal-to-apical.

2.2.2 Initial Differentiation of Hair Cells

The onset of cellular differentiation within the cochlear duct appears to coincide with the arrival of the wave of terminal mitosis in the base (Chen and Segil 1999; Lee et al. 2006) (Fig. 2.1). As discussed in Sect. 2.2.1, differentiation begins near the base and extends both apically and, to a lesser extent, basally toward the future hook region (Rubel 1978). The earliest sign of hair cell development, expression of *Atoh1*, extends from the base of the cochlea toward the apex and hook (Chen et al. 2002; Woods et al. 2004). Expression onset is rapid, with apical progenitors becoming *Atoh1*-positive roughly 24 hours after progenitors in the base. However, the subsequent differentiation of hair cells proceeds more slowly. The first signs of differentiation are observed in basal hair cells around E13, but the extreme apex of the cochlea reaches a similar stage of maturity 2–3 days later. Two aspects of this developmental wave are remarkable. First, in most developing systems, the onset of cellular differentiation is coupled with completion of terminal mitosis. In contrast, in the cochlea, terminal mitosis and differentiation are largely uncoupled. As a result, whereas hair cell differentiation in the base begins soon after terminal mitosis, progenitor cells in the apex become terminally mitotic on E12 but may not differentiate until E16. This discrepancy highlights a second remarkable aspect of this developmental wave: progenitors at the apex of the cochlea remain in an undifferentiated, but postmitotic, state for approximately 4 days. The molecular factors that regulate this inhibition of differentiation are largely unknown, but it has been shown that a secreted molecule, sonic hedgehog (Shh), produced by the spiral ganglion, acts as an inhibiter of hair cell formation (Driver et al. 2008; Bok et al. 2013). Moreover, spiral ganglion expression of *Shh* is downregulated in a basal-to-apical gradient that precedes the gradient of differentiation. Deletion of *Shh* from the spiral ganglion or elimination of the ganglion disrupts the gradient of hair cell differentiation, leading to synchronized hair cell formation along the length of the duct (Ma et al. 2000). The biological benefit of this gradient is unclear, but may play a role in region-specific patterning of hair cells in the cochlea.

2.2.2.1 *Atoh1* Is Necessary and Sufficient for Hair Cell Formation

The crucial role of Atoh1 in hair cell formation was first demonstrated through the publication of the inner ear phenotype in *Atoh1* knockout mice (Bermingham et al. 1999). Although these animals die at birth, an analysis of cochleae at E18 indicated a complete absence of hair cells. Subsequent studies in which cells within the nascent cochlear duct were forced to express *Atoh1* demonstrated strong induction of a hair cell fate (Zheng and Gao 2000; Jones et al. 2006). Perhaps two of the best examples of the critical role for Atoh1 are publications from Kelly et al. (2012) and Liu et al. (2012b) in which transgenic mouse lines were used to induce expression of *Atoh1* in the inner ears of embryos or postnatal animals in vivo. These mice showed extensive conversion of cells into hair cells throughout the organ of Corti, demonstrating that Atoh1 acts as a powerful inducer of hair cell fate.

2.2.2.2 Regulation of Hair Cell and Supporting Cell Fates Through Notch Signaling

The wave of *Atoh1* expression in the organ of Corti initiates a molecular program that ultimately leads to hair cell differentiation. However, not every cell that expresses *Atoh1* develops as a hair cell (Driver et al. 2013). These observations reveal a complex developmental regulatory mechanism that is still not fully understood (Fig. 2.2). The number of cells that maintain *Atoh1* expression and therefore go on to develop as hair cells is regulated through activation of the notch signaling pathway. This pathway includes notch receptor molecules and a family of ligands referred to as either deltas or jaggeds. Because both ligand and receptor are membrane bound, notch signaling can be activated only between cells in direct contact (reviewed in Palermo et al. 2014). Although the biological effects of notch activation are diverse, the primary outcome of notch receptor activation is inhibition of cellular differentiation. Within the cochlea, all prosensory cells express notch1 before the onset of *Atoh1* expression (Lanford et al. 1999; Murata et al. 2006). As hair cells begin to develop, they upregulate expression of two notch ligands, delta1 and jagged2 (Lanford et al. 1999; Kiernan et al. 2005a). These ligands bind notch1 and activate several known notch target genes (Lanford et al. 2000; Zine et al. 2001). Genetic deletion of any component of this signaling pathway results in an increase in the number of hair cells, demonstrating that at least one role of notch signaling is the inhibition of hair cell formation (Kiernan et al. 2005a). Moreover, lineage tracing studies have demonstrated that activated notch signaling downregulates *Atoh1* expression, explaining at a mechanistic level how this pathway can influence hair cell fate (Driver et al. 2013). Notch pathway genes are reactivated in hair cell

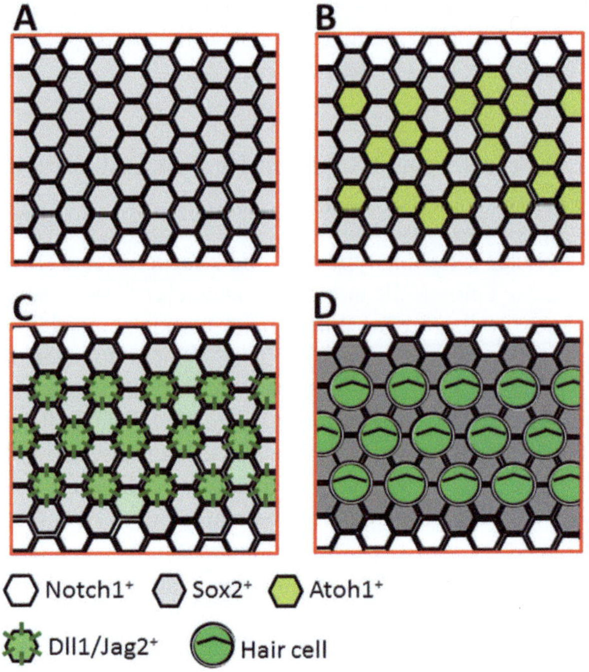

Fig. 2.2 Regulation of hair cell development. Shown is a schematic view of the surface of a hypothetical hair cell sensory epithelium. (**A**) All cells express *notch1* (black border around cells), and a subset of these cells has been specified as prosensory through expression of *Sox2* (gray). (**B**) With the prosensory cell population, a limited number of cells initiates expression of *Atoh1* (light green). (**C**) Next, some of the *Atoh1*-positive cells begin to express two notch ligands, delta-like 1 (Dll1) and jagged2 (Jag2). Binding of these ligands to notch1 in neighboring cells leads to downregulation of *Atoh1* expression and induction of a supporting cell fate. (**D**) In the mature epithelium, cells that maintain expression of *Atoh1* become hair cells, while neighboring cells that have downregulated *Atoh1* develop as supporting cells

epithelia after damage in animals that regenerate hair cells, and inhibition of notch signaling causes overproduction of new hair cells (e.g., zebrafish neuromasts, Ma et al. 2008; chicken auditory epithelium, or basilar papilla, Daudet et al. 2009; and mouse utricle, Lin et al. 2011) (see Sect 2.3.6.2).

2.2.2.3 Onset of *Atoh1* Expression

The understanding of the factors that initiate *Atoh1* expression within the prosensory domain remains limited. However, several studies have implicated the Wnt signaling pathway, the transcription factor Eya1, and its coactivator Six1 (Ahmed et al. 2012a; Jacques et al. 2012). Pharmacological activation of canonical Wnt signaling in vitro results in a significant increase in proliferation within the sensory epithelium and formation of supernumerary hair cells (Jacques et al. 2012).

Consistent with this result, genetic inactivation of the Wnt pathway in the developing cochlea beginning near E13 results in a complete absence of *Atoh1* expression and hair cell formation (Shi et al. 2013). Finally, Tcf/Lef, key Wnt targets, bind to the *Atoh1* enhancer in neural progenitor cells (Shi et al. 2010). Taken together, these results are consistent with a role for Wnt in upregulation of *Atoh1*.

BOR syndrome is a genetic disorder associated with mutations in *EYA1* and *SIX1* that includes defects in branchial arch derivatives, inner ear, and kidneys. Ahmed et al. (2012a) examined the roles of Eya1 and Six1 through forced expression in cells within Kolliker's organ. Coexpression of *Eya1/Six1* was sufficient to induce hair cell formation and expression of *Atoh1*. Moreover, Six1 was shown to bind to specific regions within the *Atoh1* enhancer, demonstrating direct regulation.

2.2.3 Hair Cell Development Beyond Atoh1

Although much attention has been focused on Atoh1 as the earliest inducer of hair cell formation, *Atoh1* expression marks only the beginning of hair cell development. Two additional transcription factors, Pou4f3 and Gfi1, are both expressed specifically in hair cells with onsets of expression that closely follow that of *Atoh1* (Erkman et al. 1996; Hertzano et al. 2004). Although hair cells still form in mice with targeted deletions of either gene, they rapidly die, demonstrating roles in hair cell survival (Wallis et al. 2003). Forced expression of all three factors drives embryonic stem cells toward a more "hair cell-like" phenotype than *Atoh1* expression alone, suggesting that Pou4f3 and Gfi1 may play roles beyond cellular survival (Costa et al. 2015).

Given the crucial role of Atoh1 in, at least, the early differentiation of hair cells, its downstream targets are of great interest. Studies in cerebellar granule neurons and dorsal spinal cord commissural neurons, two other cell types that depend on developmental expression of *Atoh1*, have used comprehensive profiling methods to define the genes that are specifically regulated by Atoh1 in those systems (Klisch et al. 2011; Lai et al. 2011). Unfortunately, similar studies have not been completed in the inner ear. However, bioinformatics has been used to identify 233 putative Atoh1 target genes in developing hair cells, identifying genes whose expression is reduced in mice lacking *Atoh1* and verifying Atoh1 binding sites in the promoters of such genes (Cai et al. 2015). Although this study did not fully define the targets of Atoh1, it provided valuable insights regarding Atoh1 regulation of hair cell formation.

The time between initial differentiation of a cell as an auditory hair cell (E14 for basal inner hair cells) and maturation of that cell (P14 or possibly later) is protracted. During that time, the cell must develop not only obvious apical and basal specializations, such as a stereociliary bundle and synaptic zones, but also the precise physiologic and anatomic structures required for mechanotransduction. The factors that mediate hair cell differentiation are poorly understood. As discussed in Sect. 2.2.3, Pou4f3 and Gfi1 may have a roles in differentiation, but the rapid death of hair cells in the absence of either gene has made it difficult to pinpoint the role. Additional studies using temporal deletion of either factor may provide clues to the roles of these factors beyond survival. Two secreted molecules that regulate hair cell maturation are thyroid hormone and retinoic acid. Deletion of thyroid hormone receptors leads to marked delays in the overall differentiation of the cochlear duct, including hair cells (Forrest et al. 1996; Rusch et al. 1998). Further, several ion channels fail to develop at the appropriate time in inner hair cells (Rusch et al. 1998; Brandt et al. 2007), while in outer hair cells, expression of *Slc26a5* (prestin) and *Kcnq4* is delayed (Weber et al. 2002; Winter et al. 2006). Unfortunately, other specific targets of thyroid hormone in hair cells have not been identified.

2.2.4 Summary of Development

At the most basic levels, hair cell progenitors appear to pass through four stages of specification, first as otic cells, next as proneurosensory cells, then as prosensory cells, and finally as hair cells. At least some of the genetic factors that regulate each step have been identified, with the most attention focused on the final decision to become a hair cell. The identification of Atoh1 as a key regulator of hair cell fate provided valuable insights regarding the developmental processes that lead to hair cell formation. However, subsequent work demonstrated that Atoh1 is just part of the answer, and more research is needed before we can fully understand hair cell development.

2.3 Hair Cell Regeneration

2.3.1 Overview

As discussed, previous studies showed that in nonmammalian vertebrates of any age, lost hair cells are replaced through conversion of supporting cells, leading to maintenance or restoration of sensory function (Fig. 2.3). However, the situation is quite different in the cochleae of mature rodents, in which the supporting cells that remain after hair cell loss converge to form a permanent scar, and hair cells are not replaced. Some vestibular hair cells are regenerated in mature rodents after damage, but only a subpopulation is replaced, and the mechanisms for hair cell replacement are different from those used by nonmammals.

This section considers what is known about hair cell replacement in nonmammals and what is currently understood about hair cell regeneration in mammals.

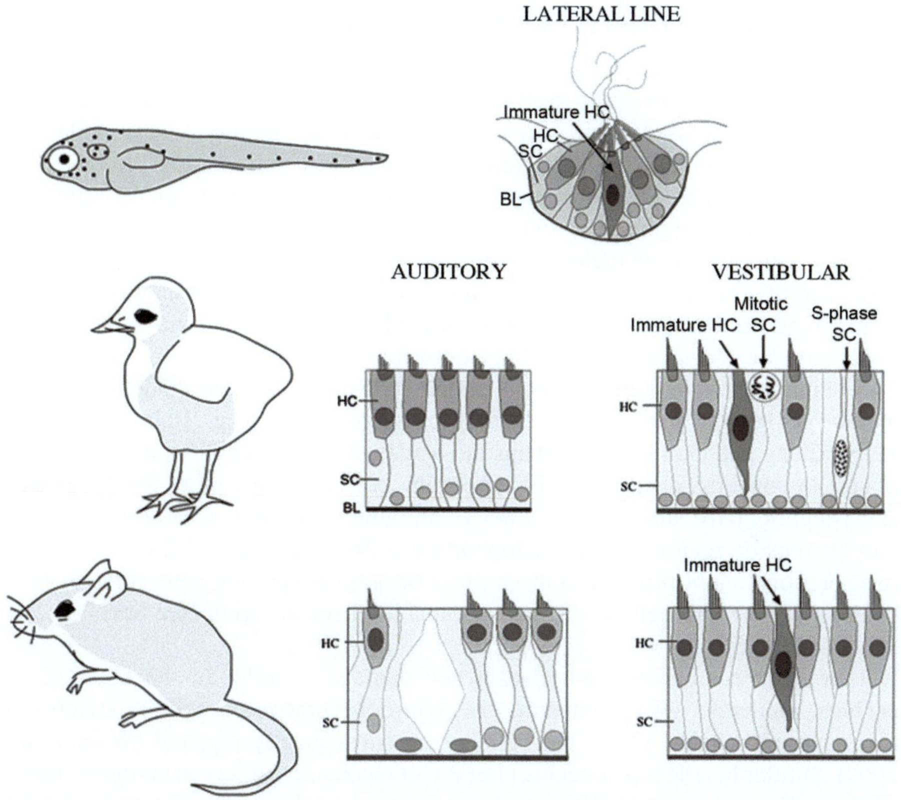

Fig. 2.3 The structure and regenerative capacity of hair cell epithelia in mammals and nonmammals. Sensory epithelia of vertebrates in the lateral line, auditory organs, and vestibular organs share a common composition of hair cells and supporting cells but vary with respect to their capacity for hair cell regeneration. Nonmammalian vertebrates—fish and birds are shown here—replace all hair cells after damage and also exhibit ongoing hair cell production, even in adulthood. By contrast, the auditory epithelium of mammals (the organ of Corti) cannot replace hair cells after damage, but vestibular epithelia regenerate a subpopulation of hair cells. HC = hair cell; SC = supporting cell; BL = basal lamina

Before proceeding, it is important to clarify a few terms. Hair cell *regeneration* refers to the replacement of hair cells in mature organs following trauma or toxic injury whereas the replacement of hair cells after injury in developing organs is not considered regeneration because organs at this time may retain the ability for repair from embryogenesis. Further, hair cell regeneration is distinct from hair cell *turnover*, which involves the natural death and addition of hair cells. This section addresses hair cell regeneration, with only minor discussions of hair cell turnover.

2.3.2 Hair Cell Injury, Death, and Repair

Hair cell damage is a first step in all studies of hair cell regeneration. Acoustic overstimulation can be used to kill hair cells specifically in auditory organs while ototoxins, such as aminoglycoside antibiotics and heavy metals, are toxic to hair cells in all sensory organs. Ototoxins are injected, delivered directly to the labyrinth, or added to cultured organs to avoid system toxicity. Diuretics such as furosemide enhance aminoglycoside damage to the cochlea. Lasers can selectively ablate hair cells, and this method is easily scaled and site directed (Balak et al. 1990; Kelley et al. 1995). Further, genetic variants of mice allow destruction of different hair cell populations (e.g., Fujioka et al. 2011; Golub et al. 2012).

The numbers, types, and positions of hair cells that are destroyed depend on the method of damage, and all methods have off-target effects (e.g., Roberto and Zito 1988; Zheng and Gao 1996). Perhaps most relevant to regeneration, ototoxins can, in certain contexts, damage or kill supporting cells (Kevetter et al. 2000; Slattery and Warchol 2010), and high-level noise can induce migration of nonsensory cells into the sensory region, displacing supporting cells (Cotanche et al. 1995). Hair cell regeneration cannot proceed in these cases because supporting cells serve as progenitors to new hair cells in all species capable of regeneration (see Sect. 2.3.3.1, 2.3.4, and 2.3.5).

Hair cells can repair themselves when damage is mild. In the mammalian cochlea, moderate noise treatments result in degenerative changes in stereocilia (Mulroy and Whaley 1984), but bundles are subsequently repaired (Wang et al. 2002). Similar bundle repair occurs after direct mechanical injury in cultured organs of Corti (Sobkowicz et al. 1996) or after aminoglycoside treatments in vestibular organs (Zheng et al. 1999a; Gale et al. 2002). By contrast, when damage is severe, hair cell corpses are ejected or decompose within the epithelium, and cellular debris is cleared by macrophages and/or supporting cells (e.g., Jones and Corwin 1996; Abrashkin et al. 2006). The factors that tip a damaged hair cell from repair to cell death remain to be determined.

2.3.3 Hair Cell Regeneration in Nonmammalian Vertebrates

The first definitive demonstrations of hair cell regeneration after damage in homeothermic animals (birds) were published in the late 1980s. Proof of regeneration came in three forms: signs of embryonic-like hair cells in the region of damage (Cotanche 1987a); recovery of hair cell numbers over time (Cruz et al. 1987); and incorporation of a nucleotide analog, tritiated thymidine, into the DNA of hair cells and supporting cells, indicating that new cells arise through mitosis (Corwin and Cotanche 1988; Ryals and Rubel 1988). Subsequent studies showed hair cell regeneration in other nonmammalian vertebrates, including inner ears of reptiles, amphibians, and fish (e.g., Baird et al. 1993; Lombarte et al. 1993; Avallone et al. 2003) and in lateral line neuromasts of amphibians and fish (Balak et al. 1990; Song et al. 1995).

Remarkably, all nonmammalian vertebrates studied to date can restore moderately damaged hair cell epithelia to original status, with only minor imperfections. Importantly, accessory structures that are also required for inner ear function, such as the tectorial membrane in the auditory organ or the otolithic membrane and cupulae in vestibular organs, are also repaired (e.g., Cotanche 1987b). As a result of this response, nonmammalian vertebrates recover function of hearing, balance, and lateral line systems throughout life (reviewed in Bermingham-McDonogh and Rubel 2003; Ryals et al. 2013).

2.3.3.1 Hair Cell Progenitors in Nonmammalian Vertebrates: Identity and Behavior

Supporting cells are the hair cell progenitors in avian inner ear epithelia (Fig. 2.4) (e.g., Hashino and Salvi 1993; Stone and Cotanche 1994). In the chicken basilar papilla, significant hair cell loss triggers supporting cell proliferation within 2 days after the onset of damage (e.g., Raphael 1992; Warchol and Corwin 1996). The extent and duration of supporting cell division are positively correlated with the amount of hair cell loss, and supporting cells stop dividing when new hair cells populate the damaged area. These findings suggest that the loss of hair cells enables supporting cells to exit growth arrest and divide, and replacement hair cells provide negative feedback to halt cell division.

In contrast to the basilar papilla, hair cells in vestibular organs and neuromasts of nonmammals undergo turnover throughout life (Jørgensen and Mathiesen 1988; Williams and Holder 2000). Therefore, there is always a small degree of supporting cell division in vestibular and neuromast organs. Upon hair cell loss, additional supporting cells are recruited to enter the cell cycle (Weisleder and Rubel 1993; Williams and Holder 2000) (Fig. 2.5). In birds, the time course of supporting cell proliferation following damage is similar in vestibular and auditory organs when lesions are comparable in scale (compare Stone et al. 1999 with Ku et al. 2014). Periods of increased supporting cell division following hair cell destruction are shorter in aquatic animals (e.g., Balak et al. 1990; Harris et al. 2003).

Although cell division is always observed in response to severe hair cell injuries, it is not required for hair cell regeneration in nonmammals. Approximately half of the hair cells regenerated in chicken basilar papillae are not labeled with a proliferation marker when it is continually infused after damage (Roberson et al. 1996, 2004), indicating new hair cells can arise without division. Moreover, hair cells are still regenerated in chickens, frogs, and salamanders when cell division is blocked (e.g., Adler et al. 1997; Baird et al. 2000; Taylor and Forge 2005). Finally, cells with phenotypes intermediate between those of supporting cells and hair cells have been observed in frogs and birds (Steyger et al. 1997; Cafaro et al. 2007). These studies indicate that some supporting cells convert into hair cells without dividing, a process called direct transdifferentiation (Fig. 2.4).

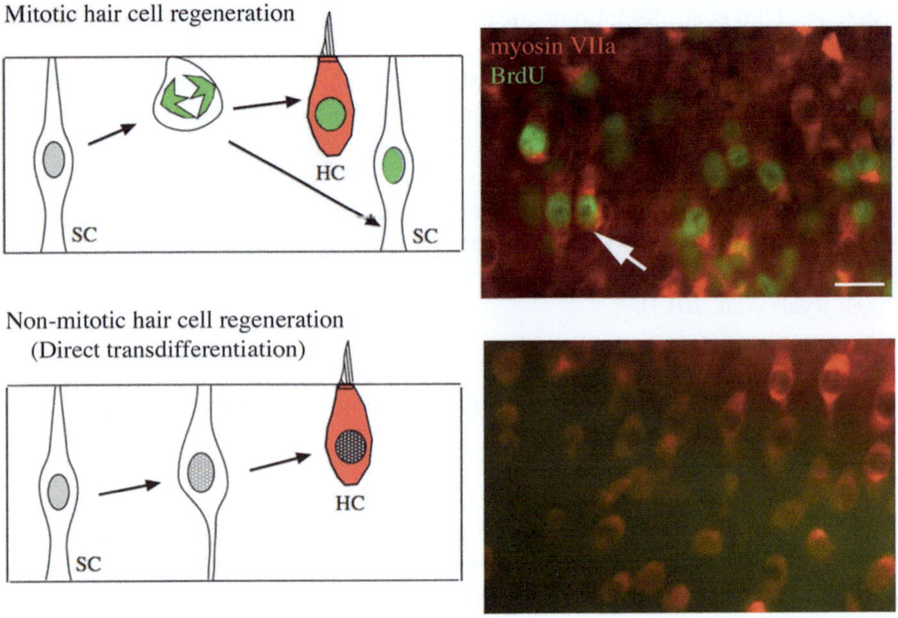

Fig. 2.4 Hair cells are regenerated using two distinct cellular processes. The two modes of hair cell regeneration in nonmammalian vertebrates are shown schematically in the *left panels* and in color digitized images from the regenerating chicken basilar papilla in the *right panels*. *Top panels*: Cells formed by mitosis are marked by BrdU (green), which is taken up by dividing supporting cells during DNA duplication and distributed to daughter cells during mitosis. Regenerated hair cells are labeled with antibodies to myosin VIIa (red). Hair cells formed by mitosis have red cytoplasm and green nuclei (arrow). *Bottom panels*: Regenerated hair cells formed by direct transdifferentiation lack BrdU because no cell division (DNA synthesis) occurs. Scale bar = 10 μm

It is not clear why a given supporting cell would undergo mitotic versus nonmitotic regeneration. In addition, it remains to be determined if all supporting cells have the capacity to form new hair cells or if there are subsets of supporting cells with different intrinsic features, such as stem cells, committed progenitors, postmitotic hair cell precursors, or terminally differentiated cells (Morest and Cotanche 2004). Several studies suggest there is a renewable progenitor cell population in mature hair cell epithelia. For instance, hair cells are regenerated after multiple successive killings in the avian basilar papilla (Niemiec et al. 1994) and in zebrafish neuromasts (Cruz et al. 2015). Further, when supporting cells are isolated and grown in culture, they can be passaged multiple times and retain the ability to differentiate into hair cells and supporting cells (Hu and Corwin 2007). However, outcomes of supporting cell divisions in birds and fishes in vivo are variable and appear to be spatially segregated in some cases, suggesting progenitors with different developmental potentials are present among supporting cells (Stone and Rubel 2000; Romero-Carvajal et al., 2015). In zebrafish neuromasts, supporting cells that divide and yield new supporting cells are enriched in the periphery of the organ, while divisions leading to two hair cells are enriched in the central zone (Romero-Carvajal

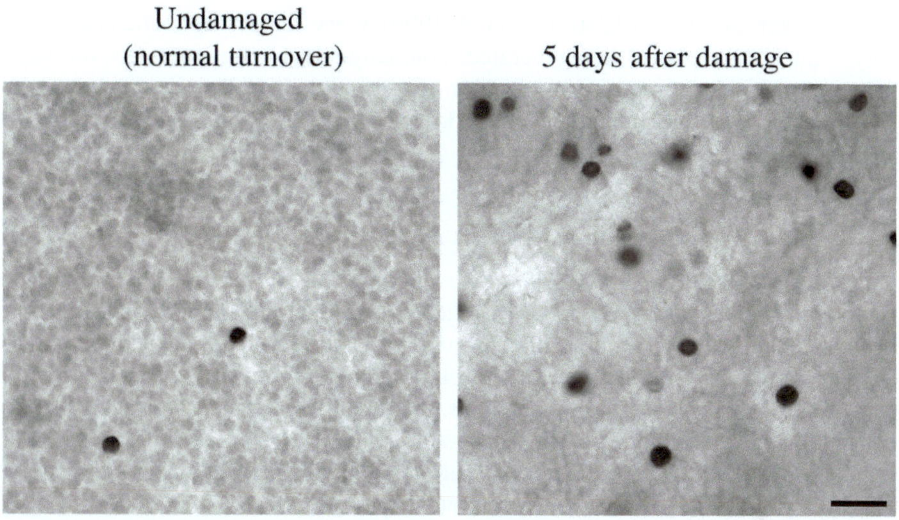

Fig. 2.5 Supporting cell division is increased in the chicken utricle following hair cell loss. Dividing supporting cells, with their nuclei marker by the proliferation marker BrdU (black), are evident in the undamaged utricle from a post-hatch chicken (*left*) due to normal hair cell turnover. Numbers of dividing supporting cells increase significantly in utricles during the 5 days after streptomycin-induced hair cell destruction (*right*). Scale bar = 10 μm

et al. 2015). Similarly, supporting cells in the neural half of the chicken basilar papilla are more likely to divide after hair cell damage, while those in the abneural half are biased toward direct transdifferentiation (Cafaro et al. 2007). Although these findings are consistent with specialized populations of supporting cells in hair cell epithelia, it is likely that spatially distributed signals also influence cell fate during regeneration.

2.3.3.2 Morphological and Functional Maturation of Regenerated Hair Cells in Nonmammalian Vertebrates

Maturation of regenerated hair cells in nonmammals has been most extensively studied in birds. Using electron microscopy and immunolabeling, investigators determined that new hair cells begin to populate the damaged auditory epithelia 3–5 days after damage (e.g., Girod et al. 1989; Ryals and Westbrook 1994). Stereociliary bundles of regenerated hair cells grow and establish proper sizes, shapes, and orientations (e.g., Cotanche and Corwin 1991; Duckert and Rubel 1993) before becoming mechanosensitive (Masetto and Correia 1997; Correia et al. 2001). Finally, synaptic connections are reestablished with peripheral nerves (Duckert and Rubel 1993; Ryals and Westbrook 1994).

In birds, several weeks are required to restore a full complement of hair cells, but it takes several months for sensory epithelia to reach full histological maturation (e.g., Duckert and Rubel 1990; Dye et al. 1999). Once sensory epithelia and accessory tissues are repaired or regenerated, nonmammalian vertebrates recover auditory or vestibular function, even after extensive hair cell loss (reviewed in Bermingham-McDonogh and Rubel 2003; Ryals et al. 2013).

2.3.4 Hair Cell Regeneration in Mammals: The Cochlea

Following the discovery of hair cell regeneration in nonmammals, investigators began to examine whether any hair cells are naturally regenerated in the mammalian cochlea. In mice and other rodents, developmental addition of cochlear hair cells ceases before or very soon after birth (e.g., Ruben 1967; Chen and Segil 1999). During the late embryonic and early postnatal periods, there is some plasticity in the organ of Corti, with small numbers of hair cells replaced after damage. For example, Kelley et al. (1995) used laser ablation to remove nascent hair cells in cultured organs of Corti from mouse embryos. They found that surrounding cells (either supporting cells or uncommitted precursor cells) differentiated into hair cells without dividing, and that a similar, but much less robust, replacement could also occur in neonatal mice. Consistent with this, a few hair cell-like cells regrow in neonatal rats after aminoglycoside-induced hair cell damage (Romand et al. 1996). Fate mapping of supporting cells in the neonatal mouse cochlea demonstrated that the immature hair cells that arise after damage are derived from supporting cells (Cox et al. 2014). Replacement potential also varies regionally, with the cochlear apex having the greatest capacity, perhaps because it is the least developmentally mature region in the organ of Corti at the time of hair cell ablation.

Even this minimal degree of regeneration is transient because cochlear supporting cells lose their ability to divide and form new hair cells within 2 weeks after birth, both in vivo and in vitro (Cox et al. 2014; White et al. 2006). As a result, auditory hair cells are not regenerated in mature mammalian cochleae after damage (e.g., Raphael and Altschuler 1991; Roberson and Rubel 1994).

2.3.5 Hair Cell Regeneration in Mammals: Vestibular Organs

Contrary to the auditory system, a limited number of hair cells are regenerated in vestibular organs of mature rodents and, possibly, humans. Using tritiated thymidine as a tracer, Warchol et al. (1993) determined that small numbers of supporting cells divide in utricles from guinea pigs and humans after aminoglycoside treatment in culture. Similar observations were made in vestibular epithelia of other rodents after hair cell damage in vivo (Rubel et al. 1995; Li and Forge 1997). These results, although significant, contrasted markedly with observations nonmammalian utricles in which hundreds of supporting cells incorporate the mitotic tracer bromodeoxyuridine (BrdU) after streptomycin treatment (Ku et al. 2014). Despite limited new cell production, cells with features typical of developing hair cells—small cell bodies or stereocilia bundles that were short and/or had a centrally located kinocilium—were detected in vestibular organs of adult rodents after damage, and numbers of these cells increased over time (Forge et al. 1993; Kawamoto et al. 2009; Golub et al. 2012). Most recently, Taylor et al. (2015) presented evidence for immature-appearing hair cells in utricles harvested from adult humans, suggesting some regeneration may also occur in primates.

One of the challenging aspects of studying vestibular hair cell regeneration in mammals is variable resistance of hair cells to damaging agents, either in vivo or in vitro. For example, in vivo injection of gentamicin to guinea pigs kills significant numbers of hair cells only in the striolar region of the utricle (Forge et al. 1993). Moreover, in some cases, hair cells may not be killed but instead may suffer apical damage that is subsequently repaired in a process that resembles the development of new hair cells (Zheng et al. 1999a), making it impossible to discriminate repaired hair cells from regenerated ones. To bypass this problem, several groups used genetic labeling to fluorescently mark supporting cells in mice and follow their fates after damage. The observation of fluorescent hair cells in utricles and cristae over time indicated some supporting cells had transdifferentiated into hair cells (Lin et al. 2011; Slowik and Bermingham-McDonogh 2013). As few postmitotic hair cells were observed in adult rodents after damage (see preceding paragraph), it is thought that mammalian vestibular hair cells are replaced predominantly by direct transdifferentiation of supporting cells (Fig. 2.4).

It is important to note that only a fraction of vestibular hair cells are regenerated after damage. In mature guinea pigs, about 70% of hair cells are replaced in the striolar region of the utricle, but this represents a small percentage of the total number of utricular hair cells (Forge et al. 1998). After nearly complete loss of hair cells was induced in utricles of transgenic *Pou4f3*^{DTR} mice, only 17% of hair cells were replaced (Golub et al. 2012). Importantly, all regenerated hair cells appear to be type II. Although these new type II hair cells develop stereociliary bundles, mechanotransduction channels, and synapses (e.g., Golub et al. 2012; Forge et al. 1998), it is not clear whether they restore vestibular function.

2.3.6 *Molecular Control of Hair Cell Regeneration in Nonmammals*

2.3.6.1 Regulators of Supporting Cell Division in Nonmammals

Because nonmammalian vertebrates naturally regenerate all types of hair cells in the inner ear and lateral line after injury, they provide excellent opportunities to study molecular signals controlling new hair cell production. Here, studies that defined the various signals regulating supporting cell division, differentiation of postmitotic cells into hair cells or supporting cells, and direct transdifferentiation of supporting cells into hair cells are discussed. However, it is important to consider that a comprehensive understanding of the genetic programs that mediate functional recovery does not exist yet.

Investigators postulated that signals derived from healthy hair cells maintain supporting cells in a state of quiescence. For instance, mature avian hair cells express fibroblast growth factors (FGFs), and FGF receptors are downregulated in supporting cells in regions where cell division occurs following hair cell loss (Bermingham-McDonogh et al. 2001; Ku et al. 2014). These findings suggest FGF signaling could, under normal conditions, maintain supporting cell quiescence. Consistent with this idea, addition of FGFs to cultured utricles or basilar papillae reduces supporting cell division (Oesterle et al. 2000; Ku et al. 2014).

A similar line of thinking suggests that supporting cell transit through the cell cycle may be stimulated by mitogens released from hair cells or other injured cells. Consistent with this idea, receptors for activins, which are secreted members of the transforming growth factor (TGF) ß family, are expressed in supporting cells (McCullar et al. 2010). Addition of activin A/B to cultured chick basilar papillae that were not treated with ototoxin increases supporting cell proliferation, suggesting release of activins from injured hair cells might trigger proliferation. Additional growth factors and growth factor–triggered enzymes have been implicated as positive regulators of supporting cell division in nonmammals, including protein kinase A (Navaratnam et al. 1996), mechanistic target of rapamycin (mTOR) and phosphoinoside-3 kinase (PI3K) (Witte et al. 2001), epidermal growth factor (White et al. 2012), insulin (Oesterle et al. 1997), TGFα, tumor necrosis factor α (Warchol 1999), and canonical Wnts (Head et al. 2013; Jacques et al. 2014).

Structural changes in the epithelium during hair cell loss may also trigger supporting cells to reenter the cell cycle. After extrusion of hair cells, supporting cells expand, presumably to reestablish the integrity of the epithelial surface. Some investigators suggest rearrangement of actin at the supporting cell surface during expansion could trigger enzymatic reactions within supporting cells to promote division (Meyers and Corwin 2007; Collado et al. 2011). Another hypothesis is that conversion of supporting cells into hair cells, and the accompanying depletion of supporting cells, stimulate supporting cell division (Roberson et al. 2004).

2.3.6.2 Regulators of Hair Cell Differentiation in Nonmammals

A critical step in hair cell regeneration is the acquisition of a hair cell phenotype in newly born precursor cells or transdifferentiating supporting cells. Cell fate is influenced by intrinsic properties of progenitor cells (e.g., transcription factors that restrict cells to a particular lineage) and local signals from neighboring cells. An excellent example of the interplay of these two mechanisms during regeneration is provided by *Atoh1* and notch signaling. During development (Sects. 2.2.1 and 2.2.2), cells in prosensory regions of the otocyst become distinguished from nonsensory regions by expressing *Sox2*. Subsequently, some prosensory cells activate *Atoh1* expression, which is sufficient to induce the hair cell fate. In chicken basilar papillae following aminoglycoside-induced hair cell damage, the earliest steps of regeneration mirror development: supporting cells reduce levels of Sox2 protein and accumulate Atoh1 protein (Cafaro et al. 2007). Many of these supporting cells progress to form hair cells, although some do not (Lewis et al. 2012). As in development, *Atoh1* transcription is modulated by notch signaling. In birds and fishes after ototoxic drug treatments, notch ligand (delta1) becomes expressed on the surface of early regenerating hair cells (Stone and Rubel 1999; Ma et al. 2008). Delta1 can bind to and activate notch receptors on neighboring supporting cells, increasing their expression of transcriptional repressors, such as Hes5. These repressors antagonize *Atoh1* transcription in some supporting cells and postmitotic precursor cells, preventing them from transdifferentiating into hair cells. Consequently, notch signaling directs cell patterning in regenerating epithelia to reestablish the precise mosaic of alternating hair cells and supporting cells. Indeed, when notch-mediated repression of *Atoh1* transcription is blocked pharmaceutically, nearly all supporting cells in the damaged area form new hair cells (Daudet et al. 2009).

2.3.6.3 Factors Limiting Hair Cell Regeneration in Mammals

There are several ideas to explain why hair cell regeneration does not occur (in the case of the cochlea) or is drastically attenuated (in the case of vestibular organs) in mature mammals. The failure of hair cell regeneration is likely due to progressive changes in supporting cells as well as signals that restrict supporting cells from responding to hair cell damage. These features of mammalian supporting cells may have evolved to ensure stability of inner ear epithelia in adulthood, which could be critical for mammalian auditory and vestibular function.

Loss of progenitor/stem cells may suppress regeneration. Oshima et al. (2007) noted a dramatic, age-related reduction in the number of stems cells that can be derived from mouse organ of Corti or utricle as defined by their ability to generate clonal colonies. This result is consistent with a loss of stem-like cells with development.

It is also possible that proregenerative signals are not activated in mammals after hair cell loss, or the signals that are turned on are not effective. To test this hypothesis, investigators treated tissues from mature mammalian inner ears with growth factors known to be mitogenic in other tissues and assessed effects on supporting cell division. In dissociated cell cultures derived from utricular epithelia of neonatal rats, Zheng et al. (1997) identified several mitogens that increased supporting cell division by two-to fourfold over untreated cultures, including fibroblast growth factor-2, TGFα, and insulin-like growth factor-1. In similar preparations, Montcouquiol and Corwin (2001) identified kinases downstream of mitogens that, when activated, trigger supporting cell division. However, the proproliferative effects of these growth factors and pathway activators were not replicated in other cellular contexts. For instance, investigators found only modest increases in supporting cell division after treatment with TGFα (with or without insulin) in organotypic cultures of mature mammalian utricles (Lambert 1994; Yamashita and Oesterle 1995) or after infusion of growth factor in vivo (Kuntz and Oesterle 1998).

Furthermore, even these modest responses are lost with age. For instance, epidermal growth factor and glial growth factor trigger supporting cell division in neonatal rat utricles (Zheng et al. 1999b; Gu et al. 2007) but have little or no effect in adult mouse utricles (Hume et al. 2003). Similarly, activation of ß-catenin (the effector of canonical Wnt signaling) triggers a transient increase in supporting cell division in the cochleae of early postnatal mice (Chai et al. 2012; Shi et al. 2013) but not in juvenile mice (Kuo et al. 2015).

Investigators have proposed mechanisms that could explain the inability of mammalian supporting cells to respond to pro-regenerative signals as they mature. Corwin and colleagues demonstrated that mammalian supporting cells develop thick apical actin networks that could prevent cell spreading, a process that may be required for cell division (Collado et al. 2011; Burns and Corwin 2014). Other groups hypothesize that natural cell cycle blockers such as retinoblastoma protein (Rb) or cyclin-dependent kinase inhibitors (CKIs) prevent supporting cells from responding to mitogenic cues that exist under normal conditions or after hair cell damage. When genes encoding Rbs or CKIs are deleted in the organ of Corti or utricular maculae of adult mammals, both hair cells and supporting cells divide, but the number of cells is limited, and many cells die during or after mitosis (e.g., Löwenheim et al. 1999; Laine et al. 2007). Supporting cell division is also triggered when potent cell cycle activators, such as cyclins or their downstream effectors, Mycs, are misexpressed (Loponen et al. 2011; Burns et al. 2012). Furthermore, misexpression of *Sox4* and *Sox11* transcription factors in vestibular supporting cells of adult mice increases mitotic hair cell regeneration, but it is unclear how long new hair cells survive (Gnedeva and Hudspeth 2015). Expression of both transcription factors in supporting cells is high embryonically and low postnatally. These studies suggest activation of negative regulators such as CKIs and suppression of positive regulators such as Sox4 and Sox11 collectively limit supporting cell division in mature mammals.

Investigators have also examined signals that control differentiation of regenerated cells. Early studies found that diffusible factors that drive new hair cell differentiation in very young cochleae are not effective in more mature cochleae (e.g., Chardin and Romand 1995, 1997). Thereafter, many researchers shifted focus away from diffusible factors to new strategies for promoting new hair cell production, namely forced expression of *Atoh1* and inhibition of the notch pathway. Investigators reasoned that forced expression of *Atoh1* in mature cochlear or vestibular epithelia after damage could promote supporting cells to convert into hair cells. In the cochlea, virally delivered *Atoh1* stimulates formation of small numbers of new hair cells in mature guinea pigs (Kawamoto et al. 2003; Atkinson et al. 2014) and mice (Kraft et al. 2013), while *Atoh1* misexpression in damaged hair cells appears to trigger bundle repair (Yang et al. 2012). *Atoh1* overexpression using transgenic mice triggers considerable new hair cell production in neonates, but it yields fewer new hair cells in adults. Moreover, in adults, the new hair cells fail to mature, and many die (Kelly et al. 2012; Liu et al., 2012a). The discrepancy between these studies underscores the challenge of defining the capacity of Atoh1 to act on its own to promote hair cell regeneration. A recent study points to the potential for activation of canonical Wnt to enhance effects of *Atoh1* misexpression in neonatal mice (Kuo et al. 2015), but further studies are needed to ascertain effectiveness in adults.

The notch signaling pathway is a potent inhibitor of *Atoh1* transcription (Sect. 2.2.2.2). Inactivation of notch with γ-secretase inhibitors (GSIs) promotes considerable conversion of supporting cells into hair cells in the neonatal cochlea but very little or no conversion in juvenile or adult cochleae (e.g., Yamamoto et al. 2006; Korrapati et al. 2013). Why do GSIs become less effective in promoting supporting cell-to-hair cell conversion as rodents mature? One explanation was offered by Maass et al. (2015), who showed that, in contrast with the avian basilar papilla (Stone and Rubel 1999), notch pathway genes are not reactivated in the mature organ of Corti after damage. However, a similar study from a different research group did show reactivation of notch signaling and even some recovery of auditory function after GSI treatment (Mizutari et al. 2013). Thus, the ability of notch signaling to modulate hair cell formation and regeneration in the adult cochlea remains controversial.

Whereas auditory and vestibular systems of adult rodents seem to have comparable brakes on supporting cell division, adult vestibular epithelia are more responsive to signals promoting direct conversion of supporting cells into hair cells. This is most likely due to the fact that some hair cells are naturally regenerated in adult rodents (Sect. 2.3.5). Virally misexpressed *Atoh1* promotes significant hair cell regeneration in adult rodent utricles in vitro (Shou et al. 2003; Staecker et al. 2007) and in vivo, where type I hair cells may also be regenerated (Schlecker et al. 2011; Xu et al. 2012). In neonatal mice, inhibition of Wnt or notch leads to increased replacement of type I and II vestibular hair cells after damage, but only type II hair cells are regenerated in adult mice (Wang et al. 2015). Inhibition of notch activity in mice leads to higher numbers of replacement hair cells in vestibular epithelia (e.g., Collado et al. 2011; Lin et al. 2011), demonstrating that notch signaling actively antagonizes vestibular hair cell regeneration at all ages.

2.3.7 Future Considerations for Hair Cell Regeneration

Hearing and balance dysfunctions are growing concerns in countries with increasing numbers of elderly individuals. To develop hair cell regeneration as a clinical therapy, researchers must determine how to more effectively drive hair cell replacement in adult mammals. Investigators are beginning to employ different scientific approaches to identify factors controlling mammalian hair cell regeneration. One of these is transcriptomics—the analysis of the set of transcripts (mRNAs) that are present in a cell at a particular time. Because the proteins that ultimately control a cell's behavior are translated from mRNAs, transcriptional profiling provides insights regarding a cell's state and potential. Several studies have published transcriptional profiles of hair cell epithelia from different species in normal conditions and after damage (e.g., Hawkins et al. 2003; Jiang et al. 2014). These studies enable comparisons of the signaling molecules, receptors, and transcription factors that are being transcribed under different conditions and at a particular time. The results will, hopefully, reveal the factors that should be tested for regulatory roles in regeneration. These sorts of analyses should also provide greater insight into properties of supporting cells in each species. For instance, are some supporting cells genetically programmed to respond in a distinct manner to damage—e.g., by remaining quiescent, dividing, or converting?

Although transcriptional profiling offers a way to model protein expression within a cell, the behavior of that cell is also controlled by other factors. For instance, posttranscriptional events regulate which proteins are generated. MicroRNAs are endogenous noncoding RNAs that direct degradation of specific transcripts. Investigation of microRNAs in hair cells is at an early stage, but it is already clear they play key roles in hair cell production (reviewed in Ushakov et al. 2013). In addition, epigenetic modifications, including DNA methylation and histone acetylation, are potent regulators of transcription and therefore represent another level at which hair cell regeneration may be modulated in mammals. Consistent with this idea, *Atoh1* transcription becomes epigenetically blocked in mouse cochlear supporting cells during postnatal maturation (Stojanova et al. 2015). Further, inhibition of histone deacetylases reduces avian hair cell regeneration (Slattery et al. 2009), although the specific mechanisms have not been defined.

Investigators hope to use these tools to drive hair cell regeneration in mature mammals by delivering drugs or manipulating gene expression to alter the protein composition and the behavior of supporting cells. It is possible, however, that such treatments will be insufficient to drive hair cell regeneration and functional recovery in adult mammals. In this case, it may be necessary to introduce another cell population to achieve recovery. Several labs are studying the capacity of mammalian stem cells, derived from young embryos or induced from mature somatic cells, to form new hair cells in a dish or on delivery to the inner ear. This scientific approach, although at an early stage, has yielded promising results in vitro (reviewed in Hu and Ulfendahl 2013).

2.3.8 Summary

Several decades of research have revealed key facets of hair cell development and regeneration, including the cellular mechanisms and some of the molecules that control those processes. In nonmammals, which regenerate hair cells after damage, the signaling molecules that direct regrowth and differentiation of hair cells mirror those employed during development, suggesting genetic programs used in embryogenesis are reactivated in adult tissues. However, there is still much work to be done before the orchestration of complete genetic and molecular pathways necessary for hair cell regeneration can be described. Nonetheless, research has proven that a deepening understanding of hair cell development hastens progress toward determining how to stimulate hair cell replacement in adult mammals. Further, as more is learned about how hair cell replacement is controlled, new ideas are generated about regulation of hair cell development. Thus, these two lines of investigation are complementary and critical for one another.

Acknowledgments The authors would like to acknowledge Edwin W Rubel as an exceptional mentor, colleague, and friend, and as one of the pioneers in the fields of both inner development and regeneration. The authors would also like to apologize for the many relevant references that were necessarily omitted because of space limitations.

Compliance with Ethics Requirements Jennifer Stone declares that she has no conflict of interest. Matthew Kelley declares that he has no conflict of interest.

References

Abrashkin, K. A., Izumikawa, M., Miyazawa, T., Wang, C. H., et al. (2006). The fate of outer hair cells after acoustic or ototoxic insults. *Hearing Research*, 218(1–2), 20–29.

Adler, H. J., Komeda, M., & Raphael, Y. (1997). Further evidence for supporting cell conversion in the damaged avian basilar papilla. *International Journal of Developmental Neuroscience*, 15(4–5), 375–385.

Ahmed, M., Wong, E. Y., Sun, J., Xu, J., et al. (2012a). *Eya1–Six1* interaction is sufficient to induce hair cell fate in the cochlea by activating *Atoh1* expression in cooperation with *Sox2*. *Developmental Cell*, 22, 377–390.

Ahmed, M., Xu, J., & Xu, P.X. (2012b). EYA1 and SIX1 drive the neuronal developmental program in cooperation with the SWI/SNF chromatin-remodeling complex and *SOX2* in the mammalian inner ear. *Development*, 139, 1965–1977.

Appler, J. M., & Goodrich, L. V. (2011). Connecting the ear to the brain: Molecular mechanisms of auditory circuit assembly. *Progress in Neurobiology*, 93, 488–508.

Atkinson, P. J., Wise, A. K., Flynn, B. O., Nayagam, B. A., & Richardson, R. T. (2014). Hair cell regeneration after *ATOH1* gene therapy in the cochlea of profoundly deaf adult guinea pigs. *PLoS ONE*, 9(7), e102077.

Avallone, B., Porritiello, M., Esposito, D., Mutone, R., et al. (2003). Evidence for hair cell regeneration in the crista ampullaris of the lizard *Podarcis sicula*. *Hearing Research*, 178(1–2), 79–88.

Baird, R. A., Torres, M. A., & Schuff, N. R. (1993). Hair cell regeneration in the bullfrog vestibular otolith organs following aminoglycoside toxicity. *Hearing Research*, 65(1–2), 164–174.

Baird, R. A., Burton, M. D., Lysakowski, A., Fashena, D. S., & Naeger, R. A. (2000). Hair cell recovery in mitotically blocked cultures of the bullfrog saccule. *Proceedings of the National Academy of Sciences of the USA*, 97(22),11722–11729.

Balak, K. J., Corwin, J. T., & Jones, J. E. (1990). Regenerated hair cells can originate from supporting cell progeny: Evidence from phototoxicity and laser ablation experiments in the lateral line system. *The Journal of Neuroscience*, 10(8), 2502–2512.

Bermingham-McDonogh, O., & Rubel, E. W. (2003). Hair cell regeneration: Winging our way towards a sound future. *Current Opinion in Neurobiology*, 13(1), 119–126.

Bermingham, N. A., Hassan, B. A., Price, S. D., Vollrath, M. A., et al. (1999). *Math1*: An essential gene for the generation of inner ear hair cells. *Science*, 284,1837–1841.

Bermingham-McDonogh, O., Stone, J. S., Reh, T. A., & Rubel, E. W. (2001). *FGFR3* expression during development and regeneration of the chick inner ear sensory epithelia. *Developmental Biology*, 238(2), 247–259.

Bok, J., Raft, S., Kong, K. A., Koo, S. K., et al. (2011). Transient retinoic acid signaling confers anterior-posterior polarity to the inner ear. *Proceedings of the National Academy of Sciences of the USA*, 108, 161–166.

Bok, J., Zenczak, C., Hwang, C. H., & Wu, D. K. (2013). Auditory ganglion source of Sonic hedgehog regulates timing of cell cycle exit and differentiation of mammalian cochlear hair cells. *Proceedings of the National Academy of Sciences of the USA*, 110, 13869–13874.

Brandt, N., Kuhn, S., Munkner, S., Braig, C., et al. (2007). Thyroid hormone deficiency affects postnatal spiking activity and expression of Ca^{2+} and K^{+} channels in rodent inner hair cells. *The Journal of Neuroscience* 27, 3174–3186.

Brooker, R., Hozumi, K., & Lewis, J. (2006). Notch ligands with contrasting functions: Jagged1 and Delta1 in the mouse inner ear. *Development*, 133, 1277–1286.

Burns, J. C., & Corwin, J. T. (2014). Responses to cell loss become restricted as the supporting cells in mammalian vestibular organs grow thick junctional actin bands that develop high stability. *The Journal of Neuroscience*, 34(5), 1998–2011.

Burns, J. C., Yoo, J. J., Atala, A., & Jackson, J. D. (2012). *MYC* gene delivery to adult mouse utricles stimulates proliferation of postmitotic supporting cells in vitro. *PLoS ONE*, 7(10), e48704.

Bylund, M., Andersson, E., Novitch, B. G., & Muhr, J. (2003). Vertebrate neurogenesis is counteracted by *Sox1–3* activity. *Nature Neuroscience*, 6,1162–1168.

Cafaro, J., Lee, G.-S., & Stone, J. S. (2007). *Atoh1* expression defines activated progenitors as well as differentiating hair cells during avian hair cell regeneration. *Developmental Dynamics*, 236, 156–170.

Cai, T., Jen, H. I., Kang, H., Klisch, T. J., et al. (2015). Characterization of the transcriptome of nascent hair cells and identification of direct targets of the *Atoh1* transcription factor. *The Journal of Neuroscience*, 35(14), 5870–5883.

Chai, R., Kuo, B., Wang, T., Liaw, E. J., et al. (2012). Wnt signaling induces proliferation of sensory precursors in the postnatal mouse cochlea. *Proceedings of the National Academy of Sciences of the USA*, 109(21), 8167–8172.

Chardin, S., & Romand, R. (1995). Regeneration and mammalian auditory hair cells. *Science*, 267(5198), 707–711.

Chardin, S., & Romand, R. (1997). Factors modulating supernumerary hair cell production in the postnatal rat cochlea in vitro. *International Journal of Developmental Neuroscience*, 15(4–5), 497–507.

Chen, P., & Segil, N. (1999). p27(Kip1) links cell proliferation to morphogenesis in the developing organ of Corti. *Development*, 126(8), 1581–1590.

Chen, P., Johnson, J. E., Zoghbi, H. Y., & Segil, N. (2002). The role of Math1 in inner ear development: Uncoupling the establishment of the sensory primordium from hair cell fate determination. *Development*, 129(10), 2495–2505.

Collado, M. S., Thiede, B. R., Baker, W., Askew, C., et al. (2011). The postnatal accumulation of junctional E-cadherin is inversely correlated with the capacity for supporting cells to convert directly into sensory hair cells in mammalian balance organs. *The Journal of Neuroscience*, 31(33), 11855–11866

Correia, M. J., Rennie, K. J., & Koo, P. (2001). Return of potassium ion channels in regenerated hair cells: Possible pathways and the role of intracellular calcium signaling. *Annals of the New York Academy of Sciences*, 942, 228–240.

Corwin, J. T., & Cotanche, D. A. (1988). Regeneration of sensory hair cells after acoustic trauma. *Science*, 240(4860), 1772–1774.

Costa, A., Sanchez-Guardado, L., Juniat, S., Gale, J. E., et al. (2015). Generation of sensory hair cells by genetic programming with a combination of transcription factors. *Development*, 142, 1948–1959.

Cotanche, D. A. (1987a). Regeneration of hair cell stereociliary bundles in the chick cochlea following severe acoustic trauma. *Hearing Research*, 30(2–3), 181–195.

Cotanche, D. A. (1987b). Regeneration of the tectorial membrane in the chick cochlea following severe acoustic trauma. *Hearing Research*, 30(2–3), 197–206.

Cotanche, D. A., & Corwin, J. T. (1991). Stereociliary bundles reorient during hair cell development and regeneration in the chick cochlea. *Hearing Research*, 52(2), 379–402.

Cotanche, D. A., Messana, E. P., & Ofsie M. S. (1995). Migration of hyaline cells into the chick basilar papilla during severe noise damage. *Hearing Research*, 91(1–2), 148–159.

Cox, B. C., Chai, R., Lenoir, A., Liu, Z., et al. (2014). Spontaneous hair cell regeneration in the neonatal mouse cochlea in vivo. *Development*, 141(4), 816–829.

Cruz, R. M., Lambert, P. R., & Rubel, E. W. (1987). Light microscopic evidence of hair cell regeneration after gentamicin toxicity in chick cochlea. *Archives of Otolaryngology – Head and Neck Surgery*, 113(10), 1058–1062.

Cruz, I., Kappedal, R., Mackenzie, S. M., Hailey, D. W., et al. (2015). Robust regeneration of adult zebrafish lateral line hair cells reflects continued precursor pool maintenance. *Developmental Biology*, 402(2), 229–238

Dabdoub, A., Puligilla, C., Jones, J. M., Fritzsch, B., et al. (2008). *Sox2* signaling in prosensory domain specification and subsequent hair cell differentiation in the developing cochlea. *Proceedings of the National Academy of Sciences of the USA*, 105, 18396–18401.

Daudet, N., Gibson, R., Shang, J., Bernard, A., et al. (2009). Notch regulation of progenitor cell behavior in quiescent and regenerating auditory epithelium of mature birds. *Developmental Biology*, 326, 86–100.

Desai, S. S., Zeh, C., & Lysakowski, A. (2005). Comparative morphology of rodent vestibular periphery. I. Saccular and utricular maculae. *Journal of Neurophysiology*, 93(1), 251–266.

Driver, E. C., Pryor, S. P., Hill, P., Turner, J., et al. (2008). Hedgehog signaling regulates sensory cell formation and auditory function in mice and humans. *The Journal of Neuroscience*, 28, 7350–7358.

Driver, E. C., Sillers, L., Coate, T. M., Rose, M. F., & Kelley, M. W. (2013). The *Atoh1*-lineage gives rise to hair cells and supporting cells within the mammalian cochlea. *Developmental Biology*, 376, 86–98.

Duckert, L. G., & Rubel, E. W. (1990). Ultrastructural observations on regenerating hair cells in the chick basilar papilla. *Hearing Research*, 48(1–2), 161–182.

Duckert, L. G., & Rubel, E. W. (1993). Morphological correlates of functional recovery in the chicken inner ear after gentamycin treatment. *The Journal of Comparative Neurology*, 331(1), 75–96.

Dye, B. J., Frank, T. C., Newlands, S. D., & Dickman, J. D. (1999). Distribution and time course of hair cell regeneration in the pigeon utricle. *Hearing Research*, 133(1–2), 17–26.

Erkman, L., McEvilly, R. J., Luo, L., Ryan, A. K., et al. (1996). Role of transcription factors *Brn-3.1* and *Brn-3.2* in auditory and visual system development. *Nature*, 381, 603–606.

Forge, A., Li, L., Corwin, J. T., & Nevill, G. (1993). Ultrastructural evidence for hair cell regeneration in the mammalian inner ear. *Science*, 259(5101), 1616–1619.

Forge, A., Li, L., & Nevill, G. (1998). Hair cell recovery in the vestibular sensory epithelia of mature guinea pigs. *The Journal of Comparative Neurology*, 397(1), 69–88.

Forrest, D., Erway, L. C., Ng, L., Altschuler, R., & Curran, T. (1996). Thyroid hormone receptor beta is essential for development of auditory function. *Nature Genetics*, 13, 354–357.

Fujioka, M., Tokano, H., Fujioka, K. S., Okano, H., & Edge, A. S. (2011). Generating mouse models of degenerative diseases using *Cre/lox*-mediated in vivo mosaic cell ablation. *The Journal of Clinical Investigation*, 121(6), 2462–2469.

Gale, J. E., Meyers, J. R., Periasamy, A., & Corwin, J. T. (2002). Survival of bundleless hair cells and subsequent bundle replacement in the bullfrog's saccule. *Journal of Neurobiology*, 50(2), 81–92.

Gnedeva, K., & Hudspeth, A. J. (2015). SoxC transcription factors are essential for the development of the inner ear. *Proceedings of the National Academy of Sciences of the USA*, 112(45), 14066–14071.

Girod, D. A., Duckert, L. G., & Rubel, E. W. (1989). Possible precursors of regenerated hair cells in the avian cochlea following acoustic trauma. *Hearing Research*, 42(2–3), 175–194.

Golub, J. S., Tong, L., Nguyen, T., Hume, C., et al. (2012). Hair cell replacement in adult mouse utricles after targeted ablation of hair cells with diphtheria toxin. *The Journal of Neuroscience*, 32(43), 15093–15105.

Goodyear, R., & Richardson, G. (1997). Pattern formation in the basilar papilla: Evidence for cell rearrangement. *The Journal of Neuroscience*, 17(16), 6289–6301.

Gu, R., Montcouquiol, M., Marchionni, M., & Corwin, J. T. (2007). Proliferative responses to growth factors decline rapidly during postnatal maturation of mammalian hair cell epithelia. *EJN European Journal of Neuroscience*, 25(5), 1363–1372.

Harris, J. A., Cheng, A. G., Cunningham, L. L., MacDonald, G., et al. (2003). Neomycin-induced hair cell death and rapid regeneration in the lateral line of zebrafish (*Danio rerio*). *Journal of the Association for Research in Otolaryngology*, 4(2), 219–234.

Hashino, E., & Salvi, R. J. (1993). Changing spatial patterns of DNA replication in the noise-damaged chick cochlea. *Journal of Cell Science*, 105(1), 23–31.

Hawkins, R. D., Bashiardes, S., Helms, C. A., Hu, L., et al. (2003). Gene expression differences in quiescent versus regenerating hair cells of avian sensory epithelia: Implications for human hearing and balance disorders. *Human Molecular Genetics*, 12(11), 1261–1272.

Head, J. R., Gacioch, L., Pennisi, M., & Meyers, J. R. (2013). Activation of canonical Wnt/β--catenin signaling stimulates proliferation in neuromasts in the zebrafish posterior lateral line. *Developmental Dynamics*, 242(7), 832–846.

Hertzano, R., Montcouquiol, M., Rashi-Elkeles, S., Elkon, R., et al. (2004). Transcription profiling of inner ears from *Pou4f3(ddl/ddl)* identifies Gfi1 as a target of the *Pou4f3* deafness gene. *Human Molecular Genetics*, 13, 2143–2153.

Hu, Z., & Corwin, J. T. (2007). Inner ear hair cells produced in vitro by a mesenchymal-to-epithelial transition. *Proceedings of the National Academy of Sciences of the USA*, 104(42), 16675–16680.

Hu, Z., & Ulfendahl, M. (2013). The potential of stem cells for the restoration of auditory function in humans. *Regenerative Medicine*, 8(3), 309–318.

Huh, S-H., Warchol, M. E., & Ornitz, D. M. (2015). Cochlear progenitor number is controlled through mesenchymal FGF receptor signaling. *Elife*, doi: 10.7554/eLife.05921.

Hume, C. R., Kirkegaard, M., & Oesterle, E. C. (2003). ErbB expression: The mouse inner ear and maturation of the mitogenic response to heregulin. *Journal of the Association for Research in Otolaryngology*, 4(3), 422–443.

Jacques, B. E., Puligilla, C., Weichert, R. M., Ferrer-Vaquer, A., et al. (2012). A dual function for canonical Wnt/beta-catenin signaling in the developing mammalian cochlea. *Development*, 139, 4395–4404.

Jacques, B. E., Montgomery, W. H. 4th, Uribe, P. M., Yatteau, A., et al. (2014). The role of Wnt/β--catenin signaling in proliferation and regeneration of the developing basilar papilla and lateral line. *Developmental Neurobiology*, 74(4), 438–456.

Jiang, L., Romero-Carvajal, A., Haug, J. S., Seidel, C. W., & Piotrowski, T. (2014). Gene-expression analysis of hair cell regeneration in the zebrafish lateral line. *Proceedings of the National Academy of Sciences of the USA*, 111(14), E1383–E1392.

Jones, J. E., & Corwin, J. T. (1996). Regeneration of sensory cells after laser ablation in the lateral line system: Hair cell lineage and macrophage behavior revealed by time-lapse video microscopy. *The Journal of Neuroscience*, 16(2), 649–662.

Jones, J. M., Montcouquiol, M., Dabdoub, A., Woods, C., & Kelley, M. W. (2006). Inhibitors of differentiation and DNA binding (Ids) regulate *Math1* and hair cell formation during the development of the organ of Corti. *The Journal of Neuroscience*, 26, 550–558.

Jørgensen, J. M., & Mathiesen, C. (1988). The avian inner ear: Continuous production of hair cells in vestibular sensory organs, but not in the auditory papilla. *Naturwissenschaften*, 75(6), 319–320.

Kawamoto, K., Ishimoto, S., Minoda, R., Brough, D. E., & Raphael, Y. (2003). *Math1* gene transfer generates new cochlear hair cells in mature guinea pigs in vivo. *The Journal of Neuroscience*, 23(11), 4395–4400.

Kawamoto, K., Izumikawa, M., Beyer, L. A., Atkin, G. M., & Raphael, Y. (2009). Spontaneous hair cell regeneration in the mouse utricle following gentamicin ototoxicity. *Hearing Research*, 247(1), 17–26.

Kelley, M. W., Talreja, D. R., & Corwin, J. T. (1995). Replacement of hair cells after laser microbeam irradiation in cultured organs of Corti from embryonic and neonatal mice. *The Journal of Neuroscience*, 15(4), 3013–3026.

Kelly, M. C., Chang, Q., Pan, A., Lin, X., & Chen, P. (2012). *Atoh1* directs the formation of sensory mosaics and induces cell proliferation in the postnatal mammalian cochlea in vivo. *The Journal of Neuroscience*, 32(19), 6699–6710.

Kempfle, J. S., Turban, J. L., & Edge, A. S. (2016). *Sox2* in the differentiation of cochlear progenitor cells. *Science Reports*, 6, 23293.

Kevetter, G. A., Blumberg, K. R., & Correia, M. J. (2000). Hair cell and supporting cell density and distribution in the normal and regenerating posterior crista ampullaris of the pigeon. *International Journal of Developmental Neuroscience*, 18(8), 855–867.

Kiernan, A. E., Cordes, R., Kopan, R., Gossler, A., & Gridley, T. (2005a). The Notch ligands *DLL1* and *JAG2* act synergistically to regulate hair cell development in the mammalian inner ear. *Development*, 132, 4353–4362.

Kiernan, A. E., Pelling, A. L., Leung, K. K., Tang, A. S., et al. (2005b). *Sox2* is required for sensory organ development in the mammalian inner ear. *Nature*, 434, 1031–1035.

Klisch, T. J., Xi, Y., Flora, A., Wang, L., Li, W., & Zoghbi, H. Y. (2011). In vivo *Atoh1* targetome reveals how a proneural transcription factor regulates cerebellar development. *Proceedings of the National Academy of Sciences of the USA*, 108, 3288–3293.

Korrapati, S., Roux, I., Glowatzki, E., & Doetzlhofer, A. (2013). Notch signaling limits supporting cell plasticity in the hair cell-damaged early postnatal murine cochlea. *PLoS ONE*, 8(8), e73276.

Kraft, S., Hsu, C., Brough, D. E., & Staecker, H. (2013). *Atoh1* induces auditory hair cell recovery in mice after ototoxic injury. *Laryngoscope*, 123(4), 992–999.

Ku, Y. C., Renaud, N. A., Veile, R. A., Helms, C., et al. (2014). The transcriptome of utricle hair cell regeneration in the avian inner ear. *The Journal of Neuroscience*, 34(10), 3523–3535.

Kuntz, A. L., & Oesterle, E. C. (1998). Transforming growth factor alpha with insulin stimulates cell proliferation in vivo in adult rat vestibular sensory epithelium. *The Journal of Comparative Neurology*, 399(3), 413–423.

Kuo, B. R., Baldwin, E. M., Layman, W. S., Taketo, M. M., & Zuo, J. (2015). In vivo cochlear hair cell generation and survival by coactivation of β-catenin and *Atoh1*. *The Journal of Neuroscience*, 35(30), 10786–10798.

Lai, H. C., Klisch, T. J., Roberts, R., Zoghbi, H. Y., & Johnson, J. E. (2011). In vivo neuronal subtype-specific targets of *atoh1* (*math1*) in dorsal spinal cord. *The Journal of Neuroscience*, 31, 10859–10871.

Laine, H., Doetzlhofer, A., Mantela, J., Ylikoski, J., et al. (2007). p19 (Ink4d) and p21 (Cip1) collaborate to maintain the postmitotic state of auditory hair cells, their codeletion leading to DNA damage and p53-mediated apoptosis. *The Journal of Neuroscience*, 27(6), 1434–1444.

Lambert, P. R. (1994). Inner ear hair cell regeneration in a mammal: Identification of a triggering factor. *Laryngoscope*, 104(6 Pt 1), 701–718.

Lanford, P. J., Lan, Y., Jiang, R., Lindsell, C., et al. (1999). Notch signalling pathway mediates hair cell development in mammalian cochlea. *Nature Genetics*, 21, 289–292.

Lanford, P. J., Shailam, R., Norton, C. R., Gridley, T., et al. (2000). Expression of Math1 and HES5 in the cochleae of wildtype and *Jag2* mutant mice. *Journal of the Association for Research in Otolaryngology*, 1(2), 161–171.

Lee, Y. S., Liu, F., & Segil, N. (2006). A morphogenetic wave of p27Kip1 transcription directs cell cycle exit during organ of Corti development. *Development*, 133, 2817–2826.

Lewis, R. M., Hume, C. R., & Stone, J. S. (2012). *Atoh1* expression and function during auditory hair cell regeneration in post-hatch chickens. *Hearing Research*, 289(1–2), 74–85.

Li, L., & Forge, A. (1997). Morphological evidence for supporting cell to hair cell conversion in the mammalian utricular macula. *International Journal of Developmental Neuroscience*, 15(4–5), 433–446.

Lin, V., Golub, J., Nguyen, T. B., Hume, C., et al. (2011). Inhibition of notch activity promotes non-mitotic regeneration of HCs in the adult mouse utricle. *The Journal of Neuroscience*, 31, 15329–15339.

Liu, Z., Dearman, J. A., Cox, B. C., Walters, B. J., et al. (2012a). Age-dependent in vivo conversion of mouse cochlear pillar and Deiters' cells to immature hair cells by *Atoh1* ectopic expression. *The Journal of Neuroscience*, 32, 6600–6610.

Liu, Z., Fang, J., Dearman, J., Zhang, L., & Zuo, J. (2012b). In vivo generation of immature inner hair cells in neonatal mouse cochleae by ectopic *Atoh1* expression. *PLoS ONE* 9(2), e89377.

Lombarte, A., Yan, H. Y., Popper, A. N., Chang, J. S., & Platt, C. (1993). Damage and regeneration of hair cell ciliary bundles in a fish ear following treatment with gentamicin. *Hearing Research*, 64(2), 166–174.

Loponen, H., Ylikoski, J., Albrecht, J. H., & Pirvola, U. (2011). Restrictions in cell cycle progression of adult vestibular supporting cells in response to ectopic cyclin D1 expression. *PLoS ONE* 6(11), e27360.

Löwenheim, H., Furness, D. N., Kil, J., Zinn, C., et al. (1999). Gene disruption of p27(Kip1) allows cell proliferation in the postnatal and adult organ of Corti. *Proceedings of the National Academy of Sciences of the USA*, 96(7), 4084–4088.

Ma, E. Y., Rubel, E. W., & Raible, D. W. (2008). Notch signaling regulates the extent of hair cell regeneration in the zebrafish lateral line. *The Journal of Neuroscience*, 28(9), 2261–2273.

Ma, Q., Anderson, D. J., & Fritzsch, B. (2000). *Neurogenin 1* null mutant ears develop fewer, morphologically normal hair cells in smaller sensory epithelia devoid of innervation. *Journal of the Association for Research in Otolaryngology*, 1, 129–143.

Maass, J. C., Gu, R., Basch, M. L., Waldhaus, J., et al. (2015). Changes in the regulation of the Notch signaling pathway are temporally correlated with regenerative failure in the mouse cochlea. *Frontiers in Cellular Neuroscience*, 9, 110.

Masetto, S., & Correia, M. J. (1997). Electrophysiological properties of vestibular sensory and supporting cells in the labyrinth slice before and during regeneration. *Journal of Neurophysiology*, 78(4), 1913–1927.

McCullar, J. S., Ty, S., Campbell, S., & Oesterle, E. C. (2010). Activin potentiates proliferation in mature avian auditory sensory epithelium. *The Journal of Neuroscience*, 30(2), 478–490.

Meyers, J. R., & Corwin, J. T. (2007). Shape change controls supporting cell proliferation in lesioned mammalian balance epithelium. *The Journal of Neuroscience*, 27(16), 4313–4325.

Mizutari, K., Fujioka, M., Hosoya, M., Bramhall, N., et al. (2013). Notch inhibition induces cochlear hair cell regeneration and recovery of hearing after acoustic trauma. *Neuron*, 77(1), 58–69.

Montcouquiol, M., & Corwin, J. T. (2001). Brief treatments with forskolin enhance s-phase entry in balance epithelia from the ears of rats. *The Journal of Neuroscience*, 21(3), 974–982.

Morest, D. K., & Cotanche, D. A. (2004). Regeneration of the inner ear as a model of neural plasticity. *Journal of Neuroscience Research*, 78(4), 455–460.

Morsli, H., Choo, D., Ryan, A., Johnson, R., & Wu, D. K. (1998). Development of the mouse inner ear and origin of its sensory organs. *The Journal of Neuroscience*, 18, 3327–3335.

Mulroy, M. J., & Whaley, E. A. (1984). Structural changes in auditory hairs during temporary deafness. *Scanning Electron Microscopy*, 1984(2), 831–840.

Murata, J., Tokunaga, A., Okano, H., & Kubo, T. (2006). Mapping of notch activation during cochlear development in mice: Implications for determination of prosensory domain and cell fate diversification. *The Journal of Comparative Neurology*, 497, 502–518.

Navaratnam, D. S., Su, H. S., Scott, S. P., & Oberholtzer, J. C. (1996). Proliferation in the auditory receptor epithelium mediated by a cyclic AMP-dependent signaling pathway. *Nature Medicine*, 2(10), 1136–1139.

Niemiec, A. J., Raphael, Y., & Moody, D. B. (1994). Return of auditory function following structural regeneration after acoustic trauma: Behavioral measures from quail. *Hearing Research*, 79(1–2), 1–16.

Oesterle, E. C., Tsue, T. T., & Rubel, E. W. (1997). Induction of cell proliferation in avian inner ear sensory epithelia by insulin-like growth factor-I and insulin. *The Journal of Comparative Neurology*, 380(2), 262–274.

Oesterle, E. C., Bhave, S. A., & Coltrera, M. D. (2000). Basic fibroblast growth factor inhibits cell proliferation in cultured avian inner ear sensory epithelia. *The Journal of Comparative Neurology*, 424(2), 307–326.

Oshima, K., Grimm, C. M., Corrales, C. E., Senn, P., et al. (2007). Differential distribution of stem cells in the auditory and vestibular organs of the inner ear. *Journal of the Association for Research in Otolaryngology*, 8(1), 18–31.

Palermo, R., Checquolo, S., Bellavia, D., Talora, C., & Screpanti, I. (2014). The molecular basis of notch signaling regulation: A complex simplicity. *Current Molecular Medicine*, 14, 34–44.

Puligilla, C., & Kelley, M. W. (2016). Dual role for *Sox2* in specification of sensory competence and regulation of Atoh1 function. *Developmental Neurobiology*, 77(1), 3–13.

Raft, S., Nowotschin, S., Liao, J., & Morrow, B. E. (2004). Suppression of neural fate and control of inner ear morphogenesis by *Tbx1*. *Development*, 131, 1801–1812.

Raft, S., Koundakjian, E. J., Quinones, H., Jayasena, C. S., et al. (2007). Cross-regulation of *Ngn1* and *Math1* coordinates the production of neurons and sensory hair cells during inner ear development. *Development*, 134, 4405–4415.

Raphael, Y. (1992). Evidence for supporting cell mitosis in response to acoustic trauma in the avian inner ear. *Journal of Neurocytology*, 21(9), 663–671.

Raphael, Y., & Altschuler, R. A. (1991). Scar formation after drug-induced cochlear insult. *Hearing Research*, 51(2), 173–183.

Riccomagno, M. M., Martinu, L., Mulheisen, M., Wu, D. K., & Epstein, D. J. (2002). Specification of the mammalian cochlea is dependent on Sonic hedgehog. *Genes and Development*, 16, 2365–2378.

Riccomagno, M. M., Takada, S., & Epstein, D. J. (2005). Wnt-dependent regulation of inner ear morphogenesis is balanced by the opposing and supporting roles of *Shh*. *Genes and Development*, 19, 1612–1623.

Roberson, D. W., & Rubel, E. W. (1994). Cell division in the gerbil cochlea after acoustic trauma. *The American Journal of Otology*, 15(1), 28–34.

Roberson, D. W., Kreig, C. S., & Rubel, E. W. (1996). Light microscopic evidence that direct transdifferentiation gives rise to new hair cells in regenerating avian auditory epithelium. *Auditory Neuroscience*, 2, 195–205.

Roberson, D. W., Alosi, J. A., & Cotanche, D. A. (2004). Direct transdifferentiation gives rise to the earliest new hair cells in regenerating avian auditory epithelium. *Journal of Neuroscience Research*, 78(4), 461–471.

Roberto, M., & Zito, F. (1988). Scar formation following impulse noise-induced mechanical damage to the organ of Corti. *The Journal of Laryngology and Otology*, 102(1), 2–9.

Romand, R., Chardin, S., & Le Calvez, S. (1996). The spontaneous appearance of hair cell-like cells in the mammalian cochlea following aminoglycoside ototoxicity. *NeuroReport*, 8(1), 133–137.

Romero-Carvajal, A., Navajas Acedo, J., Jiang, L., Kozlovskaja-Gumbrienė, A., et al. (2015). Regeneration of sensory hair cells requires localized interactions between the Notch and Wnt pathways. *Developmental Cell*, 34(3), 267–282.

Rubel, E. W. (1978). Ontogeny of structure and function in the vertebrate auditory system. In *Handbook of sensory physiology* (pp. 135–237), New York: Springer.

Rubel, E. W., Dew, L. A., & Roberson, D. W. (1995). Mammalian vestibular hair cellf regeneration. *Science*, 267(5198), 701–707.

Ruben, R. J. (1967). Development of the inner ear of the mouse: A radioautographic study of terminal mitoses. *Acta Oto-Laryngologica Supplementum*, 220, 1–44.

Rusch, A., Erway, L. C., Oliver, D., Vennstrom, B., & Forrest, D. (1998). Thyroid hormone receptor beta-dependent expression of a potassium conductance in inner hair cells at the onset of hearing. *Proceedings of the National Academy of Sciences of the USA*, 95, 15758–15762.

Ryals, B. M., & Rubel, E. W. (1988). Hair cell regeneration after acoustic trauma in adult Coturnix quail. *Science*, 240(4860), 1774–1776.
Ryals, B. M., & Westbrook, E. W. (1994). TEM analysis of neural terminals on autoradiographically identified regenerated hair cells. *Hearing Research*, 72(1–2), 81–88.
Ryals, B. M., Dent, M. L., & Dooling, R. J. (2013). Return of function after hair cell regeneration. *Hearing Research*, 297, 113–120.
Schimmang, T., & Pirvola, U. (2013). Coupling the cell cycle to development and regeneration of the inner ear. *Seminars in Cell and Developmental Biology*, 24, 507–513.
Schlecker, C., Praetorius, M., Brough, D. E., Presler, R. G. Jr., et al. (2011). Selective atonal gene delivery improves balance function in a mouse model of vestibular disease. *Gene Therapy*, 18(9), 884–890.
Shi, F., Cheng, Y. F., Wang, X. L., & Edge, A. S. (2010). Beta-catenin up-regulates *Atoh1* expression in neural progenitor cells by interaction with an *Atoh1* 3′ enhancer. *The Journal of Biological Chemistry*, 285, 392–400.
Shi, F., Hu, L., & Edge, A. S. (2013). Generation of hair cells in neonatal mice by β-catenin overexpression in Lgr5-positive cochlear progenitors. *Proceedings of the National Academy of Sciences of the USA*, 110(34), 13851–13856.
Shou, J., Zheng, J. L., & Gao, W. Q. (2003). Robust generation of new hair cells in the mature mammalian inner ear by adenoviral expression of *Hath1*. *Molecular and Cellular Neuroscience*, 23(2), 169–179.
Slattery, E. L., & Warchol, M. E. (2010). Cisplatin ototoxicity blocks sensory regeneration in the avian inner ear. *The Journal of Neuroscience*, 30(9), 3473–3481.
Slattery, E. L., Speck, J. D., & Warchol, M. E. (2009). Epigenetic influences on sensory regeneration: Histone deacetylases regulate supporting cell proliferation in the avian utricle. *Journal of the Association for Research in Otolaryngology*, 10(3), 341–353.
Slowik, A. D., & Bermingham-McDonogh, O. (2013). Hair cell generation by notch inhibition in the adult mammalian cristae. *Journal of the Association for Research in Otolaryngology*, 14(6), 813–828.
Sobkowicz, H. M., August, B. K., & Slapnick, S. M. (1996). Post-traumatic survival and recovery of the auditory sensory cells in culture. *Acta Oto-Laryngologica*, 116(2), 257–262.
Song, J., Yan, H. Y., & Popper, A. N. (1995). Damage and recovery of hair cells in fish canal (but not superficial) neuromasts after gentamicin exposure. *Hearing Research*, 91(1–2), 63–71.
Staecker, H., Praetorius, M., Baker, K., & Brough, D. E. (2007). Vestibular hair cell regeneration and restoration of balance function induced by *math1* gene transfer. *Otolaryngology and Neurotology*, 28(2), 223–231.
Steyger, P. S., Burton, M., Hawkins, J. R., Schuff, N. R., & Baird, R. A. (1997). Calbindin and parvalbumin are early markers of non-mitotically regenerating hair cells in the bullfrog vestibular otolith organs. *International Journal of Developmental Neuroscience*, 15(4–5), 417–432.
Stojanova, Z. P., Kwan, T., & Segil, N. (2015). Epigenetic regulation of *Atoh1* guides hair cell development in the mammalian cochlea. *Development*, 142(20), 3529–3536.
Stone, J. S., & Cotanche, D. A. (1994). Identification of the timing of S phase and the patterns of cell proliferation during hair cell regeneration in the chick cochlea. *The Journal of Comparative Neurology*, 341, 50–67.
Stone, J. S., & Rubel, E. W. (1999). Delta1 expression during avian hair cell regeneration. *Development*, 126, 961–973.
Stone, J. S., & Rubel, E. W. (2000). Temporal, spatial, and morphological features of hair cell regeneration in the avian basilar papilla. *The Journal of Comparative Neurology*, 417, 1–16.
Stone, J. S., Choi, Y.-S., Yamashita, H., Woolley, S. M. N., & Rubel, E. W. (1999). Progenitor cell cycling during hair cell regeneration in the vestibular and auditory epithelia of the chick. *Journal of Neurocytology*, 28, 863–876.
Takahashi, K., & Yamanaka, S. (2016). A decade of transcription factor-mediated reprogramming to pluripotency. *Nature Reviews Molecular and Cellular Biology*, 17, 183–193.

Taylor, R. R., & Forge, A. (2005). Hair cell regeneration in sensory epithelia from the inner ear of a urodele amphibian. *The Journal of Comparative Neurology*, 484(1), 105–120.

Taylor, R. R., Jagger, D. J., Saeed, S. R., Axon, P., et al. (2015). Characterizing human vestibular sensory epithelia for experimental studies: New hair bundles on old tissue and implications for therapeutic interventions in ageing. *Neurobiology of Aging*, 36(6), 2068–2084.

Ushakov, K., Rudnicki, A., & Avraham, K. B. (2013). MicroRNAs in sensorineural diseases of the ear. *Frontiers in Molecular Neuroscience*, 6, 52.

Wallis, D., Hamblen, M., Zhou, Y., Venken, K. J., et al. (2003). The zinc finger transcription factor Gfi1, implicated in lymphomagenesis, is required for inner ear hair cell differentiation and survival. *Development*, 130, 221–232.

Wang, T., Chai, R., Kim, G. S., Pham, N., et al. (2015). Lgr5+ cells regenerate hair cells via proliferation and direct transdifferentiation in damaged neonatal mouse utricle. *Nature Communications*, 6, 6613.

Wang, Y., Hirose, K., & Liberman, M. C. (2002). Dynamics of noise-induced cellular injury and repair in the mouse cochlea. *Journal of the Association for Research in Otolaryngology*, 3(3), 248–268.

Warchol, M. E. (1999). Immune cytokines and dexamethasone influence sensory regeneration in the avian vestibular periphery. *Journal of Neurocytology*, 28(10–11), 889–900.

Warchol, M. E., & Corwin, J. T. (1996). Regenerative proliferation in organ cultures of the avian cochlea: Identification of the initial progenitors and determination of the latency of the proliferative response. *The Journal of Neuroscience*, 16(17), 5466–5477.

Warchol, M. E., Lambert, P. R., Goldstein, B. J., Forge, A., & Corwin, J. T. (1993). Regenerative proliferation in inner ear sensory epithelia from adult guinea pigs and humans. *Science*, 259(5101), 1619–22.

Weber, T., Zimmermann, U., Winter, H., Mack, A., et al. (2002). Thyroid hormone is a critical determinant for the regulation of the cochlear motor protein prestin. *Proceedings of the National Academy of Sciences of the USA*, 99, 2901–2906.

Weisleder, P., & Rubel, E. W. (1993). Hair cell regeneration after streptomycin toxicity in the avian vestibular epithelium. *The Journal of Comparative Neurology*, 331(1), 97–110.

White, P. M., Doetzlhofer, A., Lee, Y. S., Groves, A. K., & Segil, N. (2006). Mammalian cochlear supporting cells can divide and trans-differentiate into hair cells. *Nature*, 441(7096), 984–987.

White, P. M., Stone, J. S., Groves, A. K., & Segil, N. (2012). EGFR signaling is required for regenerative proliferation in the cochlea: Conservation in birds and mammals. *Developmental Biology*, 363(1), 191–200.

Williams, J. A., & Holder, N. (2000). Cell turnover in neuromasts of zebrafish larvae. *Hearing Research*, 143(1–2), 171–181.

Winter, H., Braig, C., Zimmermann, U., Geisler, H. S., et al. (2006). Thyroid hormone receptors TRalpha1 and TRbeta differentially regulate gene expression of Kcnq4 and prestin during final differentiation of outer hair cells. *Journal of Cell Science*, 119, 2975–2984.

Witte, M. C., Montcouquiol, M., & Corwin, J. T. (2001). Regeneration in avian hair cell epithelia: Identification of intracellular signals required for S-phase entry. *European Journal of Neuroscience*, 14(5), 829–838.

Woods, C., Montcouquiol, M., & Kelley, M. W. (2004). *Math1* regulates development of the sensory epithelium in the mammalian cochlea. *Nature Neuroscience*, 7, 1310–1318.

Wu, D. K., & Kelley, M. W. (2012). Molecular mechanisms of inner ear development. *Cold Spring Harbor Perspectives in Biology*, 4, a008409.

Xu, J. C., Huang, D. L., Hou, Z. H., Guo, W. W., et al. (2012). Type I hair cell regeneration induced by *Math1* gene transfer following neomycin ototoxicity in rat vestibular sensory epithelium. *Acta Oto-laryngologica*, 132(8), 819–828.

Xu, P. X., Adams, J., Peters, H., Brown, M. C., et al. (1999). *Eya1*-deficient mice lack ears and kidneys and show abnormal apoptosis of organ primordia. *Nature Genetics*, 23, 113–117.

Yamamoto, N., Tanigaki, K., Tsuji, M., Yabe, D., et al. (2006). Inhibition of Notch/RBP-J signaling induces hair cell formation in neonate mouse cochleas. *Journal of Molecular Medicine* (Berlin), 84(1), 37–45.

Yamashita, H., & Oesterle, E. C. (1995). Induction of cell proliferation in mammalian inner-ear sensory epithelia by transforming growth factor alpha and epidermal growth factor. *Proceedings of the National Academy of Sciences of the USA*, 92(8), 3152–2155.

Yang, S. M., Chen, W., Guo, W. W., Jia, S., et al. (2012). Regeneration of stereocilia of hair cells by forced *Atoh1* expression in the adult mammalian cochlea. *PLoS ONE*, 7(9), e46355.

Zhang, N., Martin, G. V., Kelley, M. W., & Gridley, T. (2000). A mutation in the Lunatic fringe gene suppresses the effects of a *Jagged2* mutation on inner hair cell development in the cochlea. *Current Biology*, 10, 659–662.

Zhang, S., & Cui, W. (2014). *Sox2*, a key factor in the regulation of pluripotency and neural differentiation. *World Journal of Stem Cells*, 6, 305–311.

Zheng, J. L., & Gao, W. Q. (1996). Differential damage to auditory neurons and hair cells by ototoxins and neuroprotection by specific neurotrophins in rat cochlear organotypic cultures. *European Journal of Neuroscience*, 8(9), 1897–1905.

Zheng, J. L., & Gao, W. Q. (2000). Overexpression of *Math1* induces robust production of extra hair cells in postnatal rat inner ears. *Nature Neuroscience*, 3, 580–586.

Zheng, J. L., Helbig, C., & Gao, W. Q. (1997). Induction of cell proliferation by fibroblast and insulin-like growth factors in pure rat inner ear epithelial cell cultures. *The Journal of Neuroscience*, 17(1), 216–226.

Zheng, J. L., Keller, G., & Gao, W. Q. (1999a). Immunocytochemical and morphological evidence for intracellular self-repair as an important contributor to mammalian hair cell recovery. *The Journal of Neuroscience*, 19(6), 2161–2170.

Zheng, J. L., Frantz, G., Lewis, A. K., Sliwkowski, M., & Gao, W. Q. (1999b). Heregulin enhances regenerative proliferation in postnatal rat utricular sensory epithelium after ototoxic damage. *Journal of Neurocytology*, 28(10–11), 901–912.

Zheng, W., Huang, L., Wei, Z. B., Silvius, D., et al. (2003). The role of *Six1* in mammalian auditory system development. *Development*, 130, 3989–4000.

Zine, A., Aubert, A., Qiu, J., Therianos, S., et al. (2001). Hes1 and Hes5 activities are required for the normal development of the hair cells in the mammalian inner ear. *The Journal of Neuroscience*, 21, 4712–4720.

Chapter 3
The Molecular and Cellular Mechanisms of Zebrafish Lateral Line Development

Hillary F. McGraw, Catherine M. Drerup, Teresa Nicolson, and Alex V. Nechiporuk

Abstract In aquatic vertebrates, the lateral line mechanosensory system allows for sensation of changes in water current and mediates such behaviors as schooling, predator avoidance, and mating. The lateral line forms from placodes that arise just rostral and caudal to the otic placode. Shortly after placode formation, groups of cells will delaminate from the placodes and begin migrating either throughout the head or down the trunk of the developing embryo. These migratory groups of cells are known as the sensory ridges (head) and posterior lateral line primordium (trunk). During migration, they deposit cell clusters containing hair cell precursors. Shortly after deposition, these clusters will differentiate into mechanosensory organs called neuromasts. In larvae and adults, the lateral line system continues to elaborate; this is accomplished through a differentiation of latent precursors (larvae) as well as a cellular budding process (larvae and adults), resulting in strings of neuromasts that populate the body of aquatic vertebrates. The zebrafish (*Danio rerio*) has emerged as an exquisite model to study the formation and function of the lateral line system. This chapter describes the development of the zebrafish lateral line and the associated axonal innervations that make up the mechanosensory system.

H.F. McGraw
Division of Cell Biology and Biophysics, University of Missouri Kansas City, Kansas City, MO 64110, USA
e-mail: mcgrawh@umkc.edu

C.M. Drerup
Division of Developmental Biology, Eunice Kennedy Shriver National Institute of Child Health and Human Development, National Institutes of Health, Bethesda, MD 20892, USA
e-mail: katie.drerup@nih.gov

T. Nicolson
Vollum Institute, Oregon Health & Science University, Portland, OR 97239, USA
e-mail: nicolson@ohsu.edu

A.V. Nechiporuk (✉)
Department of Cell, Developmental and Cancer Biology, Oregon Health & Science University, Portland, OR 97239, USA
e-mail: nechipor@ohsu.edu

K.S. Cramer et al. (eds.), *Auditory Development and Plasticity*, Springer Handbook of Auditory Research 64, DOI 10.1007/978-3-319-21530-3_3

Keywords Cranial placodes • Development • Fibroblast growth factor signaling • Lateral line ganglion • Mechanosensory hair cells • Neuromast • Primordium • Wnt signaling

3.1 Introduction

Sensory systems such as touch, smell, vision, and hearing allow animals to perceive and interact with their environment. Fishes have a sensory system not found in terrestrial vertebrates, the lateral line, which mediates perception of changes in water current and provides sensory input that is used in behaviors such as schooling, predator avoidance, and mating (Parker 1904; Bleckmann 1993). The adult lateral line is composed of mechanosensory organs called neuromasts (NMs), arrayed across the surface of the body or enclosed in bony canals, and afferent and efferent nerves that innervate these sensory organs. Each NM is composed of mechanosensory hair cells and supporting cells. Hair cells contain actin-based mechanosensory stereocilia that convert mechanical vibrations into chemical signals transmitted by sensory lateral line nerves to the central nervous system (CNS) for further processing.

This chapter focuses on the development of the lateral line mechanosensory system of teleost fish, with emphasis on the formation of the posterior lateral line (pLL) in the zebrafish (*Danio rerio*), as the best-studied model of lateral line development and function. The chapter begins by describing development of embryonic lateral line placodes that give rise to all components of the lateral line system. Sections 3.2 and 3.3 focus on the molecular and cellular bases of posterior lateral line development; next, Sect. 3.4 describes the physiological properties of NM hair cells. Section 3.5 covers afferent and efferent innervation of hair cells and concludes by briefly touching on patterning of adult lateral line and lateral line variations in fish species other than zebrafish.

3.2 Early Development of the Lateral Line

3.2.1 Formation of Lateral Line Placodes

All components of the primary lateral line system are derived from transient embryonic structures called lateral line placodes. Cranial placodes are transient ectodermal thickenings that form at the end of somitogenesis in the anterior ectoderm. These structures give rise to or contribute to the sensory organs of the vertebrate head, including the olfactory epithelium; lens; inner ear; and, in aquatic vertebrates, the lateral line. Substantial molecular evidence shows that all cranial placodes are derived from a common region of embryonic ectoderm, termed pre-placodal ectoderm (PPE) (Ahrens and Schlosser 2005; Litsiou et al. 2005). Once formed, PPE cells undergo gradual and stepwise restriction of fate that ultimately leads to the formation of distinct placodes that give rise to sensory organs in the embryo.

The number of lateral line placodes varies between aquatic species. Phylogenetic analysis suggests that ancestral jawed vertebrates had six lateral line placodes: three preotic; the anterodorsal, anteroventral, and otic lateral line placodes; and three postotic; the middle, supratemporal, and posterior lateral line placodes (Northcutt 1989, 1997). All lateral line placodes give rise to neuroblasts that will differentiate into both peripheral sensory neurons and cells that will undergo extensive morphogenesis to develop into the sensory organs. The pioneer axons of neuroblasts remain in close contact with placodal cells as they migrate or elongate into the head (Northcutt et al. 1994; Northcutt and Brandle 1995) or trunk (David et al. 2002; Gilmour et al. 2004). In addition to afferent innervation, the trunk lateral line organs receive efferent innervation from the mid- and hindbrain (see Sect. 3.5 for more detail). In fishes other than teleosts, anterior lateral line placodes (but not posterior placodes) elongate to form sensory ridges that subsequently fragment to give rise to sensory organs (Schlosser 2002; Northcutt 2005). In teleost fishes, including zebrafish, both anterior and posterior placodes migrate into the head and trunk as cohesive groups of cells called the anterior and posterior lateral line primordia (aLLP and pLLP), respectively. The anterodorsal lateral line placode gives rise to supraorbital and infraorbital sensory organs (i.e., above and below the eye); the anteroventral lateral line placode gives rise to sensory organs of the cheek and lower jaw; and the otic lateral line placode (when present) forms NMs lateral to the otic vesicle (Northcutt 1997). The first two postotic lateral line placodes also form NMs near the otic vesicle: the middle lateral line placode forms middle and temporal lines of NMs, whereas the supratemporal lateral line placode forms supratemporal and posttemporal lines of NMs (Northcutt 1997). Zebrafish appear to have one pre- and one postotic placodes. However, zebrafish develop all three preotic and three postotic lateral lines. Thus, individual pre- and postotic placodes are either fused or hard to resolve using available molecular markers.

3.2.2 Molecular Mechanisms of Lateral Line Placode Formation

Pre-placodal ectoderm (PPE) is a horseshoe-shaped domain that abuts the neural plate and is lateral to the developing neural crest (Fig. 3.1A). High levels of fibroblast growth factor (FGF) in combination with low levels of Wnt and bone morphogenetic protein (BMP) signaling are necessary and sufficient to establish PPE cell fate (Ahrens and Schlosser 2005; Litsiou et al. 2005). FGF ligands are provided by the neural plate and underlying mesoderm. At the same time, prospective PPE cells express Wnt and BMP antagonists to inhibit these signaling pathways. Interestingly, high levels of Wnts and BMPs in the surrounding tissues are necessary for refining PPE boundaries. After the PPE is formed, it is further subdivided into discrete domains that give rise to individual placodes. Once formed, placodal

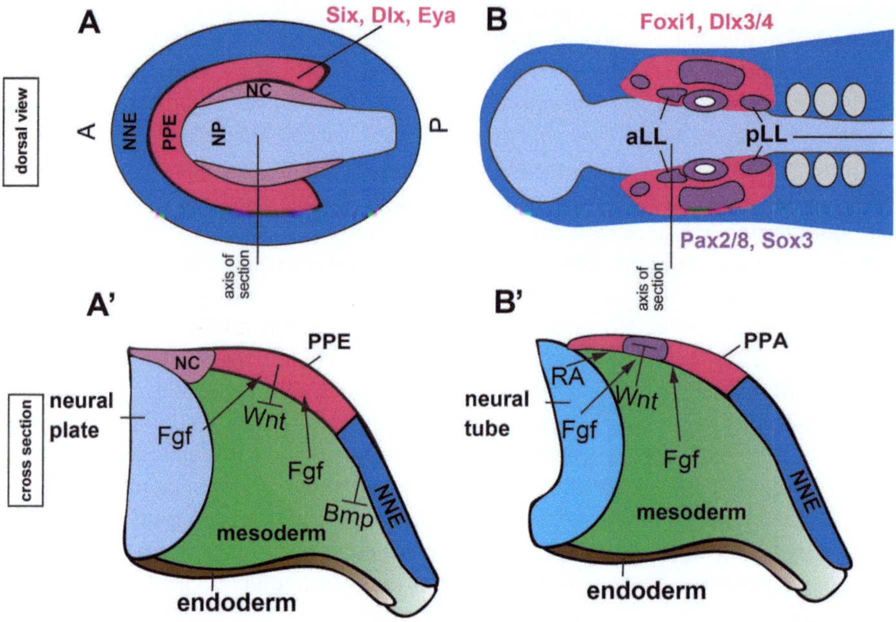

Fig. 3.1 Schematic illustration of lateral line placode development. (**A**) Dorsal view of a generalized embryo at the end of gastrulation; the horseshoe-shaped pre-placodal ectoderm (PPE) domain is depicted in magenta. This domain adjoins the neural plate (light blue) and is lateral to the forming neural crest (purple). In dark blue is the nonneural ectoderm. The PPE is defined at this stage by the expression of the factors Six, Dlx, and Eya. (**A′**) A unilateral cross section at the indicated axis and the same time point as in (**A**). The neural plate (light blue) and the head mesoderm (green) secrete fibroblast growth factors (FGFs) that induce the PPE (red); Wnt and bone morphogenetic protein (BMP) inhibitors are expressed in the PPE. (**B**) Late neurulation stage: precursors for the lateral line, otic, and epibranchial placodes (Pax2/8 and Sox3 positive; dark purple) are derived from the posterior placodal area (PPA, magenta), which is defined by the expression of Foxi1 and Dlx3/4. Inhibition of the canonical Wnt signaling in the anterior PPA biases cells toward anterior lateral line fate. (**B′**) Cross section of a late neurulation stage embryo at the indicated axis (**B**). Continued FGF signaling from the neural tube and head mesoderm together with RA from the neural plate induce the PPA domain. A = anterior; aLL = anterior lateral line precursors; NC = neural crest; NNE = nonneural ectoderm; NP = neural plate; P = posterior; pLL = posterior lateral line precursors; PPA = posterior placodal area; PPE = pan-placodal ectoderm

cells delaminate or invaginate to give rise to a variety of cell types including glia, lens fibers, mechanosensory hair cells, and neurons.

The posteriormost region of the PPE, termed the posterior placodal area (PPA), gives rise to the otic, epibranchial, and lateral line placodes (Fig. 3.1B) (Pieper et al. 2011). The same signals that promote PPE development, high FGF combined (Nechiporuk et al. 2007) with inhibition of Wnt (Freter et al. 2008), work together with retinoic acid (RA) to specify PPA cell fates (Janesick et al. 2012). Shortly after its formation, the PPA begins to express the transcription factors *foxi1* and *dlx3/4* (Lee et al. 2003; Hans et al. 2013). These transcription factors are thought to impart competence to the PPA and are necessary for PPA cells to respond to FGF signals from the neural plate and underlying mesoderm (Nissen et al. 2003; Hans et al. 2007). Specific FGF ligands vary between species; for example, in zebrafish FGF8 from the neural plate and FGF3 from the head mesoderm signal to the Foxi1-positive cells to induce two other PPA-specific transcription factors, *pax2* and *pax8* (Nechiporuk et al. 2007; Padanad and Riley 2011).

Very little is known about early steps of lateral line placode induction following formation of the Pax2/8 positive PPA. Lineage analysis in zebrafish embryos showed that the anteriormost portion of the PPA gives rise to the anterior lateral line placode (McCarroll et al. 2012; McCarroll and Nechiporuk 2013). Mosaic experiments demonstrated that cells with low or no intracellular Wnt signaling contribute to the aLL placode, while activation of the canonical Wnt signaling biases cells to the otic fate (McCarroll et al. 2012). The rostralmost cells of this region are *fgf10a*-positive and give rise to the aLLP (Nechiporuk and Raible 2008). The caudal cells express the proneural genes *neurogenin1* and *neurod* and will differentiate into neurons of the aLL ganglion (Andermann et al. 2002). Specific signals that promote development of the pLL placode are unknown. However, similar to aLL placode, anterior, *neurogenin1*- and *neurod*-positive pLL placode cells differentiate into sensory neurons of the pLL ganglia while more posterior, *fgf10a*-positive cells differentiate into pLLP (Nechiporuk and Raible 2008).

3.3 Development of the Posterior Lateral Line System

3.3.1 Primordia Migration Forms the Zebrafish Lateral Line

Although lateral line systems have been identified in species ranging from lamprey to frogs, teleost fishes, the vertebrate taxon containing the most extant members, are among the most widely studied models of aquatic mechanosensation. In the zebrafish, the lateral line is composed of two major collections of NMs, the aLL system of NMs of the head and the pLL NMs that populate the trunk. The nascent aLL and the pLL form from primordia that migrate in stereotyped manners along the surface of the embryo during the first 48 hours of development, while later migrating primordia form the more elaborate larval lateral line system (Nunez et al. 2009). A number of studies have focused on understanding development of the primary pLL, which begins to form around 20 hours post-fertilization (hpf) from the pLL placode (Sarrazin et al. 2010). The caudal portion of the pLL placode migrates toward the tail as an organized cohort of pLLP cells (Fig. 3.2A). The pLLP is a collection of approximately 100 cells made up of proliferating progenitor cells and organized epithelial cells that act as proto-NMs. As the pLLP migrates, it deposits proto-NMs every five to seven somites. Proto-NMs then differentiate into the hair cells and supporting cells of the NMs. In addition, the pLLP deposits a trail of cells between the deposited NMs, called the inter-NM cells, which serve as precursors for additional NMs that will form during larval development (Grant et al. 2005). By 48 hpf, the pLLP has reached the end of the tail, deposited approximately six NMs

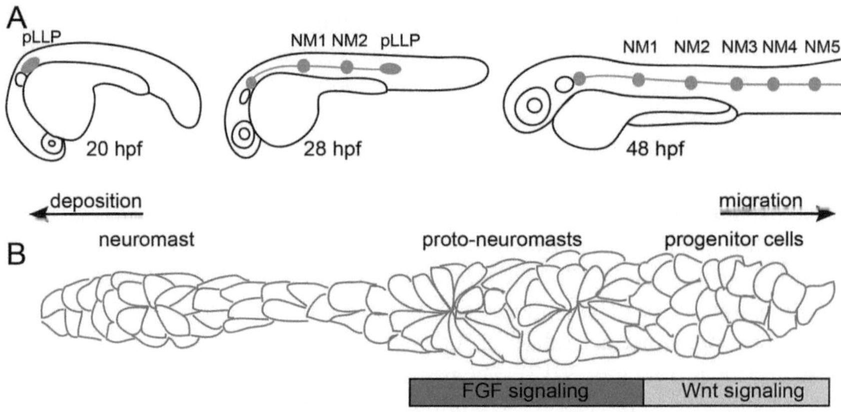

Fig. 3.2 Posterior lateral line formation in the zebrafish embryo. (**A**) Schematic showing the development of the posterior lateral line (pLL) through the migration of the pLL primordium (pLLP, gray oval). The pLLP migrates posteriorly along the trunk of the embryo between 20 and 48 hours post-fertilization (hpf), depositing approximately five neuromasts (NM 1–5, gray circles) and forming the terminal cluster (TC) at the end of the tail. (**B**) Schematic of the pLLP, a cohort of approximately 100 cells that collectively migrate to form the pLL. The leading region of the pLLP contains proliferating progenitor cells that are regulated by Wnt signaling; the trailing aspect of the pLLP contains apically constricted epithelial cells (also called proto-NMs) that are regulated by fibroblast growth factor (FGF) signaling. Mature NMs are deposited from the trailing, anteriormost aspect of the pLLP

along the trunk, and fragmented into a terminal cluster of two or three terminal NMs. As the fish increase in size and maturity, NMs bud to give rise to the elaborated adult lateral line system.

3.3.2 Organization of the pLLP

The initial NMs of the primary trunk lateral line are derived from the collectively migrating pLLP cells. During migration, mature NMs are deposited from the trailing, anteriormost region of the pLLP. Concomitantly, immature NMs are formed in the mid-region of the pLLP from polarized epithelial cells that undergo coordinated apical constriction and form the 3-4 proto-NMs within the pLLP (Ghysen and Dambly-Chaudiere 2007; Harding et al. 2014) (Fig. 3.2B). The leading, posteriormost region of the pLLP is composed of nonpolarized proliferating progenitor cells that will give rise to new proto-NMs and leader tip cells that display extensive protrusive activity and appear to direct the collective migration of the pLLP (Haas and Gilmour 2006; Nechiporuk and Raible 2008). As the pLLP migrates and deposits mature NMs, cells within the pLLP polarize and apically constrict in a reiterative manner, giving rise to the nascent trunk pLL system (Thomas et al. 2015). The cellular and molecular bases for this process are explored in more detail in Sects. 3.3.3 and 3.3.4.

3.3.3 Progenitor Cell Identity and Maintenance

Although the pLLP contains approximately 100 cells at any given time during its migration, by 48 hpf the nascent pLL comprises of approximately 300 cells (Laguerre et al. 2009), indicating that cellular proliferation is required for pLL organogenesis. Lineage analysis and cell tracing studies showed that the proliferating cells of the leading region of the pLLP give rise to the rosettes and eventually mature NMs that will form the terminal cluster (Nechiporuk and Raible 2008; McGraw et al. 2011). At the same time, proliferating cells across the pLLP contribute to formation of additional rosettes that will form NMs proximal to the terminal cluster (Nogare et al., 2017). The leading region of the pLLP is characterized by high expression of members of the canonical Wnt pathway, such as *lef1* and *axin2* (Fig. 3.2B). The formation of the pLL appears to be exquisitely sensitive to alterations in Wnt signaling. Conditional over activation of canonical Wnt signaling results in excessive proliferation and incomplete migration of the pLLP, possibly due to loss of proper pLLP patterning (Aman et al. 2010). Inhibition of the Wnt/β-catenin pathway at the level of ligand binding results in severe defects in pLLP migration, patterning, and NM deposition owing to a dramatic decrease in proliferation and an increase in cell death (McGraw et al. 2011). In particular, members of the family of secreted Wnt inhibitors, Dkk1b and Dkk2, regulate the extent of Wnt activity in the pLLP (McGraw et al. 2014). Mutants that lack Kremen1 function, a nonobligate receptor for Dkk proteins, display a dramatic decrease in cellular proliferation and survival in the pLLP resulting in truncated pLL formation. The defect can be rescued by inhibiting Dkk1b and Dkk2 function, suggesting that Kremen1 regulates the extent of Dkk activity in the pLLP (McGraw et al. 2014). In contrast, mutations in the Wnt effector protein Lef1 ultimately result in pLL truncation and incomplete NM deposition due to loss of the leading progenitor cell identity, as cells remain proliferative, but prematurely exit the leading region and are incorporated into NMs (McGraw et al. 2011). In addition, this disruption in *lef1* may also impact the NM deposition rate (Valdivia et al. 2011; Matsuda et al. 2013). The difference between global Wnt disruption at the level of ligand binding and the *lef1* mutant phenotype suggests that canonical Wnt signaling has multiple downstream effectors in the pLLP. In support of this view, Wnt signaling is required for initiation of FGF signaling (Aman and Piotrowski 2008), another molecular pathway essential for pLLP patterning (discussed in Sect. 3.3.4).

3.3.4 FGF Signaling and Proto-NM Formation

As NMs are reiteratively deposited from the trailing region of the pLLP, new proto-NMs are generated in the central portion of the pLLP rostral to the leading progenitor zone (Nechiporuk and Raible 2008). During proto-NM formation, cells undergo apicobasal polarization, concurrent with alignment of apical domains along the midline of the pLLP. The apical domains constrict in a coordinated manner, giving rise to a nascent proto-NM, which takes on a rosette-like shape (Lecaudey et al. 2008; Harding and Nechiporuk 2012). As mature NMs are deposited, newly formed proto-NMs inhabit the trailing region of the pLLP. Proto-NM renewal in the pLLP is dependent on FGF signaling (Lecaudey et al. 2008; Harding and Nechiporuk 2012) (Fig. 3.2B). In the absence of FGF signaling, rosettes fail to form and migration of the pLLP stalls. Ras–MAPK signaling is an important intracellular mediator of FGF signaling driving rosette formation, activating Rho kinase–dependent myosin II activity (Harding and Nechiporuk 2012). Activation of FGF signaling promotes transcription of the scaffolding molecule Shroom3, which is required for the apical constriction of newly formed rosettes (Ernst et al. 2012). Together, these studies suggest a model in which FGF–Ras–MAPK signaling causes transcriptional activation of Shroom3. Newly generated Shroom3 recruits Rho kinase to the apical end of columnar cells, driving activation of myosin II and subsequent apical constriction. Further rosette maturation occurs as Cadherin2 is redistributed to apical adherens junctions in a microtubule-dependent manner (Revenu et al. 2014). Recent work elucidated a novel mechanism by which FGF signals locally to rosette cells: FGF3 is sequestered to the apical lumen of the rosette for efficient FGF signaling (Durdu et al. 2014). In addition, FGF ligands interact with heparin sulfate proteoglycans to restrict FGF signaling to the pLLP (Venero Galanternik et al. 2015).

3.3.5 Regulation of NM Deposition and Spacing

NMs are deposited at regular intervals, every five to seven somites, along the trunk of the zebrafish embryo between 24 and 48 hpf. However, the molecular mechanisms that regulate this periodicity are still being investigated. Reduction of FGF signaling, chemically or genetically, results in deposition of fewer NMs farther apart along the trunk (Nechiporuk and Raible 2008). However, because FGF signaling is also required for formation of rosettes, the NM deposition defect may be secondary. Recently, several groups have suggested a direct link between the rate of proliferation in the pLLP and NM deposition (Laguerre et al. 2009; Aman and Piotrowski 2011). Wnt/β-catenin regulated proliferation in the leading region causes continual addition of cells to the pLLP, resulting in pLLP lengthening and displacement of trailing NMs (Aman et al. 2011). More recently, it has been suggested that Wnt and FGF signaling pathways coordinate NM deposition (Matsuda et al. 2013), through the Wnt target factors, *lef1* and *dusp6* (a negative regulator of the FGF pathway).

3.3.6 Directional Migration

Chemokine signaling coordinates the directional and consistent migration of the pLLP along the trunk midline (Bussmann and Raz 2015). The chemokine ligand *cxcl12a* is expressed along the horizontal myoseptum of the developing somites, while two chemokine receptors, *cxc4b and cxcr7b,* are expressed in the leading and trailing portions of the pLLP, respectively. Disruption of either the chemokine receptors or their ligand results in stalling of pLLP migration, suggesting that both receptors are necessary. Anteriorly localized Cxcr7b acts as a Cxcl12a sink that binds ligand without activation of downstream signaling pathways, accounting for directional migration of the pLLP despite homogenous *cxcl12a* expression along the trunk midline (Dona et al. 2013). Recent work suggests that while *cxcr4b* and *cxcr7b* act to direct collective migration of the pLLP, additional activity though FGF responsive cells in the trailing region is also required for proper migration (Dalle Nogare et al. 2014).

3.3.7 Hair Cell Specification and Differentiation

Hair cell specification and differentiation are directed by the interplay of the Wnt, FGF, and Notch signaling pathways (Matsuda and Chitnis 2010). Wnt signaling in the leading region of the pLLP leads to expression of *fgf10a* and *fgf3*, resulting in FGF-mediated signaling through *fgfr1*, which is expressed in the maturing rosettes (Thomas et al. 2015). As the proto-NMs mature, FGF drives expression of hair cell-specific genes, including *atoh1a* and *delta,* that are progressively restricted from several cells to one or two cells at the center of the rosette by the process of Delta/Notch-mediated lateral inhibition (Itoh and Chitnis 2001; Matsuda and Chitnis 2010). Hair cell precursor specification is FGF-dependent, as loss of FGF results in absence of *atoh1a*-positive hair cell precursors in the pLLP (Nechiporuk and Raible 2008). In contrast, disruption of the Wnt pathway in *lef1* mutants does not alter hair cell precursor specification, though it does result in a truncated pLL (McGraw et al. 2011; Valdivia et al. 2011). Once the NM is deposited, an individual hair cell precursor gives rise to two HCs of opposing polarity with respect to the orientation of the kinocilia and stereocilia. These cells arise through oriented cell division, coordinated by the planar cell polarity and Notch pathways (Lopez-Schier and Hudspeth 2006). Thus, the stereocilia and kinocilia of hair cells within individual NMs are organized in a pattern of mirror symmetry so that each NM contains two populations of cells that respond to specific directional water flow (Mirkovic et al. 2012).

3.4 Mechanosensory Hair Cells of the Lateral Line

3.4.1 Development and Physiology

Each mature NM is composed of approximately 20–30 mechanosensory hair cells, which are morphologically and functionally similar to the hair cells of the vestibular/auditory system in all vertebrates (Whitfield 2002) (Fig. 3.3A–C). After deposition of proto-NMs by the pLLP, hair cell progenitors begin dividing and postmitotic

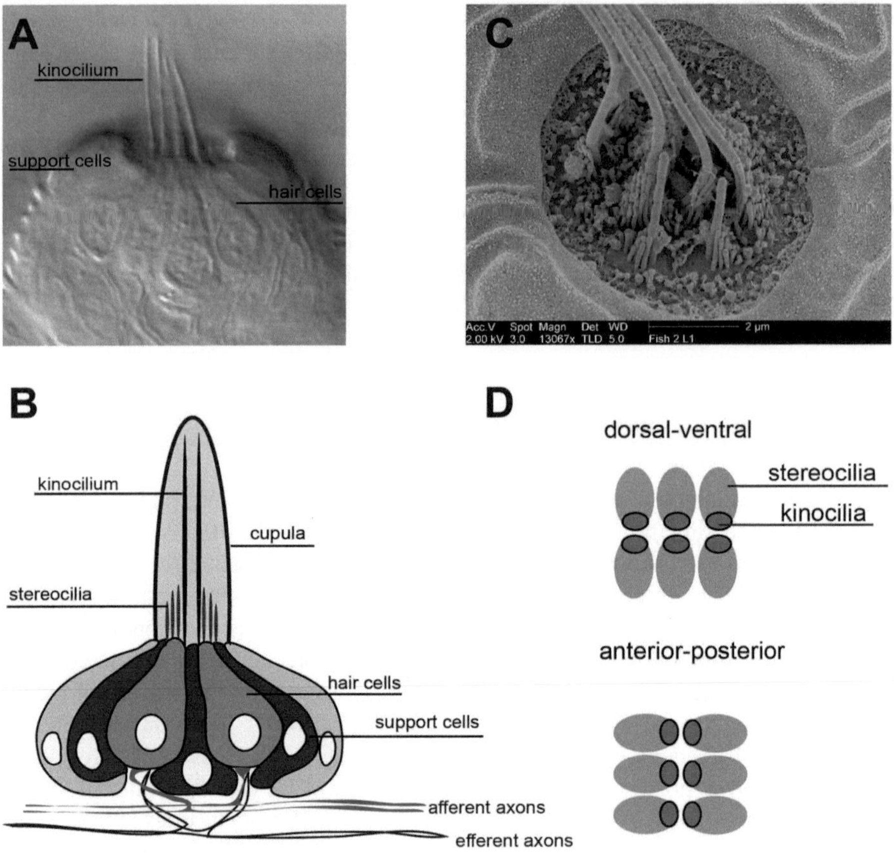

Fig. 3.3 Mechanosensory neuromast of the lateral line organ. (**A**) DIC image of a zebrafish neuromast showing the goblet-shaped hair cells, the kinocilia and supporting cells. (**B**) Representative schematic of an idealized neuromast showing hair cells extending several short stereocilia and a long kinocilium (encased in the gelatinous cupula) into the aqueous environment to sense changes of water current. Hair cells are basally innervated by afferent and efferent axons and are surrounded by supporting cells. (**C**) SEM of an apical top-down view of a neuromast showing stereocilia and longer kinocilia. Note younger hair cells in the foreground; these contain immature, shorter stereocilia and kinocilia. (**D**) Schematic showing a top-down view of the orientation of stereocilia and kinocilia in dorsoventral and anterior–posterior sensing neuromasts. (DIC image courtesy of Itallia Pacentine; SEM image courtesy of Katie Kindt)

hair cells of the lateral line organ differentiate and mature rapidly within one day. Much of the understanding of the maturation of the lateral line hair cells in zebrafish is garnered from scanning electron micrographs of hair bundles and live imaging of developing hair cells in larvae. The first step of hair cell development involves the separation of daughter cells and formation of an apical surface (Lopez-Schier and Hudspeth 2006; Mirkovic et al. 2012). Division of progenitors occurs sequentially on 2 days post-fertilization (dpf) and slows to one or two divisions per day during larval stages. Soon afterwards, a nonmotile primary cilium known as a kinocilium emerges from the center of the apical surface of each newborn cell (Kindt et al. 2012) (Fig. 3.3A, C). Unlike hair cells in the fish inner ear, the apical surface of nascent lateral line hair cells is not covered by numerous microvilli. Rather, a few (typically four to five) actin-filled processes called stereocilia arise around or to one side of the thick stub of the kinocilium (Kindt et al. 2012) (Fig. 3.3C). At this early stage of development, extracellular links spanning from the tips of these processes to the kinocilium are present. As the kinocilium grows taller, more stereocilia are added, shifting the kinocilium gradually to one side of the apical surface. The opposing planar polarity of the hair bundles of the two daughter cells can be clearly seen at this early stage (Fig. 3.3C, D). When the kinocilium is nearly twice as tall as the stereocilia, the closest and oldest group of stereocilia directly adjacent to the kinocilium begins to lengthen (Kindt et al. 2012). During these intermediate stages of development as the kinocilium quickly outpaces the growth of the stereocilia, a staircase arrangement of the initial rows of stereocilia forms. Tip links running from shorter to taller stereocilia become apparent during this phase. Eventually, each hair cell generates up to five rows of stereocilia, all of which appear to be interconnected with tip links (Fig. 3.4). The kinocilium grows to a final height of around 15 μm, whereas the tallest stereocilia are 1–2 μm tall in lateral line hair cells of larvae.

The entire process of hair bundle development requires less than 24 hours (Lopez-Schier and Hudspeth 2006; Kindt et al. 2012). A long kinocilium relative to the bundle of stereocilia is reminiscent of the ampullary hair cells in semicircular canals, which also sense the motion of the overlying membranous cupula caused by the movement of fluid inside the canals. The cupula consists of a gelatinous material that completely covers the NM hair bundles, extending several microns beyond the tips of the kinocilia (Fig. 3.3B).

Mechanotransduction, which is the conversion of mechanical energy into an electrical signal, can be measured in hair cells of the zebrafish lateral line organ using several different methods. A recent study characterized individual hair cells using a combination of in vivo imaging and scanning electron microscopy, and was able to correlate morphology to the onset of evoked calcium transients (Kindt et al. 2012). Nascent hair cells of the zebrafish lateral line organ begin transducing mechanical stimuli at very early stages of development (within several hours of forming an apical surface), when only kinocilial links are visible. Surprisingly, the initial responses to stimuli are inhibitory, that is, when the fledgling structure is deflected toward the stereocilia. This response is in contrast to mature hair cells, where deflection of the hair bundle towards the stereocilia results in hyperpolarization. This early reversed response in nascent hair cells requires kinocilial links and transduction channels as evoked calcium transients are absent in *cdh23* or *pcdh15a* mutants that lack both kinocilial and tip links and no longer detectable in hair cells treated with known transduction channel blockers (Kindt et al. 2012) (Fig. 3.4). At a later stage in development, when more rows of stereocilia are present, the immature bundle responds not only to an inhibitory stimulus, but also to a positive one. The hair cells are bidirectional and can change their response pattern in a matter of minutes. Eventually the response to an inhibitory deflection disappears as the cells mature (Kindt et al. 2012).

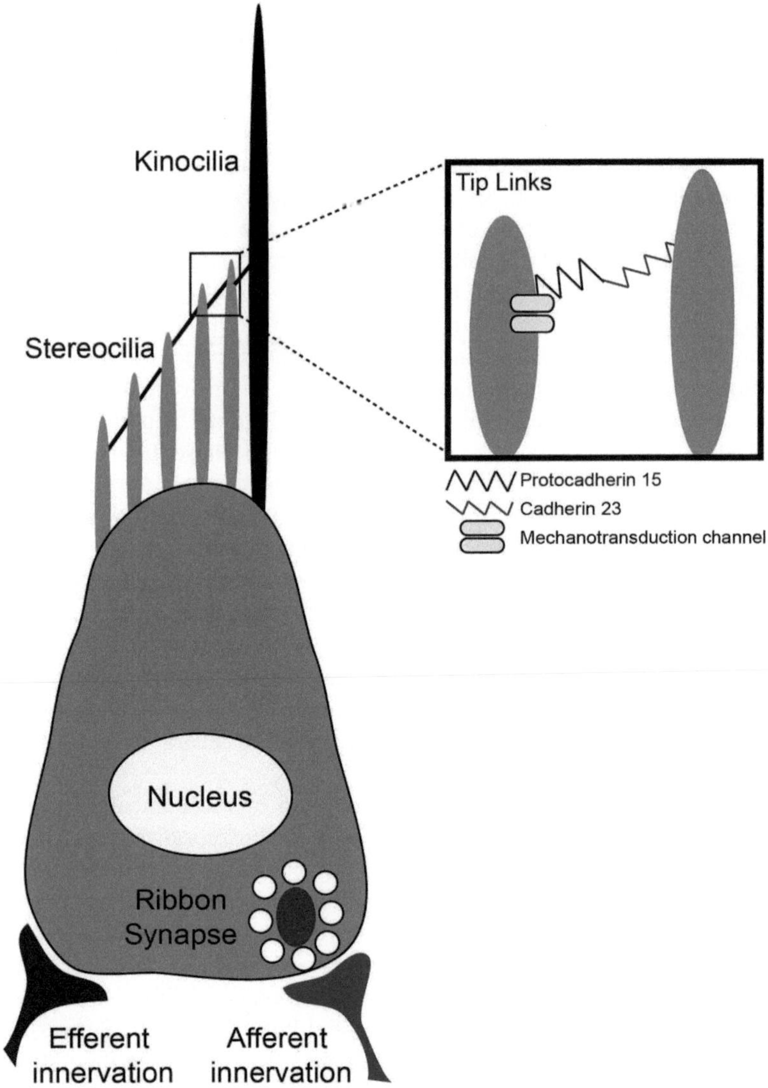

Fig. 3.4 Morphology of mechanosensory hair cells. Schematic of a generalized hair cell showing the short apical stereocilia and the long kinocilium. The nucleus localizes on the basal side of the cell, as do ribbon synapses and afferent/efferent innervating axons. **Inset**: Schematic showing the tip links between the stereocilia and kinocilum. Protocadherin 15 and cadherin 23 form the tip links. When the hair cell bundles are displaced, the mechanotransduction channel is activated, thus mediating hair cell depolarization

As the apical hair bundles mature, the basolateral surface of the hair cells receive innervation by the lateral line nerve (Ghysen and Dambly-Chaudiere 2004). Like photoreceptors and bipolar cells of the retina and photoreceptors of the pineal gland, lateral line hair cells possess ribbon synapses (Fig. 3.4). These excitatory synapses consist of a dense body surrounded by synaptic vesicles that is attached to the presynaptic membrane opposite of a postsynaptic density in the afferent fiber (Sidi et al. 2004; Sheets et al. 2012). Initially, the major protein constituent of the ribbon, Ribeye a and b (the gene is duplicated in zebrafish), is detectable as cytoplasmic particles throughout the soma of inner ear hair cells as early as 22 hpf. Within several hours, the particles coalesce and attach to various sites at the basolateral plasma membrane (Sheets et al. 2011). The afferent fibers penetrate into the neuroepithelium during this phase and make stable contacts with hair cells around 12 hours post-mitosis (Dow et al. 2015). During the early phase of innervation, hair cells appear to create basal projections that reach out to nearby afferent fibers, which in turn extend toward the projections (Dow et al. 2015). The hair cell projections then contract once a stable contact has been made. Eventually each NM obtains an extensive web of afferent fibers just below the hair cell layer, with each hair cell making up to three afferent synapses (Nagiel et al. 2009; Sheets et al. 2011; Pujol-Marti et al. 2014).

Release of neurotransmitter at ribbon synapses occurs as a result of depolarization of the soma. As with other ribbon synapses, calcium influx at presynaptic sites is mediated by voltage-gated calcium channels (Matthews and Fuchs 2010). The particular type of channel in lateral line hair cells is the L-type Cav1.3a channel, which is closely associated with the ribbon (Sidi et al. 2004; Sheets et al. 2012). Interestingly, there appears to be an interdependent relationship between the activity of Cav1.3a channels and the ribbon itself (Sheets et al. 2011, 2012). On attachment of the ribbon to the plasma membrane, Cav1.3a channels appear to cluster within the active zones (Sheets et al. 2011). When Ribeye b is exogenously overexpressed in hair cells, ectopic ribbons can be found inside the soma, either near a plasma membrane or some other membrane, such as the outer membrane layer of the nucleus or mitochondria (Sheets et al. 2011). Wherever a large aggregate of Ribeye is visible, clusters of Cav1.3a are also detectable. Thus, ribbons appear to target and cluster Cav1.3a channels at membrane interfaces.

Conversely, Cav1.3a channels can affect the size, shape, and number of ribbons (Sheets et al. 2012). In the absence of Cav1.3a channels, the ribbons initially increase in size (3 dpf), but at 5 dpf ribbon size is not significantly different. In contrast, if the mutant Cav1.3a protein is stable and present at the active zone, as is the case of the R284C mutation that renders the channel inactive, then the ribbons remain enlarged in mature hair cells. In both the null and "channel dead" *cav1.3a* mutants, most of the ribbons take on a more amorphous and less spherical shape than wild-type ribbons. During early development, ribbon synapses also contain higher numbers of ribbons in *cav1.3a* mutants. These results can be mimicked with the Cav1.3 antagonist isradipine (Sheets et al. 2012), suggesting that calcium influx can influence the assembly of Ribeye proteins.

Synaptic transmission can be assessed by recording the spiking activity of the afferent neurons (Obholzer et al. 2008; Trapani and Nicolson 2010). Loose patch recordings of spiking activity of afferent neurons have been extremely helpful in determining how release of neurotransmitter at the ribbon synapses is affected in various mutants. In some mutants such as *cav1.3a* and *vesicular glutamate transporter 3* (*vglut3*) mutants, the deficits are quite severe; the mutant synapses are essentially silent (Obholzer et al. 2008; Sheets et al. 2012). Despite the lack of synaptic transmission in *vglut3* mutants, the ribbon synapses are relatively normal, showing a small increase in the postsynaptic density (visualized with pan-MAGUK antibodies; Sheets et al. 2012). Indeed, innervation appears to be normal in various mutants that disrupt synaptic transmission in lateral line hair cells (Sidi et al. 2004; Nagiel et al. 2009). This phenotype suggests that the development of ribbon synapses largely does not depend on the release of neurotransmitter. However, there is evidence to support the role of hair cell activity in synaptogenesis (Faucherre et al. 2009), and also evidence for nerve fiber interactions (Pujol-Marti et al. 2014). Thus, synaptogenesis may be in part hard wired, but also dependent on interaction with or between the afferent neurons. The molecular mechanisms of synapse formation in lateral line hair cells remain to be determined.

3.5 Posterior Lateral Line Innervation

3.5.1 Pioneer Axon Extension and Pathfinding

The posterior lateral line ganglion (pLLg) is a collection of neuronal cell bodies situated caudal to the developing ear in embryonic and larval zebrafish. This group of cells is derived from the rostral part of the pLL placode while the pLLP discussed in Sect. 3.2.1 arises from the caudal portion of this ectodermal structure. As the pLLP migrates away to form the sensory organs, a group of approximately 20 cells is left behind that differentiate into the sensory neurons of the primary pLL system. Afferent axons extend from the pLLg and form the pLL nerve. This nerve develops in a somewhat unusual fashion: unlike other systems in which axons grow toward targets sometimes a great distance away using landmarks, the axons of the pLL nerve use guidance cues they receive directly from cells of the pLLP (Metcalfe 1985). Consequently, full axon extension in this system relies on complete pLLP migration (Fig. 3.5A, B).

The first afferent axons to extend from the pLLg are referred to as pioneer axons. These axons extend as the pLLP begins migrating, with initial extension visible at 22 hpf and full extension completed when the pLLP reaches the tail at 48 hpf (Fig. 3.5B). The growth cones of these axons maintain contact with the central and trailing portion of the pLLP (Metcalfe 1985), which expresses various growth factors and cell adhesion molecules. Several neurotrophin receptors are expressed by the pLLg neurons including TrkB (Martin et al. 1995), Ret (Schuster et al. 2010),

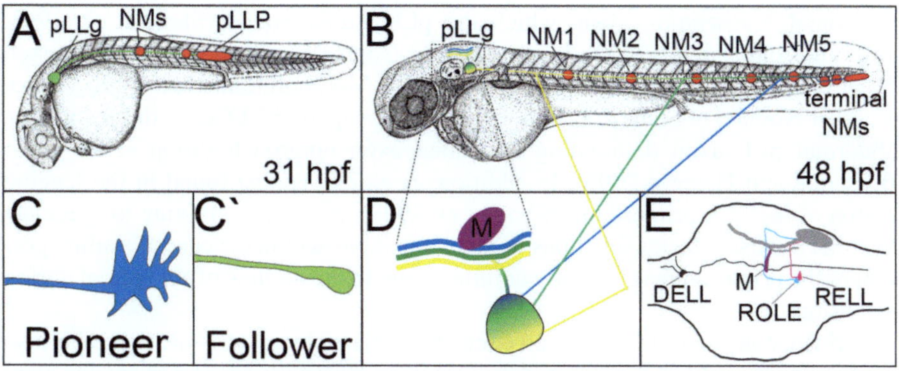

Fig. 3.5 Topography and somatotopy of posterior lateral line (pLL) innervation. (**A**) Schematic of a 31 hours post-fertilization (hpf) embryo in which the pLL pioneer axons have extended approximately halfway through the trunk. The pLL axons and posterior lateral line ganglion (pLLg) are shown in green in this image and neuromasts (NM) and the migrating pLL primordium (pLLP) are shown in red. (**B**) By 48 hpf, pioneer axons have reached the tail and central projections have reached the medial octavolateralis nucleus (MON). Gradient of yellow to blue in the pLLg, axons, and CNS illustrate the innervation of posterior neuromasts (NMs) by dorsal pLLg neurons that project into the dorsal MON (**B, D**). (**C**) During initial axon extension, axons with growth cones in the pLL primordium (pioneer axons) have elaborated growth cones (**C**). They will go on to innervate caudal NMs. (**C′**) Follower axon growth cones are more bulbous and these projections will innervate more rostral NMs. These central projections are more likely to contact the Mauthner neuron dendrite (purple in **D**). (**E**) The efferent nuclei (diencephalic efferent neurons to the lateral line [DELL], rhombencephalic octavolateral efferent neurons [ROLE], and rhombencephalic efferent neurons to the lateral line [RELL]) are depicted in this dorsal view. The DELL (black) lies in the diencephalon while the ROLE (blue) and RELL (pink) lie in the hindbrain. The Mauthner neuron (M) is shown in purple. The pLLg and the central projections from this ganglion are depicted in gray

GFRα (GDNF family receptor alpha; Shepherd et al. 2004) and p75 (Brosamle and Halpern 2009) receptor tyrosine kinases. Brain-derived neurotrophic factor (BDNF), glial cell line derived neurotrophic factor (GDNF), and nerve growth factor (NGF), their respective ligands, are present in the migrating pLLP (Shepherd et al. 2004; Germana et al. 2010). Work from Schuster and colleagues demonstrated GDNF-Ret signaling is necessary for axon extension during both development and regeneration (Schuster et al. 2010). Loss of this neurotrophin–receptor pair by a combined morpholino knockdown resulted in pLL axon truncation and failed reinnervation following nerve injury (Schuster et al. 2010). Anecdotal evidence also exists for BDNF–TrkB signaling in pLL axon extension: loss of TrkB in growth cones of pLL axons correlates with axon truncation (Drerup and Nechiporuk 2013). A direct causal relationship between TrkB–BDNF signaling and axon extension in this system has, however, not been investigated. NGF–p75 signaling during pLL axon outgrowth has also not been explored.

Similar to other systems, repulsive factors and cell–cell contacts also contribute to guidance of extending pLL axons. In the case of the pLL, two repulsive factors, Nogo and Semaphorins, have been identified in the musculature surrounding the horizontal myoseptum, along which the pLL axons extend (Shoji et al. 1998; Brosamle and Halpern 2009). In addition, the Nogo receptor, Nogo66, is expressed by pLL neurons (Brosamle and Halpern 2009). Morpholino mediated knockdown of either Nogo66 or SemaZ1a (orthologous to Semaphorin III/Collapsin) resulted in abnormal pLL axon pathfinding or stalled axon outgrowth (Shoji et al. 1998; Brosamle and Halpern 2009). In addition, SemaZ1a is also found in the leading region of the pLL primordium (Shoji et al. 1998), making it tempting to speculate that this repulsive factor functions to maintain axons within the central/trailing portion of the primordium as well as confining them to the midline of the animal during extension.

Evidence indicates that cell–cell contacts mediated by glycosylated proteins also mediate proper pLL axon outgrowth. One such modification, important for axon outgrowth in many systems, is the glycoepitope Human Natural Killer 1 (HNK-1). Injection of an antibody raised against HNK-1 that hinders interaction with this molecule results in severe truncation of pLL axon outgrowth with no effect on pLL primordium migration or pLL system patterning (Becker et al. 2001). As HNK-1 coats the surface of pLL axons (Metcalfe et al. 1990), and modulates homophilic interactions with NCAM (Cole and Schachner 1987), an adhesion molecule found in the pLLP (Becker et al. 2001), it is likely that a third mechanism of pLL axon pathfinding is through regulated homophilic interactions of glycosylated adhesion proteins during axon outgrowth.

3.5.2 Peripheral Topography and Central Somatotopy of the pLL

Neurons that compose the pLLg are highly similar in their general projection patterns and neurotransmission. They innervate planar polarized hair cells along the trunk, extend central projections into the medial octavolateralis nucleus (MON), and receive glutamatergic neurotransmission from sensory hair cells (Fig. 3.5D). However, in order for the system to functionally distinguish between stimuli that require a rapid escape response versus those necessary for schooling, prey, or mate proximity detection, differential activity among lateral line NMs must be interpreted. This sensory processing is at least partially achieved through differential input from NMs along the body axis to the CNS, leading to disparate responses.

A number of elegant studies have mapped both the topography of the pLLg innervation of NMs along the rostral–caudal axis as well as their central projections. The interpretation of these pLLg mapping studies is complicated by there being progressive neurogenesis in this ganglion between 24 and 72 hpf. During early stages (before 72 hpf), topographical mapping of sensory innervation of the pLLg is not clear (Gompel et al. 2001; Sato et al. 2010), whereas after early neurogenesis and subsequent axon extension is complete, rostral–caudal NM innervation can be mapped onto the pLLg (Pujol-Marti et al. 2010, 2012). These studies, which have been compiled over the last 30 years, show that pLLg organization correlates with the localization of the growth cone in relation to the migrating pLLP, the complexity of the growth cone of that axon and the NM that will eventually be innervated.

Pioneer axons of the pLL nerve have elaborate growth cones (Gompel et al. 2001; Sato et al. 2010). These axons extend into the caudal trunk and tail to innervate the distal NMs (Gompel et al. 2001; Pujol-Marti et al. 2010). The neurons that extend these axons are some of the first born, and their cell bodies are located dorsally in the pLLg at 6 dpf (Pujol-Marti et al. 2010). Structurally, dorsal neurons have much larger cell somas than neurons that innervate more rostral sensory organs that differentiate later in development (Pujol-Marti et al. 2010). Neurons that extend pioneer axons project centrally along the path taken by the ninth cranial nerve (Dorsky et al. 2000) and bifurcate in the MON. Compared to the central projections of other pLLg neurons, the pioneer neuron projections lie dorsalmost in the lateral column of the MON (Alexandre and Ghysen 1999; Pujol-Marti et al. 2012) (Fig. 3.5D). Interestingly, these central axons, which relay signals from distal NMs, are more likely to be in close contact with the dendrite of the Mauthner neuron (Alexandre and Ghysen 1999; Pujol-Marti et al. 2012), whose activity directly dictates the rapid C-bend escape response (Eaton et al. 1988). The fast escape response, which does not rely on central processing but rather an immediate motor response, may thus be predominantly driven by signals received by the large, dorsal pLLg neurons from the caudal region of the animal.

As mentioned in Sect. 3.5.1, rostral NMs are more likely to be innervated by pLLg neurons that differentiate later and lie in the ventral portion of the ganglion (Pujol-Marti et al. 2010, 2012) (Fig. 3.5D). During extension, their axons have much simpler, bulbous growth cones and are typically referred to as follower axons (Fig. 3.5C′). Rather than responding to primordium cues in order to extend, these axons likely use the pioneer axons as a scaffold for elongation and pathfinding. The central projections of these later born neurons extend ventral to those of the earlier born neurons in the MON, making them less likely to contact the Mauthner neuron (Pujol-Marti et al. 2012).

3.5.3 The Unexplored Role of Efferents

Whereas much attention has been paid to the mechanisms of afferent axon synaptogenesis and function, efferent innervation of the pLL has gone largely unexplored. Early work on this system revealed the presence of three nuclei in the CNS that contain efferent neurons innervating pLL NMs (Fig. 3.5E). Two of these nuclei are situated in the hindbrain and are referred to either as the rhombencephalic octavolateral efferent neurons (ROLE) and rhombencephalic efferent neurons to the lateral line (RELL) or the rostral efferent nucleus (REN) and caudal efferent nucleus (CEN) respectively (Metcalfe 1985; Bricaud et al. 2001). These nuclei are cholinergic and act to suppress afferent activity. The axons of the RELL neurons project rostrally and then turn to exit the CNS at roughly the same axial level as the pLL afferents enter. ROLE neurons project axons laterally, and they exit the CNS along the eighth cranial nerve. Neurons in both the RELL and ROLE innervate all ipsilateral NMs (Metcalfe 1985). The third efferent nucleus that innervates the NMs of the lateral line is the diencephalic efferent neurons to the lateral line (DELL). As their name suggests, these neurons lie in the ventral diencephalon (Metcalfe et al. 1985). DELL neurons are tyrosine hydroxylase positive, and are likely dopaminergic (Bricaud et al. 2001). Axons of the DELL neurons exit the CNS and innervate all ipsilateral NMs by 72 hpf. The function of efferent NM innervation is unclear but, as each type of neuron innervates all NMs, it is thought that these nuclei function as global regulators of afferent excitability in the pLL, fine tuning responses.

3.5.4 Innervation of Planar Polarized Hair Cells

As described in Sect. 3.3.7, in each NM hair cells are planar polarized across the sensory organ midline. Half of the hair cells are sensitive to movement of water in one direction, while the opposite half respond to polar opposite currents (Nagiel et al. 2008, 2009; Faucherre et al. 2009). Whereas primary NMs are sensitive to water movement in the rostral–caudal axis, secondary NMs (those laid down by the secondary primordium; see Sect. 3.6) are polarized to respond to water movement in the dorsoventral water column. Innervation of hair cells begins soon after the deposition of NMs. Swollen regions corresponding to axon terminal boutons are apparent morphologically in NMs as early as 2.5 dpf, soon after hair cells are identifiable by expression of the ET4 transgene (Nagiel et al. 2008). A single afferent axon can innervate several adjacent NMs but will make stable contacts only with hair cells that are similarly polarized (Faucherre et al. 2009; Nagiel et al. 2009). Amazingly, this polarized innervation is maintained on hair cell regeneration after ablation, arguing for strict control of afferent contacts within subregions of each NM (Faucherre et al. 2009). How afferent axons make unique contacts with hair cells of a particular polarity remains an open question.

One common mechanism of synapse formation and stabilization is synaptic activity. In contrast to this model, afferent innervation of polarized hair cells in the pLL does not require evoked or spontaneous synaptic release (Nagiel et al. 2009). If not activity, what controls the polarized sensory organ innervation? Molecular differences between hair cells of opposite polarities must exist that allow axons to differentiate between them before innervation. As hair cells and the afferent axons that innervate them arise from the same tissue, the posterior lateral line placode, it was tempting to speculate that a lineage relationship may play a role in differential synaptogenesis. Fate mapping of placodal offspring, however, revealed that unlike some insect mechanosensory organs, sensory neurons and the sensory hair cells they innervate are not related by lineage (Gompel et al. 2001).

Although the molecular differences are as of yet unidentified, the mechanism of cell contact sorting is beginning to be unraveled. Recent work by Dow and colleagues discovered the presence of contacts early in synapse formation between afferent axons and hair cells in NMs (Dow et al. 2015). Using serial block face microscopy, they revealed polarized afferent innervation as early as 2–15 hours after hair cell division, indicating that polarity specificity arises early after hair cell differentiation. Though unable to shed light on how polarized innervation originates early in pLL development, that is, when the first hair cells are born, this work beautifully illustrates how polarity of innervation can be perpetuated: once newly born hair cells are sorted into the appropriate side of the NM, they extend actin and microtubule filled projections to contact the afferent axon that is innervating sister hair cells. These projections elicit filopodial outgrowth from the afferent axon, stimulating the formation of a new axon–hair cell synapse on the newly differentiated hair cell (Dow et al. 2015). With the now rapid evolution of new imaging and cell biological techniques, the identification of the molecular address of polarized hair cells will likely be elucidated in the near future.

3.6 Postembryonic Lateral Line Development in the Zebrafish

Collective migration of the pLLP gives rise to the primary trunk NMs that contain hair cells with anterior-posterior oriented stereocilia (Thomas et al. 2015) (Fig. 3.3D). In addition to the nascent NMs of the primary pLL, the pLLP deposits a trail of inter-NMs cells that will proliferate and give rise to additional trunk NMs (Grant et al. 2005). Recent work has demonstrated that ErbB/neuregulin signaling regulates Wnt and FGF activity to direct proliferation and differentiation of inter-NM cells (Lush and Piotrowski 2014). During early larval development, a secondary primordium (prim II) follows the path of the primary primordium and gives rise to NMs with dorsoventrally oriented hair cells (Fig. 3.3D). In addition, a dorsal primordium (primD) migrates along the dorsal aspect of the trunk, also depositing NMs with dorsoventrally oriented hair cells. The anterior lateral line forms from primordia that are less well defined. Together, these primordia give rise to the initial lateral line systems that will elaborate to form the adult lateral line system (Fig. 3.6A) (Nunez et al. 2009).

Following metamorphosis, the adult zebrafish trunk NMs are arranged in "stitches" that run dorsoventrally from the posterior lateral line (Fig. 3.6B). To give rise to the additional NMs that make up a stitch, cells bud off from the primary NMs, migrate, proliferate, and differentiate into hair and supporting cells. Recent work suggests that afferent innervation of NMs is required for proper stitch formation during larval and adult development (Wada et al. 2013). The adult anterior or cranial lateral line system NMs are enclosed in fluid-filled dermal bone canals that form during larval development in zebrafish (Webb and Shirey 2003). Canals are suggested to be an adaptation that improves the "signal-to-noise" ratio of mechanotransduction, thus allowing better detection of prey (Ghysen & Dambly-Chaudiere, 2007).

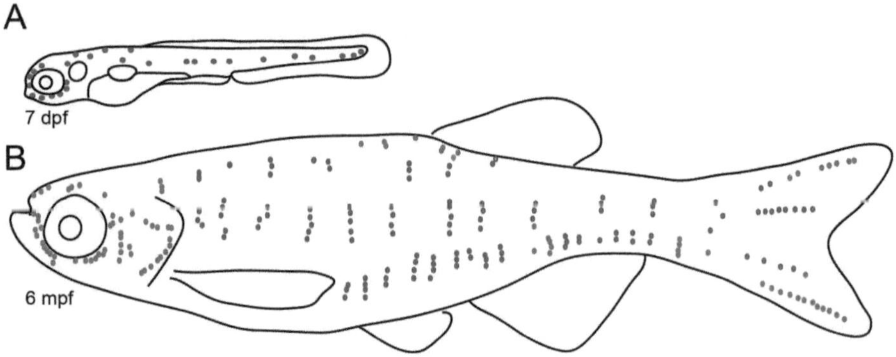

Fig. 3.6 Neuromast patterning in larval and adult zebrafish. (**A**) Schematic of a 7 days post-fertilization (dpf) zebrafish larva showing the nascent anterior and posterior lateral line (gray circles). (**B**) Schematic of the 6 months post-fertilization (mpf) adult zebrafish showing the elaborate pattern of neuromasts across the surface of the body (gray circles). Schematics are not to scale

3.7 Summary

The zebrafish lateral line system has emerged as an excellent model with which to study a variety of subjects from basic organogenesis to large-scale chemical screens aimed at identifying potential therapeutics to protect against deafness or treat cancer. This chapter has covered the early formation of the lateral line system from its placodal origins, to the collective migration of the lateral line primordium, hair cell specification and physiology, and finally the afferent and efferent innervation that mediates lateral line mechanosensation. Development of the lateral line requires precise coordination of several molecular signaling pathways that drive complex cellular interactions to regulate organ formation and innervation. With the advent of new technologies such as genome editing and advanced imaging, future work on the lateral line will elucidate the mechanisms underlying hair cell damage, regeneration, and axonal maintenance. In addition, the collective migration of the lateral line primordium has been suggested to be morphologically and genetically similar to collective cell invasion in epithelial cancers (Friedl and Gilmour 2009), making the zebrafish an attractive model in which to study potential targets for treating such diseases.

Compliance with Ethics Requirements
Catherine M. Drerup declares no competing financial or ethical interests.
Hillary F. McGraw declares no competing financial or ethical interests.
Alex V. Nechiporuk declares no competing financial or ethical interests.
Teresa Nicolson declares no competing financial or ethical interests.

References

Ahrens, K., & Schlosser, G. (2005). Tissues and signals involved in the induction of placodal Six1 expression in *Xenopus laevis*. *Developmental Biology*, 288(1), 40–59.

Alexandre, D., & Ghysen, A. (1999). Somatotopy of the lateral line projection in larval zebrafish. *Proceedings of the National Academy of Sciences of the USA*, 96(13), 7558–7562.

Aman, A., & Piotrowski, T. (2008). Wnt/beta-catenin and Fgf signaling control collective cell migration by restricting chemokine receptor expression. *Developmental Cell*, 15(5), 749–761.

Aman, A., & Piotrowski, T. (2011). Cell–cell signaling interactions coordinate multiple cell behaviors that drive morphogenesis of the lateral line. *Cell Adhesion and Migration*, 5(6), 499–508.

Aman, A., Nguyen, M., & Piotrowski, T. (2010). Wnt/β-catenin dependent cell proliferation underlies segmented lateral line morphogenesis. *Developmental Biology*, doi:10.1016/j.ydbio.2010.10.022.

Aman, A., Nguyen, M., & Piotrowski, T. (2011). Wnt/beta-catenin dependent cell proliferation underlies segmented lateral line morphogenesis. *Developmental Biology*, 349(2), 470–482.

Andermann, P., Ungos, J., & Raible, D. W. (2002). Neurogenin1 defines zebrafish cranial sensory ganglia precursors. *Developmental Biology*, 251(1), 45–58.

Becker, T., Becker, C. G., Schachner, M., & Bernhardt, R. R. (2001). Antibody to the HNK-1 glycoepitope affects fasciculation and axonal pathfinding in the developing posterior lateral line nerve of embryonic zebrafish. *Mechanisms of Development*, 109(1), 37–49.

Bleckmann, H. (1993). Role of the lateral line and fish behavior. In T. J. Pitcher (ed.), *Behaviour of Teleost Fishes*, 2nd ed. (pp. 201–246). New York: Springer.

Bricaud, O., Chaar, V., Dambly-Chaudiere, C., & Ghysen, A. (2001). Early efferent innervation of the zebrafish lateral line. *The Journal of Comparative Neurology*, 434(3), 253–261.

Brosamle, C., & Halpern, M. E. (2009). Nogo-Nogo receptor signalling in PNS axon outgrowth and pathfinding. *Molecular and Cellular Neuroscience*, 40(4), 401–409.

Bussmann, J., & Raz, E. (2015). Chemokine-guided cell migration and motility in zebrafish development. *The EMBO Journal*, 34(10), 1309–1318.

Cole, G. J., & Schachner, M. (1987). Localization of the L2 monoclonal antibody binding site on chicken neural cell adhesion molecule (NCAM) and evidence for its role in NCAM-mediated cell adhesion. *Neuroscience Letters*, 78(2), 227–232.

Dalle Nogare, D., Somers, K., Rao, S., Matsuda, M., et al.(2014). Leading and trailing cells cooperate in collective migration of the zebrafish posterior lateral line primordium. *Development*, 141(16), 3188–3196.

David, N. B., Sapede, D., Saint-Etienne, L., Thisse, C., et al. (2002). Molecular basis of cell migration in the fish lateral line: Role of the chemokine receptor CXCR4 and of its ligand, SDF1. *Proceedings of the National Academy of Sciences of the USA*, 99(25), 16297–16302.

Dona, E., Barry, J. D., Valentin, G., Quirin, C., et al. (2013). Directional tissue migration through a self-generated chemokine gradient. *Nature*, 503(7475), 285–289.

Dorsky, R. I., Moon, R. T., & Raible, D. W. (2000). Environmental signals and cell fate specification in premigratory neural crest. *Bioessays*, 22(8), 708–716.

Dow, E., Siletti, K., & Hudspeth, A. J. (2015). Cellular projections from sensory hair cells form polarity-specific scaffolds during synaptogenesis. *Genes and Development*, 29(10), 1087–1094.

Drerup, C. M., & Nechiporuk, A. V. (2013). JNK-interacting protein 3 mediates the retrograde transport of activated c-Jun N-terminal kinase and lysosomes. *PLoS Genetics*, doi:10.1371/journal.pgen.1003303.

Durdu, S., Iskar, M., Revenu, C., Schieber, N., et al. (2014). Luminal signalling links cell communication to tissue architecture during organogenesis. *Nature*, 515(7525), 120–124.

Eaton, R. C., DiDomenico, R., & Nissanov, J. (1988). Flexible body dynamics of the goldfish C-start: Implications for reticulospinal command mechanisms. *The Journal of Neuroscience*, 8(8), 2758–2768.

Ernst, S., Liu, K., Agarwala, S., Moratscheck, N., et al. (2012). Shroom3 is required downstream of FGF signalling to mediate proneuromast assembly in zebrafish. *Development*, 139(24), 4571–4581.

Faucherre, A., Pujol-Marti, J., Kawakami, K., & Lopez-Schier, H. (2009). Afferent neurons of the zebrafish lateral line are strict selectors of hair-cell orientation. *PLoS ONE*, doi:10.1371/journal.pone.0004477.

Freter, S., Muta, Y., Mak, S. S., Rinkwitz, S., & Ladher, R. K. (2008). Progressive restriction of otic fate: The role of FGF and Wnt in resolving inner ear potential. *Development*, 135(20), 3415–3424.

Friedl, P., & Gilmour, D. (2009). Collective cell migration in morphogenesis, regeneration and cancer. *Nature Review of Molecular Cell Biology*, 10(7), 445–457.

Germana, A., Laura, R., Montalbano, G., Guerrera, M. C., et al. (2010). Expression of brain-derived neurotrophic factor and TrkB in the lateral line system of zebrafish during development. *Cellular and Molecular Neurobioloy*, 30(5), 787–793.

Ghysen, A., & Dambly-Chaudiere, C. (2004). Development of the zebrafish lateral line. *Current Opinion in Neurobiology*, 14(1), 67–73.

Ghysen, A., & Dambly-Chaudiere, C. (2007). The lateral line microcosmos. *Genes and Development*, 21(17), 2118–2130.

Gilmour, D., Knaut, H., Maischein, H. M., & Nusslein-Volhard, C. (2004). Towing of sensory axons by their migrating target cells in vivo. *Nature Neuroscience*, 7(5), 491–492.

Gompel, N., Dambly-Chaudiere, C., & Ghysen, A. (2001). Neuronal differences prefigure somatotopy in the zebrafish lateral line. *Development*, 128(3), 387–393.

Grant, K. A., Raible, D. W., & Piotrowski, T. (2005). Regulation of latent sensory hair cell precursors by glia in the zebrafish lateral line. *Neuron*, 45(1), 69–80.

Haas, P., & Gilmour, D. (2006). Chemokine signaling mediates self-organizing tissue migration in the zebrafish lateral line. *Developmental Cell*, 10(5), 673–680.

Hans, S., Christison, J., Liu, D., & Westerfield, M. (2007). Fgf-dependent otic induction requires competence provided by Foxi1 and Dlx3b. BMC *Developmental Biology*, doi:10.1186/1471-213X-7-5.

Hans, S., Irmscher, A., & Brand, M. (2013). Zebrafish Foxi1 provides a neuronal ground state during inner ear induction preceding the Dlx3b/4b-regulated sensory lineage. *Development*, 140(9), 1936–1945.

Harding, M. J., & Nechiporuk, A. V. (2012). Fgfr-Ras-MAPK signaling is required for apical constriction via apical positioning of Rho-associated kinase during mechanosensory organ formation. *Development*, 139(17), 3130–3135.

Harding, M. J., McGraw, H. F., & Nechiporuk, A. (2014). The roles and regulation of multicellular rosette structures during morphogenesis. *Development*, 141(13), 2549–2558.

Itoh, M., & Chitnis, A. B. (2001). Expression of proneural and neurogenic genes in the zebrafish lateral line primordium correlates with selection of hair cell fate in neuromasts. *Mechanisms of Development*, 102(1–2), 263–266.

Janesick, A., Shiotsugu, J., Taketani, M., & Blumberg, B. (2012). RIPPLY3 is a retinoic acid-inducible repressor required for setting the borders of the pre-placodal ectoderm. *Development*, 139(6), 1213–1224.

Kindt, K. S., Finch, G., & Nicolson, T. (2012). Kinocilia mediate mechanosensitivity in developing zebrafish hair cells. *Developmental Cell*, 23(2), 329–341.

Laguerre, L., Ghysen, A., & Dambly-Chaudiere, C. (2009). Mitotic patterns in the migrating lateral line cells of zebrafish embryos. *Developmental Dynamics*, 238(5), 1042–1051.

Lecaudey, V., Cakan-Akdogan, G., Norton, W. H., & Gilmour, D. (2008). Dynamic Fgf signaling couples morphogenesis and migration in the zebrafish lateral line primordium. *Development*, 135(16), 2695–2705.

Lee, S. A., Shen, E. L., Fiser, A., Sali, A., & Guo, S. (2003). The zebrafish forkhead transcription factor Foxi1 specifies epibranchial placode-derived sensory neurons. *Development*, 130(12), 2669–2679.

Litsiou, A., Hanson, S., & Streit, A. (2005). A balance of FGF, BMP and WNT signalling positions the future placode territory in the head. *Development*, 132(18), 4051–4062

Lopez-Schier, H., & Hudspeth, A. J. (2006). A two-step mechanism underlies the planar polarization of regenerating sensory hair cells. *Proceedings of the National Academy of Sciences of the USA*, 103(49), 18615–18620.

Lush, M. E., & Piotrowski, T. (2014). Sensory hair cell regeneration in the zebrafish lateral line. *Developmental Dynamics*, 243(10), 1187–1202.

Martin, S. C., Marazzi, G., Sandell, J. H., & Heinrich, G. (1995). Five Trk receptors in the zebrafish. *Developmental Biology*, 169(2), 745–758.

Matsuda, M., & Chitnis, A. B. (2010). *Atoh1a* expression must be restricted by Notch signaling for effective morphogenesis of the posterior lateral line primordium in zebrafish. *Development*, 137(20), 3477–3487.

Matsuda, M., Nogare, D. D., Somers, K., Martin, K., et al. (2013). *Lef1* regulates *Dusp6* to influence neuromast formation and spacing in the zebrafish posterior lateral line primordium. *Development*, 140(11), 2387–2397.

Matthews, G., & Fuchs, P. (2010). The diverse roles of ribbon synapses in sensory neurotransmission. *Nature Reviews of Neuroscience*, 11(12), 812–822.

McCarroll, M. N., & Nechiporuk, A. V. (2013). *Fgf3* and *Fgf10a* work in concert to promote maturation of the epibranchial placodes in zebrafish. *PLoS ONE*, 8(12), e85087.

McCarroll, M. N., Lewis, Z. R., Culbertson, M. D., Martin, B. L., et al. (2012). Graded levels of Pax2a and Pax8 regulate cell differentiation during sensory placode formation. *Development*, 139(15), 2740–2750.

McGraw, H. F., Drerup, C. M., Culbertson, M. D., Linbo, T., et al. (2011). *Lef1* is required for progenitor cell identity in the zebrafish lateral line primordium. *Development*, 138(18), 3921–3930.

McGraw, H. F., Culbertson, M. D., & Nechiporuk, A. V. (2014). Kremen1 restricts Dkk activity during posterior lateral line development in zebrafish. *Development*, 141(16), 3212–3221.

Metcalfe, W. K. (1985). Sensory neuron growth cones comigrate with posterior lateral line primordial cells in zebrafish. *The Journal of Comparative Neurology*, 238(2), 218–224.

Metcalfe, W. K., Kimmel, C. B., & Schabtach, E. (1985). Anatomy of the posterior lateral line system in young larvae of the zebrafish. *The Journal of Comparative Neurology*, 233(3), 377–389.

Metcalfe, W. K., Myers, P. Z., Trevarrow, B., Bass, M. B., & Kimmel, C. B. (1990). Primary neurons that express the L2/HNK-1 carbohydrate during early development in the zebrafish. *Development*, 110(2), 491–504.

Mirkovic, I., Pylawka, S., & Hudspeth, A. J. (2012). Rearrangements between differentiating hair cells coordinate planar polarity and the establishment of mirror symmetry in lateral-line neuromasts. *Biology Open*, 1(5), 498–505.

Nagiel, A., Andor-Ardo, D., & Hudspeth, A. J. (2008). Specificity of afferent synapses onto plane-polarized hair cells in the posterior lateral line of the zebrafish. *The Journal of Neuroscience*, 28(34), 8442–8453.

Nagiel, A., Patel, S. H., Andor-Ardo, D., & Hudspeth, A. J. (2009). Activity-independent specification of synaptic targets in the posterior lateral line of the larval zebrafish. *Proceedings of the National Academy of Sciences of the USA*, 106(51), 21948–21953.

Nechiporuk, A., & Raible, D. W. (2008). FGF-dependent mechanosensory organ patterning in zebrafish. *Science*, 320(5884), 1774–1777.

Nechiporuk, A., Linbo, T., Poss, K. D., & Raible, D. W. (2007). Specification of epibranchial placodes in zebrafish. *Development*, 134(3), 611–623.

Nissen, R. M., Yan, J., Amsterdam, A., Hopkins, N., & Burgess, S. M. (2003). Zebrafish foxi one modulates cellular responses to Fgf signaling required for the integrity of ear and jaw patterning. *Development*, 130(11), 2543–2554.

Nogare DD, Nikaido M, Somers K, Head J, Piotrowski T, Chitnis AB. In toto imaging of the migrating Zebrafish lateral line primordium at single cell resolution. Dev Biol. 2017 Feb 1;422(1):14–23.

Northcutt, R. G. (1989). The phylogenetic distribution and innervation of craniate mechanoreceptive lateral lines. In P. G. S. Coombs & H. Münz (Eds.), *Mechanosensory lateral line: Neurobiology and evolution* (pp. 17–78). New York: Springer.

Northcutt, R. G. (1997). Evolution of gnathostome lateral line ontogenies. *Brain, Behavior and Evolution*, 50(1), 25–37.

Northcutt, R. G. (2005). Ontogeny of electroreceptors and their neural circuitry. In T. H. Bullock (Ed.), *Electroreception* (pp. 112–131). New York: Springer Science+Business Media.

Northcutt, R. G., & Brandle, K. (1995). Development of branchiomeric and lateral line nerves in the axolotl. *The Journal of Comparative Neurology*, 355(3), 427–454.

Northcutt, R. G., Catania, K. C., & Criley, B. B. (1994). Development of lateral line organs in the axolotl. *The Journal of Comparative Neurology*, 340(4), 480–514.

Nunez, V. A., Sarrazin, A. F., Cubedo, N., Allende, M. L., et al. (2009). Postembryonic development of the posterior lateral line in the zebrafish. *Evolution & Development*, 11(4), 391–404.

Obholzer, N., Wolfson, S., Trapani, J. G., Mo, W., et al. (2008). Vesicular glutamate transporter 3 is required for synaptic transmission in zebrafish hair cells. *The Journal of Neuroscience*, 28(9), 2110–2118.

Padanad, M. S., & Riley, B. B. (2011). Pax2/8 proteins coordinate sequential induction of otic and epibranchial placodes through differential regulation of foxi1, sox3 and fgf24. *Developmental Biology*, 351(1), 90–98.

Parker, G. H. (1904). The function of the lateral-line organs in fishes. *Bulletin of U.S. Burrow of Fish*, 24, 185–207.

Pieper, M., Eagleson, G. W., Wosniok, W., & Schlosser, G. (2011). Origin and segregation of cranial placodes in *Xenopus laevis*. *Developmental Biology*, 360(2), 257–275.

Pujol-Marti, J., Baudoin, J. P., Faucherre, A., Kawakami, K., & Lopez-Schier, H. (2010). Progressive neurogenesis defines lateralis somatotopy. *Developmental Dynamics*, 239(7), 1919–1930.

Pujol-Marti, J., Zecca, A., Baudoin, J. P., Faucherre, A., et al. (2012). Neuronal birth order identifies a dimorphic sensorineural map. *The Journal of Neuroscience*, 32(9), 2976–2987.

Pujol-Marti, J., Faucherre, A., Aziz-Bose, R., Asgharsharghi, A., et al. (2014). Converging axons collectively initiate and maintain synaptic selectivity in a constantly remodeling sensory organ. *Current Biology*, 24(24), 2968–2974.

Revenu, C., Streichan, S., Dona, E., Lecaudey, V., et al. (2014). Quantitative cell polarity imaging defines leader-to-follower transitions during collective migration and the key role of microtubule-dependent adherens junction formation. *Development*, 141(6), 1282–1291.

Sarrazin, A. F., Nunez, V. A., Sapede, D., Tassin, V., et al. (2010). Origin and early development of the posterior lateral line system of zebrafish. *The Journal of Neuroscience*, 30(24), 8234–8244.

Sato, A., Koshida, S., & Takeda, H. (2010). Single-cell analysis of somatotopic map formation in the zebrafish lateral line system. *Developmental Dynamics*, 239(7), 2058–2065.

Schlosser, G. (2002). Development and evolution of lateral line placodes in amphibians I. Development. *Zoology*, 105(2), 119–146.

Schuster, K., Dambly-Chaudiere, C., & Ghysen, A. (2010). Glial cell line-derived neurotrophic factor defines the path of developing and regenerating axons in the lateral line system of zebrafish. *Proceedings of the National Academy of Sciences of the USA*, 107(45), 19531–19536.

Sheets, L., Trapani, J. G., Mo, W., Obholzer, N., & Nicolson, T. (2011). Ribeye is required for presynaptic Ca(V)1.3a channel localization and afferent innervation of sensory hair cells. *Development*, 138(7), 1309–1319.

Sheets, L., Kindt, K. S., & Nicolson, T. (2012). Presynaptic CaV1.3 channels regulate synaptic ribbon size and are required for synaptic maintenance in sensory hair cells. *The Journal of Neuroscience*, 32(48), 17273–17286.

Shepherd, I. T., Pietsch, J., Elworthy, S., Kelsh, R. N., & Raible, D. W. (2004). Roles for GFRalpha1 receptors in zebrafish enteric nervous system development. *Development*, 131(1), 241–249.

Shoji, W., Yee, C. S., & Kuwada, J. Y. (1998). Zebrafish semaphorin Z1a collapses specific growth cones and alters their pathway in vivo. *Development*, 125(7), 1275–1283.

Sidi, S., Busch-Nentwich, E., Friedrich, R., Schoenberger, U., & Nicolson, T. (2004). gemini encodes a zebrafish L-type calcium channel that localizes at sensory hair cell ribbon synapses. *The Journal of Neuroscience*, 24(17), 4213–4223.

Thomas, E. D., Cruz, I. A., Hailey, D. W., & Raible, D. W. (2015). There and back again: development and regeneration of the zebrafish lateral line system. *Wiley Interdisciplinary Reviews: Developmental Biology*, 4(1), 1–16.

Trapani, J. G., & Nicolson, T. (2010). Physiological recordings from zebrafish lateral-line hair cells and afferent neurons. *Methods Cellular Biology*, 100, 219–231.

Valdivia, L. E., Young, R. M., Hawkins, T. A., Stickney, H. L., et al.(2011). Lef1-dependent Wnt/beta-catenin signalling drives the proliferative engine that maintains tissue homeostasis during lateral line development. *Development*, 138(18), 3931–3941.

Venero Galanternik, M., Kramer, K. L., & Piotrowski, T. (2015). Heparan sulfate proteoglycans regulate Fgf signaling and cell polarity during collective cell migration. *Cell Reports*, doi:10.1016/j.celrep.2014.12.043.

Wada, H., Ghysen, A., Asakawa, K., Abe, G., et al. (2013). Wnt/Dkk negative feedback regulates sensory organ size in zebrafish. *Current Biology*, 23(16), 1559–1565.

Webb, J. F., & Shirey, J. E. (2003). Postembryonic development of the cranial lateral line canals and neuromasts in zebrafish. *Developmental Dynamics*, 228(3), 370–385.

Whitfield, T. T. (2002). Zebrafish as a model for hearing and deafness. *Journal of Neurobiology*, 53(2), 157–171.

Chapter 4
Glutamate Signaling in the Auditory Brainstem

Jason Tait Sanchez and Yong Lu

Abstract Glutamate signaling in the auditory brainstem is a dynamic process. It involves the perfect coordination of pre- and postsynaptic factors that reliably permit the transfer of information between neurons. When compared to other brain regions, glutamate signaling in the auditory brainstem is unique in that it provides the initial mechanisms underlying the perception of behaviorally relevant communication signals. Disruptions in such mechanisms are thought to underlie several hearing-related disorders. Relying on the foundation of work by Dr. Edwin Rubel and colleagues, this chapter offers a conceptual overview of factors that regulate glutamate signaling in the auditory brainstem, as well as determinants responsible for development, activity-dependent regulation, pathophysiology, and neuromodulation.

Keywords AMPA receptor • Cochlear nucleus • Heterosynaptic modulation • Homeostasis • Homosynaptic modulation • Inferior colliculus • Intracellular calcium concentration • Long-term synaptic plasticity • Metabotropic glutamate receptor • Neural development • NMDA receptor • Short-term synaptic plasticity • Superior olivary complex • Synaptic transmission

J.T. Sanchez (✉)
Roxelyn and Richard Pepper Department of Communication Sciences and Disorders, The Hugh Knowles Hearing Research Center, and Department of Neurobiology and the Interdepartmental Neuroscience Program, Northwestern University, Evanston, IL 60208, USA
e-mail: jason.sanchez@northwestern.edu

Y. Lu
Department of Anatomy and Neurobiology, Northeast Ohio Medical University, 4209 State Rt. 44, Rootstown, OH 44272, USA
e-mail: ylu@neomed.edu

K.S. Cramer et al. (eds.), *Auditory Development and Plasticity*, Springer Handbook of Auditory Research 64, DOI 10.1007/978-3-319-21530-3_4

Abbreviations

$[Ca^{2+}]_i$	Intracellular free calcium concentration
AMPA-R	α-Amino-3-hydroxy-5-methyl-4-isoxazolepropionic acid receptor
AP	Action potential
AVCN	Anteroventral cochlear nucleus
CN	Cochlear nucleus
CNS	Central nervous system
DCN	Dorsal cochlear nucleus
EPSC	Excitatory postsynaptic current
EPSP	Excitatory postsynaptic potential
$GABA_BR$	γ-Aminobutyric acid receptor
GPCR	G protein–coupled receptor
IC	Inferior colliculus
iGluR	Ionotropic glutamate receptors
ILD	Interaural level difference
IPSC	Inhibitory postsynaptic current
ITD	Interaural time difference
LSO	Lateral superior olive
LTD	Long-term depression
LTP	Long-term potentiation
mGluR	Metabotropic glutamate receptor
MNTB	Medial nucleus of trapezoid body
MSO	Medial superior olive
NL	Nucleus laminaris
NM	Cochlear nucleus magnocellularis
NMDA-R	*N*-Methyl-D-aspartate receptor
SOC	Superior olivary complex
TTX	Tetrodotoxin
VCN	Ventral cochlear nucleus
VGCC	Voltage-gated calcium channel

4.1 Introduction to Glutamate Signaling

Glutamate is an excitatory neurotransmitter in the central nervous system (CNS). When released from presynaptic axon terminals it activates receptors embedded within the plasma membrane of postsynaptic neurons and provides the primary mode of information transfer between neurons in the brain of all vertebrates. As such, glutamate signaling is essential for neural excitation and the formation and regulation of synapses (Traynelis et al. 2010). Glutamate signaling is also implicated in numerous biological processes, from learning and memory to neurological disorders (Paoletti et al. 2013).

Unlike in most other brain regions, however, glutamate signaling in the vertebrate auditory system is highly specialized and unique, properties thought critical for mechanisms underlying the perception of behaviorally relevant communication signals. How auditory glutamate signaling develops its functional specializations is a remarkable feat of biological engineering. Decades of research in both avians and mammals clearly show that specific molecular, chemical, and cellular events must be perfectly orchestrated for proper maturation of auditory glutamate signaling. Disruptions in these processes are thought to underlie several hearing-related disorders. To validate the hypothesis that abnormal glutamate signaling leads to auditory pathophysiology, an appreciation for normal glutamate signaling in mature and developing auditory brainstem nuclei is warranted. The purpose of this chapter is to provide a conceptual overview of glutamate signaling in the auditory system. Particular attention will focus on general factors that regulate glutamate signaling in the auditory brainstem, as well as determinants responsible for activity-dependent regulation, pathophysiology, and neuromodulation. When appropriate, interested readers will be referred to more detailed studies on glutamate signaling in other regions of the auditory system and elsewhere in the brain.

4.2 Synaptic Excitation

Synaptic excitation is the process by which glutamate is released by a presynaptic axon terminal and binds to and activates glutamate receptors located within the plasma membrane of a neighboring neuron (Fig. 4.1). It relies profoundly on

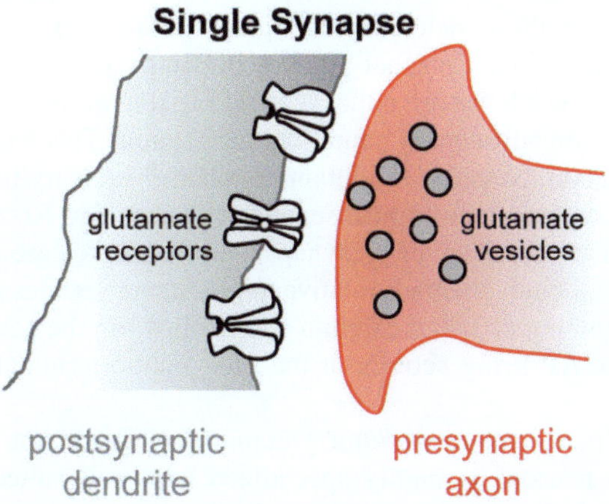

Fig. 4.1 Schematic representation of a single synapse in the avian nucleus laminaris. A single synapse that consists of a presynaptic axon (red) and a postsynaptic dendrite (gray). The presynaptic axon has vesicles (gray circles) that contain the neurotransmitter glutamate. The postsynaptic dendrite contains receptors of glutamate (white structures). The arrival of an action potential to the presynaptic axon bouton releases glutamate into the synaptic cleft (space between pre- and postsynaptic processes). Glutamate binds to the postsynaptic receptors, permitting a chemical (and ionic) transfer of information between the two neurons. [Modified from Sanchez (2013), with permission]

the availability of presynaptic glutamate vesicles, the release of glutamate by exocytosis, the binding of glutamate to postsynaptic receptors, and the subsequent removal of glutamate from the synaptic cleft. As mentioned in Sect. 4.1, synaptic excitation is essential for the process of communication between pre- and postsynaptic neurons.

4.2.1 Presynaptic Release

In general, presynaptic release of glutamate can be described as a five-step process: (1) anchoring of glutamate vesicles via specialized proteins; (2) mobilization of glutamate vesicles; (3) exogenous calcium influx and the initiation of glutamate release; (4) glutamate vesicle docking, fusion, and exocytosis; and (5) the recycling of glutamate and presynaptic glutamate vesicles.

Glutamate-containing vesicles are stored at the axon terminal in two different locations when a neuron is at rest. A relatively small number of glutamate vesicles are positioned in the membrane at "active zones," where presynaptic glutamate release occurs, while most other glutamate vesicles are contained near active zones until needed. These nearby glutamate vesicles are held into place by calcium sensitive vesicle membrane proteins that bind to actin filaments, microtubules, and various other elements of the cytoskeleton. On arrival of an action potential (AP) to the presynaptic axon terminal, voltage-gated calcium channels (VGCCs) located at these terminals open and permit the entry of exogenous calcium. The influx of exogenous calcium triggers numerous events that result in the movement, docking, fusion, and exocytosis of glutamate from its stored vesicles into the synaptic cleft. Following exocytosis, the vesicle, now acting as a continuous extension of the presynaptic membrane, forms a pit that pinches off into the terminal to form a new, vacant vesicle. The vesicle is then either recycled or refilled with glutamate and the process is repeated on subsequent neural depolarization. This entire process, from the opening of calcium channels to glutamate release, can vary in time from a few hundred microseconds to tens of milliseconds (Denker and Rizzoli 2010). In the auditory system, the properties of presynaptic glutamate release are dependent on (1) the location of calcium channels relative to glutamate vesicles in the presynaptic terminal, (2) the amount of initial calcium entry following the generation of an AP, and (3) the rate of AP firing activity at the presynaptic terminal (Fioravante and Regehr 2011).

The relationship between presynaptic glutamate release and information transfer between auditory neurons is highly specialized and well suited for the precise processing of sound. For example, presynaptic glutamate release is activity dependent and synapses can act as dynamic filters. Auditory synapses that have a high probability of presynaptic glutamate release (i.e., more glutamate is released upon the initial part of sustained depolarization) act as low-pass filters and respond best to relatively low-frequency signals. Conversely, auditory synapses that have a low-probability of release (i.e., less glutamate is released on sustained depolarization) act as high-pass filters and respond optimally to higher-frequency signals (Fortune and Rose 2001). These forms of presynaptic dynamic filtering contribute to short-term synaptic plasticities and are prominent in the avian nucleus laminaris (NL) and the mammalian superior olivary complex (SOC) (Kuba et al. 2002; Schneggenburger et al. 2002). Both are binaural circuits important for behaviorally relevant auditory functions such as sound localization and discrimination (discussed in Sect. 4.4.1).

4.2.2 Ionotropic Glutamate Receptors

Fast glutamate signaling is mediated by ionotropic glutamate receptors (iGluRs) in the CNS, including the auditory brainstem. IGluRs are ligand-gated ion channels that are divided into two primary functional subtypes named after their agonist compounds: the AMPA (α-amino-3-hydroxy-5-methyl-4-isoxazolepropionic acid receptor type and the NMDA (*N*-methyl-D-aspartate) type. Both receptor subtypes present with diverse biophysical properties based largely on their subunit composition, splice variants, and mRNA editing (Dingledine et al. 1999). This diversity is especially evident in the auditory brainstem. Such receptor properties help regulate information transfer between auditory neurons and assist in establishing specific functional specializations of both subtypes, a process thought critical for their individual roles in hearing. For example, AMPA-type glutamate receptors (AMPA-R) are defined by the connection of four subunits, GluA1-4 (Fig. 4.2). The distinctive subunit composition of AMPA-Rs found in the auditory brainstem (discussed in Sects. 4.2.4 and 4.3.1) permits unusually high permeability to divalent cations and conveys extremely rapid synaptic responses; activation and deactivation rates can occur on the order of a few hundred of microseconds (Parks 2000). It is well established that ultrafast and highly reliable synaptic responses mediated by AMPA-Rs are essential for normal auditory temporal processing in the brainstem, and disruptions in this process are thought to underlie various hearing-related problems (Trussell 1999; Yang et al. 2011).

The NMDA-type glutamate receptor (NMDA-R) is also characterized by its subunit composition, containing the mandatory GluN1 subunit and a complement of GluN2 and GluN3 subunits (Fig. 4.2). Similar to those of AMPA-Rs, the biophysical properties of NMDA-Rs are defined by their subunit composition. Unlike AMPA-Rs however, the GluN2 and GluN3 subunits impart much slower synaptic response kinetics (often lasting hundreds of milliseconds) and larger divalent cation sensitivity and permeability. Although it is currently not understood how "slower" NMDA-R responses contribute to precise auditory temporal processing, it is clear that they trigger distinct intracellular signaling pathways important for regulating numerous neuronal properties in the brain, such as synaptic plasticity and memory formation (Luscher et al. 2000; Sanz-Clemente et al. 2013).

Whether located elsewhere in the brain or in the auditory brainstem, NMDA-Rs are distinct from other glutamate receptors in two ways. First, they are voltage-dependent and ligand gated; second, activation requires cobinding of glutamate and

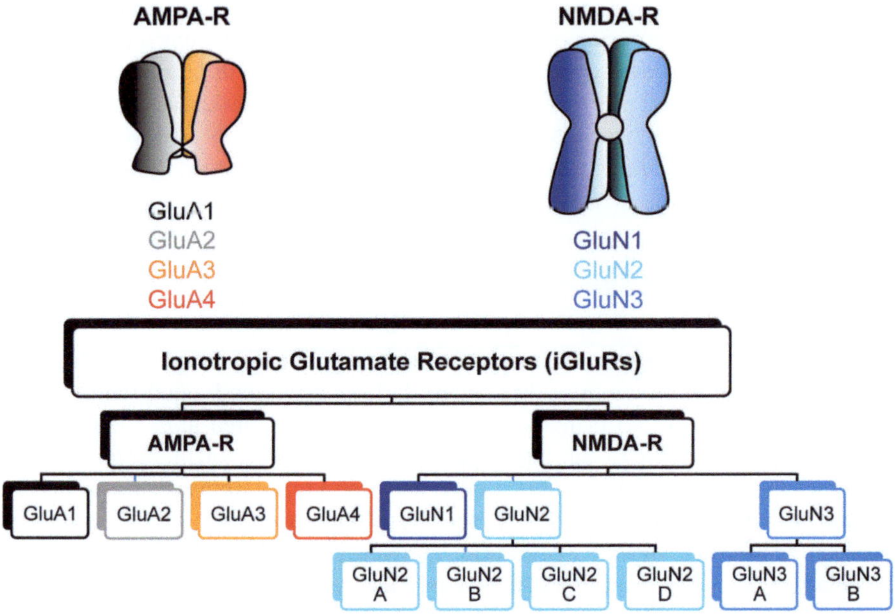

Fig. 4.2 Ionotropic glutamate receptors and their subunit content. Schematic representation of ionotropic glutamate receptors (iGluRs) found in the avian and mammalian auditory system. AMPA-type glutamate receptors contain four subunits, termed GluA1, A2, A3, and A4. NMDA-type glutamate receptors contain GluN1, GluN2, and GluN3. The GluN2 subunit consists of a subfamily termed GluN2A, 2B, 2C, and 2D. The GluN3 subunit consists of a subfamily termed GluN3A and 3B

the inhibitory neurotransmitter glycine (Sanz-Clemente et al. 2013). Typically, at negative membrane voltages, NMDA-Rs are blocked by extracellular magnesium and zinc ions and often require depolarization mediated by the AMPA-R to relieve this block. Once NMDA-Rs are relieved from the blockade, activation of the channel permits the permeability of sodium, potassium, and calcium ions to flow across the neurons membrane in a voltage-dependent manner (Bartlett and Wang, 2013). Responses mediated by NMDA-Rs during normal and pathophysiological auditory conditions have been extensively investigated (Sanchez et al. 2015), and, similarly to drug development targeting metabotropic glutamate receptors (mGluRs; see Sect 4.2.3), drug development targeting NMDA-Rs for treatment of specific auditory problems is currently underway (Sahley et al. 2013).

In both avians and mammals, specialized AMPA-Rs and NMDA-Rs are expressed in numerous auditory nuclei and show noticeable morphological and functional specializations that can be related to their roles in hearing. As first suggested by Parks (2000), auditory glutamate receptors present with (1) a unique but characteristic structural profile, (2) a tightly controlled regulation of their expression patterns throughout development, and (3) specific pharmacological targets to manipulate functional properties during auditory pathophysiological states (Parks 2000). Research supporting such hypotheses has become evident in the last two decades and is discussed further in Sects. 4.3 and 4.4.

4.2.3 Metabotropic Glutamate Receptors

Glutamate signaling in the CNS is executed not only through fast iGluRs, but also through relatively slow but equally important mGluRs. These are G protein–coupled receptors (GPCRs) categorized in the same subfamily as type B receptors for γ-aminobutyric acid ($GABA_BR$), another GPCR that is ubiquitously expressed in the CNS and exerts modulatory functions in the CNS. Two independent research groups reported the discovery of mGluRs about three decades ago (Sladeczek et al. 1985; Nicoletti et al. 1986). Since then, eight members of mGluRs have been cloned and divided into three groups, based on their molecular homology, pharmacology, and signaling transduction pathways (Niswender and Conn 2010). The neuromodulatory functions of these receptors under normal and pathophysiological conditions have been extensively investigated and drug development targeting mGluRs for treatment of CNS disorders has been rigorously pursued (Nicoletti et al. 2015).

Rubel and colleagues, using the chicken (*Gallus gallus domesticus*) as the animal model system, pioneered anatomical and physiological studies of mGluRs in the auditory brainstem. About 10 years after the discovery of mGluRs, Rubel and colleagues (Zirpel et al. 1994) reported metabolic functions of glutamate in the chicken cochlear nucleus magnocellularis (NM), a brainstem structure analogous to the mammalian anteroventral cochlear nucleus (AVCN). A series of mechanistic studies from this research group and others in the following years investigated in-depth the regulation of calcium signaling by mGluRs under normal and abnormal hearing conditions (e.g., deafferentation and deprivation; see Sects. 4.2.4 and 4.4.3). In parallel, the anatomical distribution and physiological roles of mGluRs in the mammalian auditory brainstem were explored. For example, the expression of mGluRs has been well studied in the cochlear nucleus (CN), where all three groups of mGluRs are detected in various subdivisions (Petralia et al. 2000; Lu 2014). The physiology of mGluRs functioning as autoreceptors modulating glutamatergic transmission and as heteroreceptors modulating nonglutamatergic transmission in the auditory brainstem have been extensively investigated primarily using in vitro preparations (see Sects. 4.5.1 and 4.5.2).

4.2.4 Calcium Signaling

Calcium is one of the most important signaling molecules in biological systems. In the CNS, calcium is involved in (1) triggering chemical transmission, (2) participating in the generation of APs, (3) signal transduction as a second messenger, (4) neural plasticities, and (5) regulating neuronal survival and death.

As such, the regulation of calcium signaling and maintenance of calcium homeostasis becomes a routine—yet extremely critical—task for neurons to function properly and to survive; excessively high intracellular free calcium ($[Ca^{2+}]_i$) is as detrimental to neural functions as is too little calcium (Berridge 2014). There are many pathways for calcium signaling to occur within a neuron, arising from both intra- and extracellular sources. Both iGluRs and mGluRs participate significantly in these extracellular processes. As previously mentioned, it is well established that NMDA-Rs are highly permeable to calcium, and indeed calcium influx via NMDA-Rs is usually a necessary signaling step in most forms of short- and long-term plasticities (see Sects. 4.2.2, 4.4.1, and 4.4.2). Although AMPA-Rs permit no or minimal influx of calcium elsewhere in the CNS, the AMPA-Rs in the auditory brainstem are most unusual in that they are highly permeable to calcium (Parks 2000). This feature exists especially in the timing-coding neurons where the high speed of information processing is critical, for example, in the mammalian AVCN, superior olivary complex, and in the avian NM and NL (Trussell 1999). For example, in NM (Otis et al. 1995) and NL neurons (Raman et al. 1994), owing to their relative lack of the GluA2 subunit, AMPA-Rs are highly calcium permeable (Ravindranathan et al. 2000). In fact, calcium signaling triggered by synaptic activation of the auditory nerve afferents to NM is mediated primarily by AMPA-Rs (Wang et al. 2012). AMPA-Rs may similarly dominate synaptically released glutamate-induced calcium signaling in NL neurons, because subthreshold glutamatergic excitatory postsynaptic potentials (EPSPs) are able to trigger calcium signaling in NL dendrites (Blackmer et al. 2009).

Through a number of signaling pathways, mGluRs also mediate calcium signaling in the CNS by releasing calcium from internal stores such as endoplasmic reticulum. Calcium release from internal stores has been thoroughly characterized in the avian NM by Rubel and colleagues (Rubel and Fritzsch 2002) (Fig. 4.3A). Zirpel et al. (1995) reported that activation of mGluRs with bath-applied agonists caused an increase in $[Ca^{2+}]_i$ in NM neurons. Paradoxically, loss of afferent glutamatergic inputs to NM neurons led to a threefold increase in $[Ca^{2+}]_i$ preceding neuronal degeneration (Zirpel et al. 1995), which was prevented by activation of mGluRs using orthodromic stimulation of the auditory nerve afferents or by bath application of mGluR agonists (Zirpel et al. 1998, 2000). It is also reported that group I mGluRs, when activated by exogenous bath application of agonists, increase $[Ca^{2+}]_i$ via release of calcium from internal stores (Lachica et al. 1998; Kato and Rubel 1999). Although subsequent work using exogenous application of mGluR agonists confirmed the phenomenon of mGluR-induced increase in $[Ca^{2+}]_i$ (Zirpel and Parks 2001; Wang et al. 2012), synaptic activation of the afferents to NM demonstrated that the calcium increase caused by synaptically released glutamate is largely due to iGluRs but not mGluRs (Wang et al. 2012) (Fig. 4.3B–D). This is not, however, because synaptically released glutamate did not activate mGluRs, as it is known that synaptically released glutamate via spillover does activate mGluRs, at least group II mGluRs located on the presynaptic inhibitory terminals in NM neurons (Tang et al. 2013).

Perhaps the paradoxical results observed by Zirpel et al. (1995, 1998) can be interpreted that under physiological conditions, mGluRs activated by synaptically

Fig. 4.3 (continued) increase. (**C, D**) The Ca^{2+} signaling induced by synaptically released glutamate is mediated primarily by ionotropic glutamate receptors because antagonists for NMDA-R (APV) and AMPA-R (NAS) largely eliminate the responses. [From Wang et al. (2012), with permission]

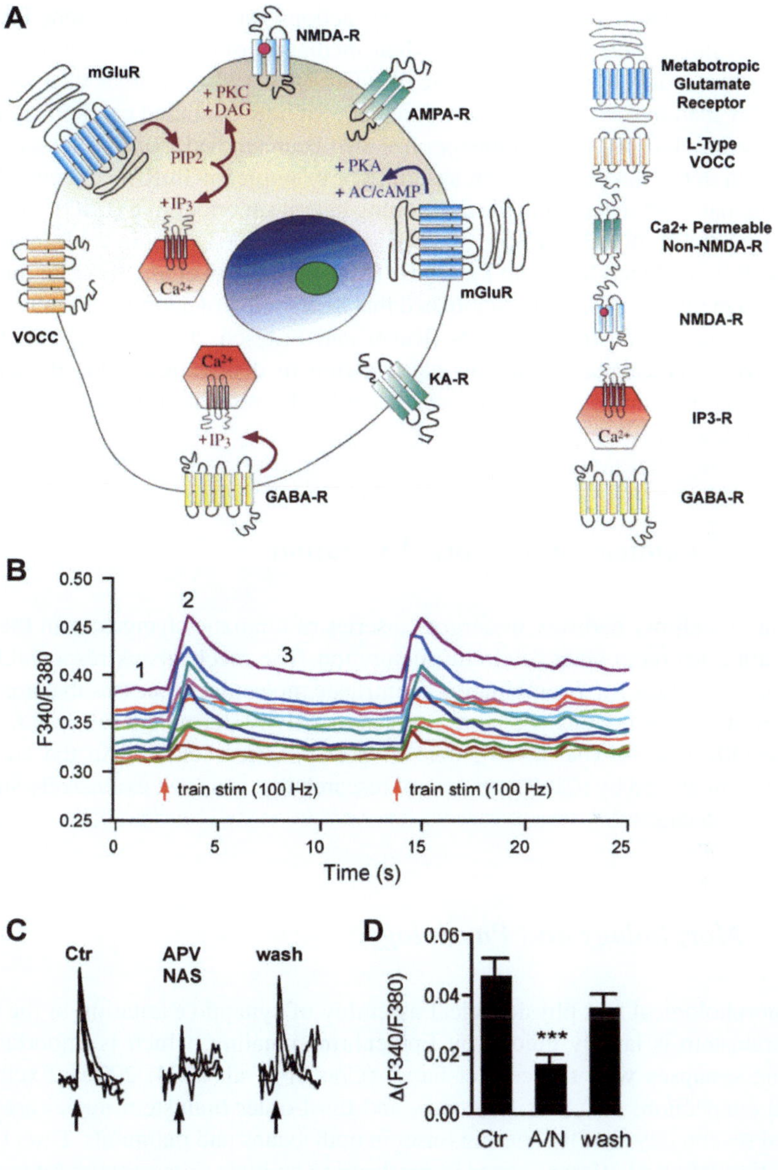

Fig. 4.3 Calcium signaling pathways in avian cochlear nucleus magnocellularis (NM) neurons. (**A**) Schematic drawings showing that ionotropic glutamate receptors (AMPA-R, NMDA-R, KA-R) in NM neurons are Ca^{2+} permeable, allowing Ca^{2+} influx upon receptor activation. Metabotropic glutamate receptors (mGluR) display bidirectional regulation of calcium signaling. On one hand, they increase intracellular Ca^{2+} concentration by causing Ca^{2+} release through internal stores. On the other hand, they suppress activity of voltage-operated Ca^{2+} channels, reducing Ca^{2+} influx. The latter role may be more important in maintaining Ca^{2+} homeostasis in these highly active neurons. [From Lachica et al. (1998), with permission.] (**B**) Synaptically released glutamate, evoked by electrical activation of the auditory nerve afferents, induces transient Ca^{2+}

released glutamate exert their modulatory actions on limiting calcium increase through multiple pathways. However, their mediation of calcium increase is minimal. In other words, the dominant role of mGluRs in NM neurons is their modulation rather than initiation of calcium signaling. MGluR-mediated calcium signaling in the mammalian auditory brainstem has also been studied, but to a lesser extent and less in depth than in the avian system. For example, mGluR-mediated calcium signaling has been detected in the developing lateral superior olive (LSO), mediated by group I and II mGluRs (Ene et al. 2003; Kim et al. 2003), and in the inferior colliculus (IC), mediated by group I mGluRs (Martinez-Galan et al. 2012). In multiple mammalian models, it is well established that deprivation of auditory afferent inputs leads to neuronal death in the CN (Rubel and Fritzsch 2002; Harris and Rubel 2006). Whether calcium signaling plays a role in deprivation-induced neuronal death in the mammalian CN, and whether mGluRs regulate this process, remain largely unknown.

4.3 Development of Synaptic Excitation

The entire auditory pathway undergoes a series of dramatic changes from the time of synaptic neural contact to circuit maturation. The mechanisms responsible for this transformation are a combination of intrinsic molecular processes that are independent of the sensory environment, coupled to activity-dependent refinement. This is especially true for the development of postsynaptic excitation in the auditory brainstem mediated by iGluRs, an area of research that has been extensively studied for several decades.

4.3.1 Morphology and Physiology

The morphological and physiological assembly of synaptic excitation in the auditory brainstem is largely guided by Eph/ephrin signaling, which is important for forming synapses with the correct target (Cramer et al. 2004, 2006). Excitatory axonal connections into first-, second-, and third-order brainstem nuclei are fully formed several days before hearing onset in both avians and mammals. Direct electrical stimulation of afferent axons in pre-hearing animals shows that these connections form functional excitatory synapses shortly after they innervate their target nuclei. Within days to a week after this initial establishment, the developmental regulation and maturation of glutamate signaling ensures rapid and reliable signal transmission between auditory neurons and it is clear that both AMPA-Rs and NMDA-Rs mediate excitatory events (Hoffpauir et al. 2006; Steinert et al. 2010).

At nearly all developing auditory synapses in the brainstem, excitatory responses have two pharmacologically and kinetically distinct components: an early AMPA-R response lasting a few milliseconds and a late NMDA-R response lasting several tens of milliseconds (Zhang and Trussell 1994; Pliss et al. 2009) (Fig. 4.4A). As a result, an extensive amount of work has been conducted over the last several decades studying the developmental profile of ionotropic glutamate signaling in the auditory brainstem of both avians and mammals (Trussell 1998; Parks 2000). From these studies, several common themes emerge regarding structure and function of iGluR signaling. First, both receptors' subunit compositions are developmentally regulated, changing dramatically before, during, and after the onset of hearing (Fig. 4.4B). Second, both types of receptors are upregulated during a prehearing time period that functionally peaks at the onset of hearing, as measured by excitatory postsynaptic currents (EPSCs). Third, AMPA-R–mediated EPSCs continue to increase in magnitude after hearing onset while the NMDA-R–mediated responses are significantly reduced but remain relatively

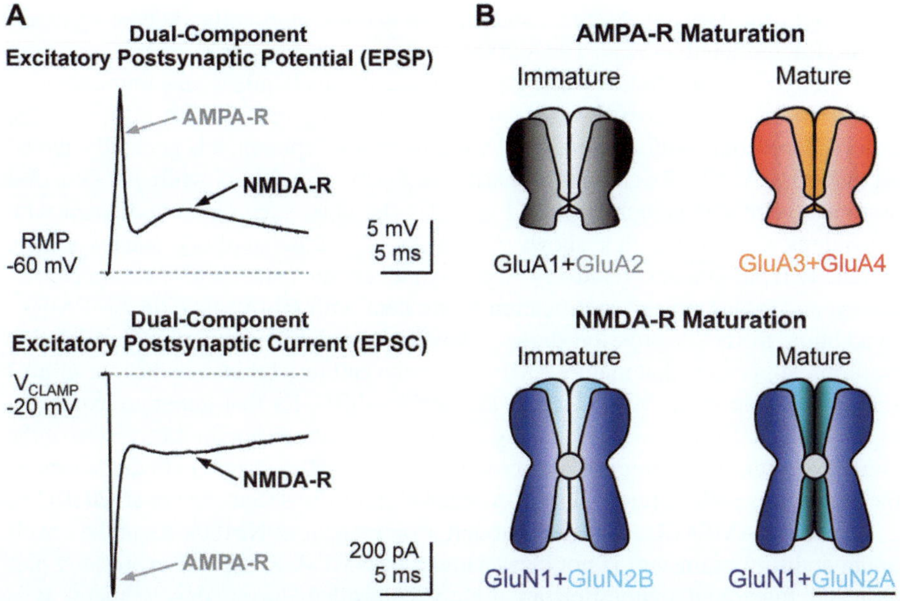

Fig. 4.4 Dual component glutamate signaling and schematic maturation of ionotropic glutamate receptors. (**A**) Representative traces of dual component excitatory responses show fast and slow phases mediated by AMPA-Rs and NMDA-Rs, respectively. *Top trace* represents in vitro excitatory postsynaptic potential (EPSP) recorded in current-clamp configuration following afferent stimulation. *Bottom trace* represents in vitro excitatory postsynaptic current (EPSC) recorded in voltage-clamp configuration following afferent stimulation. RMP = resting membrane potential; V_{CLAMP} = membrane voltage at which the neuron was held. (**B**) AMPA-Rs and NMDA-Rs change their subunit content during development. Early in development, immature synapses typically present with AMPA-Rs containing GluA1, A2 subunits and NMDA-Rs containing the GluN2B subunit (*left*). Conversely, late in development, more mature synapses have AMPA-Rs containing the GluA3, A4 subunits and NMDA-Rs containing the GluN2A subunits (*right*). These developmental changes in glutamate receptor subunit content are biological hallmarks for synaptic maturation and are essential for normal glutamate signaling. [Modified from Sanchez et al. 2015, with permission]

constant long after the maturation of hearing (Fig. 4.5A). Finally, a notable quickening in both receptors' EPSC kinetics occurs throughout development, owing in large part to modifications in their subunit compositions (Fig. 4.5B).

With respect to the AMPA-R, in the developing auditory brainstem the number of subunits that make up the channel is not completely clear. There is evidence that elsewhere in the brain most native AMPA-Rs include more than one subunit type (i.e., heterotetramers) but there is also credible support for AMPA-Rs in the brain stem that are homotetramers (Petralia et al. 1997; Tomita et al. 2003). Regardless of the exact subunit composition, it is now clear that all four of the AMPA-R homologous receptor subunit proteins (GluA1-4) occur in two alternatively spliced isoforms, termed flip and flop (Niedzielski and Wenthold 1995). The flip isoform results in slower activation and desensitization while the flop isoform results in much faster activation and desensitization on glutamate activation. The flip and flop alternative splicing is also developmentally regulated, with the flip isoform prevailing early in auditory development and the flop isoform dominating at hearing maturation (Ravindranathan et al. 1997, 2000).

Although several studies—using physiological, biochemical and immunohistological approaches—have shown somewhat conflicting results with respect to the exact subunit composition of the AMPA-R during development, it is generally agreed on that early AMPA-Rs contain primarily the GluA1-2 subunits while the GluA3-4 dominate later in development (see Fig. 4.4). The latter scenario is consistent with physiological evidence that shows AMPA-Rs at mature auditory synapses have extremely rapid kinetics, relatively large channel conductance, high calcium permeability, and strong inward rectification (consistent with GluA2-lacking AMPA-Rs). In addition, mRNA expression shows a developmental increase in the GluA3-4 flop isoforms, suggesting that mature AMPA-Rs in the auditory brainstem lack the GluA2 subunit (Raman et al. 1994; Otis et al. 1995). AMPA-Rs that generate extremely rapid and large EPSCs and have strong inward rectification lack the expression of the GluA2 subunit, which imparts slower kinetics and smaller channel conductances on the receptor as well as linear current–voltage relationships (Sanchez et al. 2010).

Similar to AMPA-Rs, the exact subunit arrangement of NMDA-Rs in the developing auditory brainstem is not clear. However, NMDA-R subunit expression pattern and functional maturation are tightly controlled during development. It is generally agreed on that NMDA-Rs form heterotetramers between two mandatory GluN1 subunits and two GluN2 subunits. Multiple receptor isoforms with distinct auditory brainstem distribution patterns arise by selective splicing of the GluN1 transcript and differential expression of the GluN2 subunits. For example, NMDA-Rs composed of the GluN1 and GluN2B subunits predominate at virtually all early

Fig. 4.5 (continued) hearing onset (*right*). GluA2 Lacking $_{\text{BLOCKADE}}$ = application of IEM-1460 (100 μM). (**B**) *Bottom*: Representative traces of NMDA-R–mediated EPSCs recorded in voltage clamp configuration from the avian NL (*left*). NMDA-R responses become faster with development relative to hearing onset (before hearing onset = blue trace; after hearing onset = black trace). NMDA-R responses in NL are more sensitive to GluN2B-specific antagonist before hearing onset (*right*). +2B$_{\text{BLOCK}}$ = application of Ro25-6981 (1 μM). [Modified from Sanchez et al. (2010, 2015), with permission]

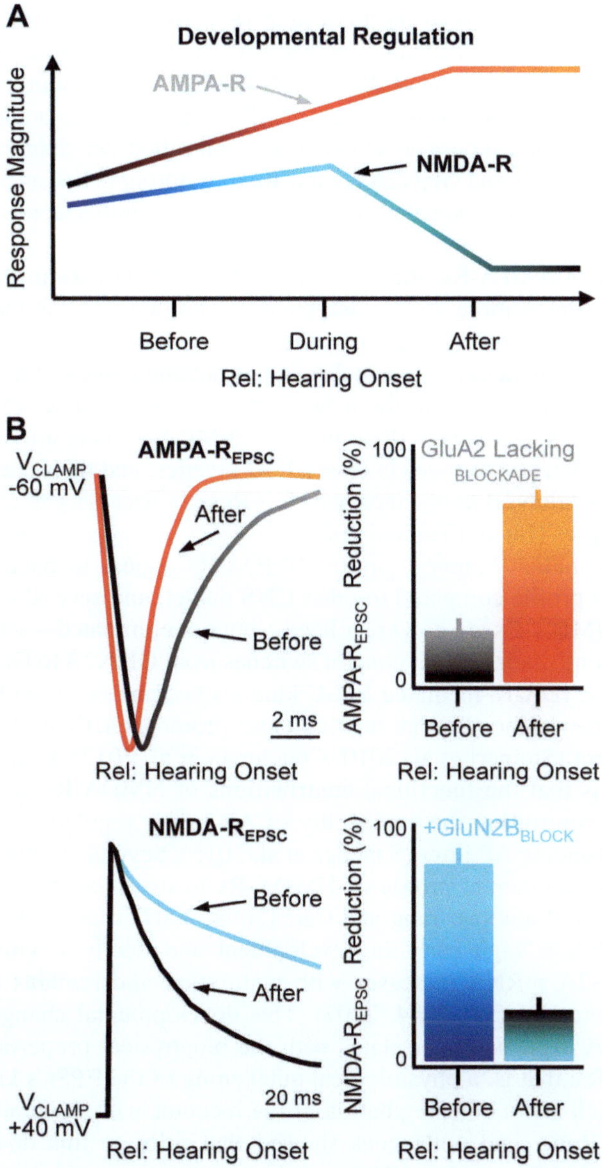

Fig. 4.5 Developmental profile of ionotropic glutamate receptors in the avian brainstem. (**A**) Schematic of the developmental regulation of iGluR response magnitude relative to hearing onset in the avian auditory brainstem. Similar to that of AMPA-Rs, NMDA-R response magnitude is upregulated and peaks during hearing onset. Unlike AMPA-Rs, which continue to increase in magnitude and stabilize, NMDA-Rs are downregulated after hearing onset but maintain into maturation. (**B**) *Top*: Representative traces of isolated AMPA-R–mediated EPSCs recorded in voltage clamp configuration from the avian nucleus laminaris (NL) (*left*). AMPA-R responses become faster with development relative to hearing onset (before hearing onset = black trace; after hearing onset = red trace). AMPA-R responses in NL are more sensitive to GluA2-lacking antagonist after

forming auditory synapses in the brainstem (Sanchez et al. 2015). This subunit arrangement confers high calcium permeability and slow response kinetics on the channel, with the decay phase of their EPSC lasting several hundred milliseconds (Sanz-Clemente et al. 2013). In contrast, NMDA-Rs at mature auditory synapses are downregulated and show a developmentally controlled switch in subunit content comprising the GluN1 and GluN2A composition, resulting in lower calcium permeability and faster response kinetics lasting several tens of milliseconds (see Fig. 4.4) (Flint et al. 1997).

Unlike that of AMPA-Rs, the functional role of NMDA-Rs in the developing auditory brainstem is more diverse and complex. For example, in the mammalian CN, NMDA-Rs are downregulated soon after hearing onset and their EPSCs become faster, consistent with a developmental switch in subunit content (Pliss et al. 2009). However, functional studies in the avian NM—the analogous counterpart to the mammalian AVCN—challenge these findings. NMDA-Rs here are downregulated but lack developmental changes in their EPSC kinetics, and their pharmacological profile suggests GluN2B dominance, which is thought necessary to eliminate inappropriate synapses (Lu and Trussell 2007).

In the mammalian binaural circuit, NMDA-Rs appear to have a traditional developmental profile compared to other CNS nuclei, and several related themes occur. First, NMDA-Rs are downregulated—but not eliminated—soon after hearing onset. Second, their subunit content switches from GluN2B to GluN2A. Third, the resulting NMDA-R–mediated EPSC kinetics become faster with maturation. Finally, their persistence beyond hearing onset promotes activity-dependent synaptic refinement (Steinert et al. 2010; Couchman et al. 2012). Consistent among these studies is that the functional contributions of NMDA-Rs during development include controlling the probability of AP firing, regulating first-spike AP latency, and reducing AP jitter (Sanchez et al. 2015). Several studies have characterized the developmental profile of NMDA-Rs in the avian binaural circuit as well. In a series of studies, Tang and Carr (2004, 2007) showed that the level of GluN2B mRNA is high early in development and declines with maturation, whereas GluN2A mRNA increases with maturation and remains elevated into adulthood (Tang and Carr 2004, 2007). This developmental change in receptor subunit mRNA expression correlates with the biophysical properties associated with NMDA-Rs: that is, a physiological quickening of the EPSCs kinetics due to a subunit switch and a downregulation of the receptor with maturation (Sanchez et al. 2010). Rubel and colleagues showed that prior to this downregulation, NMDA-Rs limit AP output of NL neurons by summating responses that prevent sodium channel deactivation, owing to the slow time course of GluN2B-containing NMDA-Rs (Sanchez et al. 2012). When considering binaural input, summating NMDA-R responses act like a temporal filter by controlling AP output only when strong and appropriately timed inputs arrive, discarding other inputs that arrive outside a coincidence window. It has been suggested that NMDA-Rs act in an activity-dependent manner to assist in the establishment of appropriate connections between binaural neurons, sculpting specialization for accurate binaural circuit computations (discussed next).

NMDA-Rs in the mammalian lateral lemniscus (LL) and IC are also developmentally downregulated and response kinetics become faster, indicating the classical change in subunit content. Studies from a variety of species reinforce the view that NMDA-Rs mediate later and slower responses compared to the AMPA-Rs during a developmental period that includes hearing onset (Zhang and Kelly 2001; Ma et al. 2002). At mature synapses, however, functional NMDA-Rs are unique compared to other brain regions. Their long-lasting conductances amplify postsynaptic responses more efficiently than AMPA-Rs and provide a mode for synaptic summation and integration. They can also generate APs independent of AMPA-Rs and control first-spike latencies (Sanchez et al. 2007). A possible explanation for such diversity is likely due to the numerous subunit arrangements and biophysical properties associated with NMDA-Rs. Although the exact subunit composition is not known, significant levels of GluN2D-containing NMDA-Rs are abundant in the mature IC (Wenzel et al. 1996). GluN2D-containing NMDA-Rs have a weaker magnesium block of the channel and may be associated with AMPA-R–independent NMDA-R activity (Hollmann and Heinemann 1994).

During developmental time periods that consist of initial neuronal contact, hearing onset, synaptic refinement, and maturation, glutamate signaling is dependent on interactions between AMPA-Rs and NMDA-Rs. As the auditory brainstem develops, both receptors show significant changes in their response to glutamate activation, suggesting a reciprocal relationship between receptor function and circuit maturation. As mentioned earlier in this section, both receptors develop relatively fast response kinetics as a result of changes in their subunit content. This quickening of responses and changes in subunit content contribute to the reliable encoding of auditory information in both avians and mammals and are considered critical hallmarks for the normal development of synaptic excitation.

4.3.2 Cotransmission: Excitation Mediated by Inhibitory Neurotransmitters

A noteworthy aspect of synaptic excitation found in the developing auditory brainstem is cotransmission. Cotransmission is defined as the release of more than one neurotransmitter from a single neuron. In the mammalian LSO, principal neurons exhibit cotransmission. This promotes excitatory signaling by the release of inhibitory neurotransmitters that activate postsynaptic NMDA-Rs.

Typically, cotransmission involves two comparable neurotransmitters (i.e., GABA and glycine) that impose similar effects (e.g., inhibition) on the postsynaptic target. In the developing LSO—a nucleus responsible for encoding intensity differences between the two ears used for sound localization—immature inputs from the medial nucleus of the trapezoid body (MNTB) corelease the inhibitory neurotransmitters GABA and glycine. Normally inhibitory at mature synapses, these neurotransmitters are excitatory at immature synapses owing to high intracellular chloride during CNS development (Kandler and Friauf 1995). Interestingly, MNTB inputs to the LSO also corelease glutamate (Gillespie et al. 2005). The excitatory effects of immature GABA and glycine released at MNTB–LSO synapses—along with coreleased glutamate—relieve magnesium blockade and activate NMDA-Rs. Corelease of glutamate permits compartmentalized calcium influx through NMDA-Rs located on LSO dendrites (Kalmbach et al. 2010). This NMDA-R activation is prominent during a period of early synapse strengthening and is downregulated by hearing onset, corresponding with a specific time window of synaptic refinement when inhibitory MNTB inputs are also excitatory. It has been hypothesized that these inhibitory synapses utilize well-established mechanisms of excitatory synaptic plasticity by modulating properties of GABA and glycine receptors, rather than traditional AMPA-R changes (Kandler and Gillespie 2005).

4.4 Activity-Dependent Regulation of Glutamate Signaling

One remarkable feature of auditory brainstem nuclei is the ability to undergo functional changes in response to normal and aberrant synaptic activity. Specific types of activity-dependent changes in glutamate signaling can be either short term or long term and result in normal or pathophysiological functions.

4.4.1 Short-Term Synaptic Plasticity

With respect to short-term synaptic plasticity, functional changes occur on the order of seconds to minutes and can be facilitated (enhanced synaptic strength) or depressed (reduced synaptic strength). Factors that regulate short-term synaptic plasticity include pre- and postsynaptic properties, which when combined play an important computational role in auditory processing by acting as a filter to encode temporal cues of sound (Xu-Friedman and Regehr 2004).

Excitatory presynaptic terminals influence the firing of postsynaptic neurons by releasing different amounts of glutamate in an activity-dependent manner. At the presynaptic terminal, repetitive APs that are separated by only a few milliseconds can transiently decrease the response of postsynaptic targets. This synaptic depression can last from hundreds of milliseconds to seconds. It is clear that in the auditory brainstem—and elsewhere in the brain—numerous presynaptic factors account for short-term synaptic depression. They include (1) the depletion of available glutamate, (2) inactivation of glutamate release sites, and (3) endogenous/exogenous calcium concentration levels (Zucker and Regehr 2002).

The depletion of presynaptically released glutamate is often dependent on the number of glutamate vesicles ready to be released and their subsequent release following sustained AP depolarization. This is referred to as the readily releasable pool (RRP) and release probability, respectively. APs arriving at the presynaptic terminal before the RRP is replenished result in fewer released vesicles. Even if the RRP is not completely depleted following initial AP arrival at the axon terminal, fusion of ensuing glutamate vesicles at the release site is limited owing to the inactivation at that site. Recovery from synaptic depression is thought to reflect the time it takes to clear fused vesicles from inactivated release sites, as well as the time it takes for the RRP to be replenished.

The probability of glutamate vesicle release is also dependent on endogenous and exogenous concentration levels of calcium and its subsequent influx, resulting in varying short-term synaptic plasticities. Variations in endogenous calcium levels and exogenous calcium influx are thought to account for differences in short-term synaptic depression and facilitation. For example, at low exogenous calcium levels, very little calcium enters the presynaptic axon terminal on AP depolarization and the probability of glutamate release is reduced. In this low-calcium environment, the buildup of residual endogenous calcium from an initial AP can have a significant effect by contributing to more glutamate release on subsequent APs. This scenario results in short-term synaptic facilitation. By experimentally altering exogenous calcium concentration from low to high levels, auditory brainstem neurons can exhibit synaptic dynamics that change from postsynaptic facilitation to depression, respectively. Thus, both types of short-term synaptic dynamics can occur at the same synapse, and exogenous calcium concentration levels regulate which one predominates.

Working in concert with presynaptic mechanisms, postsynaptic factors also contribute to short-term synaptic plasticity. Postsynaptic factors often lead to synaptic depression, which occurs largely through desensitization of postsynaptic AMPA-Rs in both avians and mammals (Trussell and Fischbach 1989; Chanda and Xu-Friedman 2010). During periods of repetitive AP activity, a majority of AMPA-Rs become nonresponsive to the glutamate ligand (i.e., desensitized) and close their channel pores. It can take several milliseconds to tens of seconds before AMPA-Rs recover from their desensitized state. The time course of desensitization is therefore related to the receptors' ability to close and enter a nonconducting state, and as such the accumulation of excessive glutamate in the synaptic cleft can impose differential effects on AMPA-R desensitization. This differential desensitization to accumulated glutamate is activity dependent, synapse specific, and receptor subunit dependent; resulting in varying degrees of postsynaptic depression observed in the auditory brainstem. Another factor to consider is the structure of the synapse itself. Because of many closely spaced glutamate release sites and the high probability of its release at calyceal synapses (e.g., endbulb and calyx of Held in the CN and MNTB, respectively), glutamate emanating from one release site can diffuse, activate, and desensitize neighboring receptors and thereby increase AMPA-R desensitization (Isaacson and Walmsley 1996; Walmsley et al. 2006).

4.4.2 Long-Term Synaptic Plasticity

Long-term synaptic plasticity is defined as a persistent change in postsynaptic strength based on recent activity patterns of either excitatory or inhibitory presynaptic signaling. Synaptic plasticity at excitatory synapses can either strengthen or weaken synapses over long periods of time in response to either increases or decreases in their activity, respectively. These synaptic dynamics are referred to as long-term potentiation (LTP; increase in postsynaptic strength) and long-term depression (LTD; decrease in postsynaptic strength). In contrast to short-term synaptic plasticity, long-term synaptic plasticity can persist for hours to weeks. Both LTP and LTD are thought to underlie cellular mechanisms associated with learning and memory (Malenka 1994). Activity dependent regulation of glutamate signaling is found in the developing and mature auditory brainstem. Its functional contribution is still under investigation but it has been hypothesized to play a role in maintaining a balance between excitation and inhibition as well as auditory learning in the brainstem (Feldman and Knudsen 1998; Wu et al. 2002).

Glutamate signaling plays a role in long-term synaptic plasticity within the dorsal cochlear nucleus (DCN), a subnucleus in the mammalian CN. DCN fusiform neurons and cartwheel interneurons express AMPA-R and NMDA-Rs mainly at their apical dendrites, where parallel fibers arising from granule neurons of the DCN make contact. In contrast, very few (if any) NMDA-Rs are expressed at fusiform basal dendrites where auditory nerve inputs are located (Bilak et al. 1996).

A clear function exists for NMDA-Rs on apical dendrites in the DCN. Here, they contribute to both LTP and LTD, which are partially blocked by receptor-specific antagonists, indicating an NMDA-R dependence (Fujino and Oertel 2003; Tzounopoulos et al. 2004). Buffering internal calcium levels completely eliminated long-term plasticity in all cases, indicating calcium influx through NMDA-Rs is a likely mechanism for their contribution to long-term plasticity in the DCN. Interestingly, NMDA-R–dependent potentiation cannot be induced by stimulation of auditory nerve inputs, consistent with studies indicating low expression of NMDA-Rs on basal dendrites (Bilak et al. 1996). The functional role of synaptic plasticity in the DCN has yet to be determined but one suggested possibility is to maintain a balance between excitation and inhibition, as activity-dependent mechanisms alter synaptic strength in the DCN (Tzounopoulos 2008).

Most neurons in the IC exhibit synaptic enhancement that resembles LTP, and glutamate signaling appears to play a partial role. However, this synaptic enhancement is expressed through both AMPA-Rs and NMDA-Rs and, in a subset of neurons, the magnitude of synaptic enhancement is dependent on stimulus frequency as well as internal calcium levels. LTP is more prominent using higher tetanus stimulus frequencies and with internal calcium levels buffered, indicating synaptic strengthening is partially dependent on calcium influx through VGCCs rather than NMDA-Rs (Wu et al. 2002). It has been hypothesized that synaptic enhancement in the IC may provide a substrate for auditory learning, and NMDA-Rs have been implicated in this process (Feldman and Knudsen 1994, 1998). For example, in the IC of the barn owl (*Tyto alba*), differences in arrival time of sound from two ears are mapped across frequency-specific channels. The output of these signals converges in the external nucleus of the IC (ICX) and creates auditory space maps. Information is then communicated to the auditory thalamus and optic tectum, where space-specific neurons have overlapping auditory and visual receptive fields, creating a multimodal pathway. Evidence of auditory learning is well documented in this pathway. Adaptive adjustments in localization abilities are learned by juvenile barn owls when auditory or visual inputs are deprived and traces of this early learning persist into adulthood. Physiological studies show that blocking NMDA-Rs in the ICX inhibits the expression of learned responses, and anatomical studies support a reorganization of the IC-to-ICX projection that reflects changes in learned localization abilities (Feldman and Knudsen 1997).

4.4.3 Pathophysiology and Glutamate Signaling

As mentioned in Sect. 4.3.1, auditory inputs from the periphery are essential for the development and maintenance of auditory nuclei in the CNS. Deprivation of afferent inputs, by genetic or surgical deafness, changes the structure and function of central auditory nuclei. The most dramatically affected central auditory structure is the CN, neurons of which receive direct glutamatergic excitatory input from the auditory nerve. As a continued and expanded theme of work based on Levi-Montalcini (1949), Rubel and colleagues painstakingly detailed the changes in cell survival and cell death in both the avian and mammalian CN. Through a diligent series of studies covering nearly four decades, they examined a multitude of important proteins and molecules in the avian NM after various manipulations of the activity levels from the auditory periphery to produce different degrees of hearing loss. Multiple review articles by Rubel and colleagues are devoted to these issues (Rubel and Fritzsch 2002; Harris and Rubel 2006). A summary of these findings follows, with a focus on the effects of deafness on glutamate signaling in the CN and a few other prominent auditory brainstem nuclei.

Cochlear ablation or application of tetrodotoxin (a potent voltage-gated sodium channel blocker) to the cochlea immediately ceased spiking activity in the avian NM (Born and Rubel 1988) (Fig. 4.6A). Interestingly, 7 days after cochlear ablation, the excitatory transmission at NM still remained, measured as the amplitude of EPSCs in response to single-pulse electrical shocks to the auditory nerve afferents in acute brain slice preparations, or as the frequency of spontaneous EPSCs (Kuba et al. 2010). This observation is somewhat surprising, because 7 days after cochlear ablation, about 30% of NM neurons degenerated (Born and Rubel, 1985), and the surviving neurons were expected to have diminishing excitatory synapses and thus reduced EPSCs. Plausible interpretations are that the surviving neurons retain the synaptic activity after passing a critical time point for them to degenerate, or more likely, they re-obtain glutamatergic inputs from sources other than the ipsilateral auditory nerve. The latter hypothesis is indeed supported by a number of studies in which the excitatory transmission in mammalian CN was examined after deafness. For example, cochlear ablation in developing mice caused newly emerged excitatory inputs onto the dendrites of AVCN bushy cells (Lu et al. 2007). An increase in the glutamatergic excitatory inputs to the AVCN from the somatosensory system after deafening has been demonstrated as a compensatory mechanism to upregulate excitation in neurons that lost their auditory inputs (Shore 2011). Consistently, in

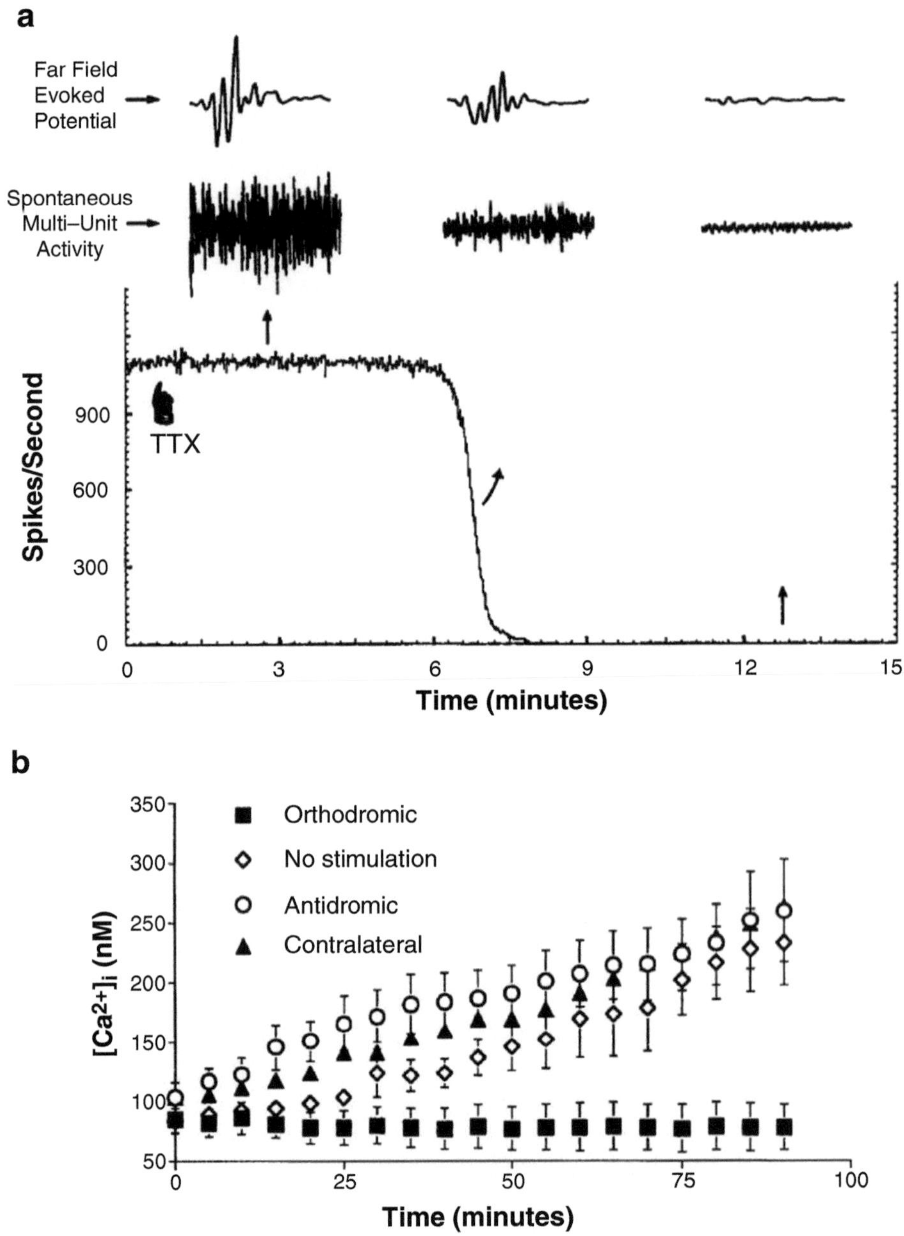

Fig. 4.6 Regulation of calcium signaling by afferent inputs in avian cochlear nucleus magnocellularis (NM) neurons. (**A**) Injection of tetrodotoxin (TTX, a voltage-gated sodium channel blocker) into the cochlea eliminated spiking activity completely, an effect similar to cochlea removal. [From Born and Rubel (1988), with permission.] (**B**) In a brain slice preparation, intracellular Ca^{2+} concentration of NM neurons increased steadily (no stimulation) to more than twofold of the basal level. Orthodromic stimulation of the auditory nerve to NM neurons maintained the Ca^{2+} level. Antidromic stimulation or stimulation of the contralateral auditory nerve failed to do so. [From Zirpel and Rubel (1996), with permission]

congenitally deaf mice, the excitatory synapses onto AVCN bushy cells are strengthened, although with an altered synaptic release profile toward greater asynchronous release, likely due to weakened calcium buffering capability in presynaptic terminals (Oleskevich and Walmsley 2002). The influence of auditory deprivation goes beyond the CN in the auditory brainstem, and the changes in glutamatergic transmission after auditory deprivation differ among varying auditory nuclei (Walmsley et al. 2006). For example, in the LSO, glutamate signaling decreases in strength after cochlear ablation (Kotak and Sanes 1997), whereas in the IC, glutamate signaling increases (Vale and Sanes 2002). In contrast, there are no changes in glutamate signaling in MNTB neurons between genetically deaf and normal mice (Oleskevich et al. 2004; Youssoufian et al. 2005).

Rubel and colleagues' work did not stop at just observing the anatomical and physiological changes in neurons in the central auditory system after deafness. Instead, more important questions were addressed to determine the neuronal mechanisms involved, which may ultimately lead to strategies to prevent further neuronal degeneration from occurring. As such, a series of studies using calcium imaging established a working model that contains two hypotheses, defined as (1) "the calcium hypothesis" for the interpretation of calcium-overload-induced cell degeneration and (2) "the mGluR hypothesis" for the interpretation of mGluR-mediated regulation of calcium signaling and possible neuroprotective function of mGluRs. The calcium hypothesis stated that calcium dysregulation after sensory deprivation led to overload of intracellular calcium, reaching a hypercalcemic condition for initiation of neuronal degeneration processes in the avian NM (Zirpel et al. 1998; Zirpel and Parks 2001) (Fig. 4.6B). Furthermore, afferent deprivation decreased plasma membrane calcium ATPase 2, a calcium efflux protein, leading to dysregulation of calcium signaling (Wang et al. 2009). The mGluR hypothesis stated that under normal hearing conditions, mGluRs regulate calcium signaling and maintain calcium homeostasis. Calcium dysregulation therefore was caused by the reduction of glutamate signaling to NM, which silenced mGluR activity (Kato et al. 1996; Zirpel and Rubel 1996). Such a critical regulation of calcium signaling and homeostasis by mGluRs is achieved via multiple pathways including activation of protein kinase A and protein kinase C (Zirpel et al. 1998), suppression of VGCCs (Lachica et al. 1995; Lu and Rubel 2005), and regulation of internal stores in the endoplasmic reticulum (Kato and Rubel 1999).

In addition to calcium signaling, structure and function of a variety of proteins are also regulated by afferent inputs, and mGluRs play a significant role in this regulation. Hyson and Rubel (1989) showed that deprivation of the auditory nerve input disrupts protein synthesis in NM neurons, and this activity-dependent regulation of protein synthesis does not involve iGluRs (Hyson and Rubel 1989; Hyson 1997). Rather, the process relies on activity of mGluRs (Hyson 1998). Blocking either group I or group II mGluRs eliminates activity-dependent regulation of protein synthesis machinery (Nicholas and Hyson 2004; Carzoli and Hyson 2011), suggesting that mGluRs are required to maintain protein synthesis in NM neurons. These studies using the same avian model further support the "mGluR hypothesis" proposed by Rubel and colleagues. Evidence supporting this hypothesis has also been demonstrated in mammals, to a lesser extent (Harris and Rubel 2006; Harris et al. 2008). Future studies are needed to pursue the treatment potential of mGluRs in preventing or reducing cell death in the central auditory system after hearing loss or deafness.

4.5 Neuromodulation of Glutamate Signaling

Every synapse in the CNS is subject to neuromodulation, a broad term defined here as the regulation of synaptic and intrinsic neural properties by GPCRs activated by neurotransmitters. As autoreceptors, mGluRs regulate synaptic strength of glutamate signaling in the auditory brainstem. Glutamate signaling is also subject to neuromodulation mediated by a multitude of other GPCRs. Neuromodulation plays an important role in maintaining a balance between synaptic excitation and inhibition and thus normal neural function.

4.5.1 Homosynaptic Modulation

Homosynaptic modulation is defined here as mGluR-mediated regulation of glutamate signaling (autoreceptor-mediated neuromodulation). An almost universally existing feedback mechanism in nearly all the glutamatergic synapses in the CNS, this type of modulation controls the amount of glutamate release to prevent overactivation of the system, which could lead to excitotoxicity. Via different signaling pathways, multiple groups of mGluRs may be involved in this modulation depending on the location of the synapse, and they are likely to be recruited under different levels of neural activity (Rondard and Pin 2015). Interestingly and somewhat surprisingly, in the avian NM, such a seemingly universal autoreceptor-mediated modulation was not detected (Otis and Trussell 1996; Tang et al. 2013). One plausible interpretation for the lack of mGluR modulation of the excitatory input to NM is to maintain high fidelity in signaling transmission in the intensity domain at this first central auditory processing stage. In NL, although the modulation was detected, it was functioning mainly in low-frequency tuned neurons (Okuda et al. 2013) (Fig. 4.7A).

In the mammalian VCN, group I mGluRs facilitate excitatory responses by depolarizing bushy cells, counteracting inhibition caused by presynaptic $GABA_BRs$

Fig. 4.7 (continued) Postsynaptic mGluRs regulate excitability of neurons in the cochlear nucleus, MNTB, LSO, and IC. Presynaptic $GABA_BRs$ exert an inhibitory action on the glutamatergic transmission in all the synapses tested. [Modified from Lu (2014), with permission.] Modulation of glutamatergic transmission by other G protein–coupled receptors (e.g., cannabinoid receptors in NM, and adenosine A1 receptors in MNTB) is not shown

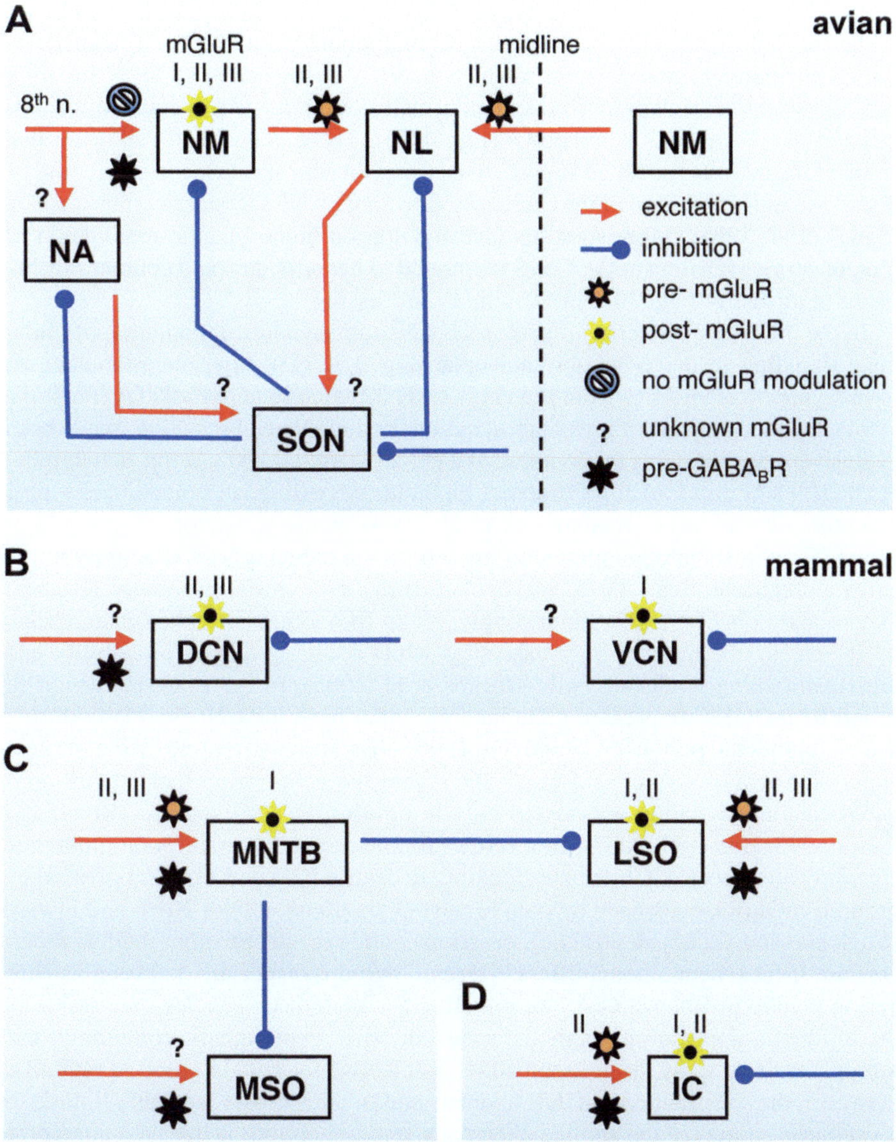

Fig. 4.7 Neuromodulation of glutamatergic transmission in auditory brainstem. (**A**) In the avian auditory brainstem, presynaptic group II and III mGluRs suppress glutamatergic transmission in nucleus laminaris in a coding frequency-dependent manner. Presynaptic γ-aminobutyric acid receptors ($GABA_BRs$) but not metabotropic glutamate receptor (mGluRs) modulate the excitatory inputs to cochlear nucleus magnocellularis (NM). Postsynaptic mGluRs suppress voltage-gated Ca^{2+} channels in NM neurons. [Modified from Lu (2014), with Permission.] (**B–D**) In the mammalian auditory brainstem, presynaptic group II and III mGluRs modulate glutamatergic transmission in medial nucleus of trapezoid body (MNTB) and lateral superior olive (LSO) neurons, whereas group II mGluRs suppress glutamatergic transmission in the inferior colliculus (IC).

(Chanda and Xu-Friedman 2011). In addition, mGluRs regulate neuronal excitability of DCN neurons bi-directionally, participating in the induction of LTP and LTD at the parallel fiber multisensory pathway (Fujino and Oertel 2003). These effects are all attributed to postsynaptic mGluRs in VCN bushy cells and DCN fusiform and cartwheel cells. Whether presynaptic mGluRs modulate glutamatergic transmission in CN neurons is unknown (Fig. 4.7B). There is one study showing that some of the mGluR-mediated suppressive effects on neuronal firing in vivo could be due to synaptic inhibition of the excitatory (presumably glutamatergic) transmission (Sanes et al. 1998). More rigorous investigation on homosynaptic modulation of glutamatergic signaling at the CN is warranted to better understand cellular mechanisms at this early stage of central auditory processing.

In the brainstem nuclei upstream of the CN, homosynaptic modulation of glutamate signaling seems to be more universal (Fig. 4.7C, D). After the establishment of presynaptic recordings from the calyx of Held synapse at the MNTB (Forsythe 1994), the mechanisms of presynaptic modulation mediated by mGluRs have been extensively studied using this synapse as a model (Trussell 2002; Kopp-Scheinpflug et al. 2011). Group I mGluRs suppress glutamate signaling via cannabinoid type 1 receptors on the calyx (Kushmerick et al. 2004), while activation of group II or group III mGluRs inhibits glutamate signaling, via reducing VGCC activity in the calyx (Takahashi et al. 1996) or regulating the exocytosis machinery (Barnes-Davies and Forsythe 1995). Interestingly, group III mGluRs alter the synaptic state by increasing the number of release sites while reducing release probability and thus maintaining EPSC strength (Billups et al. 2005). In the LSO, activation of group III mGluRs suppressed glutamatergic input from the CN, especially during a developmental period (Wu and Fu 1998) that is consistent with the protein expression results (Nishimaki et al. 2007) and the observation that mGluR-triggered calcium signaling is detected only in developing LSO neurons (Kotak and Sanes 1995; Ene et al. 2007). In the IC, variable results of mGluRs on glutamate signaling are reported. Glutamate signaling in central IC seems to be suppressed by group II mGluRs, evidenced by both in vitro (Farazifard and Wu 2010) and in vivo (Voytenko and Galazyuk 2011) observations, while results on other mGluR members are inconsistent among different research groups, especially between in vitro and in vivo experiments. One interpretation for such discrepancies may reside in the different application methods used. In vivo experiments commonly use iontophoresis to apply drugs assumed to be selective for certain group of mGluRs. However, the specificity of mGluR agonists and/or antagonists is usually limited to low concentrations of the applied chemicals, and iontophoresis does not guarantee low concentrations at the recorded neurons. Alternative methods such as pressure ejection may give rise to concentration-independent specific effects (Lu 2014). Indeed, a recent in vivo study using both methods revealed that blocking glycinergic inhibition with pressure ejection of strychnine at low concentration produced a specific effect on ITD coding in MSO neurons, whereas iontophoresis application of strychnine generated a time-dependent mixed complex effect owing to nonspecific drug actions (Franken et al. 2015). Overall, more research needs to be done to investigate the cellular mechanisms in vitro, as well as the physiological roles in vivo on the homosynaptic modulation of glutamate signaling in the auditory brainstem nuclei. Together, this may advance our understanding of the auditory modulation that occurs before information transfers to auditory subcortical and cortical structures.

4.5.2 Heterosynaptic Modulation

In contrast to homosynaptic modulation, heterosynaptic modulation is defined as regulation of glutamate signaling mediated by non-mGluR GPCRs, a very broad topic owing to the large number of GPCRs in the CNS. Here, the focus is primarily on one very important GPCR, $GABA_BR$. GABA, as the most abundant inhibitory neurotransmitter in vertebrate brains, activates $GABA_BRs$, leading to neuromodulation of a variety of ion channels in both pre- and postsynaptic neurons. One feature for $GABA_BR$-mediated neuromodulation is that its effect on glutamate signaling is universal at all synapses studied (Chalifoux and Carter 2011; Kantamneni 2015). While $GABA_BR$ modulation represents a powerful potential mechanism to regulate the activity level of the whole brain, it may also pose a challenge to achieve specific effects using drugs targeting $GABA_BRs$ in a particular neural circuit.

Being universal, $GABA_BR$-mediated modulation nonetheless serves different functions in varying auditory brainstem structures. In avian NM neurons, $GABA_BR$ modulation produces stable steady-state excitatory responses by enhancing synaptic efficacy and reducing synaptic depression of glutamate signaling (Brenowitz et al. 1998; Brenowitz and Trussell 2001). In the mammalian DCN, $GABA_BRs$ suppress the excitatory inputs to fusiform cells regardless of the origin of the inputs, enhancing synaptic facilitation of the parallel fiber input while preventing synaptic depression of the auditory fiber input (Irie and Ohmori 2008). In the mammalian VCN, expression of $GABA_BRs$ was detected (Juiz et al. 1994), but auditory responses from VCN neurons were not affected by agonists of $GABA_BRs$ (Ebert and Ostwald 1995).

In the MSO, $GABA_BRs$ modulate glutamate signaling to adjust the neural gain and achieve fast adaption for ITD processing (Stange et al. 2013). This modulation on EPSCs diminished after maturation of hearing (Hassfurth et al. 2010), suggesting its role in the development of neural circuits encoding ITDs. $GABA_BRs$ also regulate binaural sensitivity of LSO neurons to ILDs by suppressing both EPSCs and IPSCs with a stronger modulatory efficacy on excitation than on inhibition (Magnusson et al. 2008). Similar modulation of EPSCs by presynaptic $GABA_BRs$ was observed in the IC (Sun et al. 2006). In the MNTB, $GABA_BRs$ reduce amplitude of EPSCs (Barnes-Davies and Forsythe 1995). However, although the inhibitory inputs to MNTB neurons seem to be strong (Mayer et al. 2014), the source of endogenous GABA input to the MNTB is still unclear and thus the physiological significance of this modulation remains elusive. At the same synapse, presynaptic glycine receptors enhance glutamate signaling via a slight depolarization of the presynaptic membrane (Turecek and Trussell 2001), counteracting the inhibitory effect of presynaptic $GABA_BRs$ (Otis and Trussell 1996). This modulation is an excellent example of bidirectional regulation of neurotransmitter release. Different neurotransmitter receptors work in a push–pull fashion to maintain physiological dynamics of the system.

Besides neuromodulation of glutamate signaling by mGluRs and $GABA_BRs$, other GPCRs play additional roles in regulating the strength of excitation in various auditory nuclei. In the avian NM, activation of cannabinoid receptors reduces glutamate release (Stincic and Hyson 2011), and this modulation may be linked to group I mGluR signaling as shown in the MNTB (Kushmerick et al. 2004). In NM, cholinergic inputs are present (Code and Carr 1994) but their functional roles have not been explored. In MNTB, under high-frequency input conditions, adenosine A1 receptors are activated and suppress EPSCs (Wong et al. 2006). In addition to presynaptic modulation, plasticity of postsynaptic properties caused by GPCRs and other modulatory molecules inevitably affects synaptic integration of glutamatergic inputs and thus affects neuronal output. For example, in the mammalian DCN, activation of metabotropic acetylcholine receptors increases glutamatergic responses in cartwheel cells via suppression of Ca^{2+}-activated K^+ channels (He et al. 2014). Zinc, released from glutamatergic vesicles, also inhibits NMDA-R responses in cartwheel cells (Anderson et al. 2015). In the VCN, purinergic modulation via P2X receptors increases spike activity as well as Ca^{2+} signaling in bushy cells (Dietz et al. 2012). As in other brain areas, heterosynaptic modulation of glutamate signaling by GPCRs in the auditory brainstem is broad and evolving, and relevant research is actively ongoing.

4.6 Summary

As in other sensory systems, glutamate is the main excitatory neurotransmitter in the auditory brainstem. Soon after the discovery of glutamate as a chemical transmitter, excitotoxicity caused by excessive glutamate release in the brain was also revealed. Both normal glutamate signaling and abnormal pathological processes are mediated by primarily iGluRs. Although the receptor subtypes and their specific functions in the auditory brainstem have been well studied, much is unknown about their structure and function during auditory pathophysiologies. Moreover, the developmental regulation of iGluRs properties can provide insight into how individual neurons establish biophysical properties associated with normal glutamate signaling.

Relying on work from other brain regions (Henley et al. 2011), several scenarios are possible. First, NMDA-Rs control the number of AMPA-Rs by restricting their insertion into early developing synapses (Hall et al. 2007; Hall and Ghosh 2008). Conversely, as the system becomes refined following the onset of hearing, NMDA-Rs switch their subunit content (from GluN2B to GluN2A) and actively recruit AMPA-Rs into the synapse. It has been shown that the genetic removal of NMDA-Rs negatively affects AMPA-R trafficking in other brain regions, modifying neural excitability (Gambrill and Barria 2011). The functional consequence of this genetic manipulation in the developing auditory system could result in degradation of temporal processing essential for normal hearing. Indeed, research has shown that alterations in AMPA-R subunit content result in abnormal auditory brainstem responses and late evoked potentials (Yang et al. 2011). The mechanism(s) by which variation in AMPA-R trafficking occurs is beyond the scope of this chapter but include the regulation of transcription, translation, and protein degradation factors.

In the avian auditory brainstem, mGluRs are shown to be neuroprotective based on their roles in regulating intracellular calcium concentration and protein synthesis and in maintaining intracellular homeostasis. To determine fully the signaling pathways involved, as well as to test alternative hypotheses, further studies regarding specific mechanisms are clearly warranted. In addition, more in vitro physiology studies on mGluRs in the mammalian auditory brainstem are needed to establish a foundational basis for mGluR modulation in normal-hearing animals. Protein expression of various members of mGluRs in the auditory brainstem needs to be determined at both light and electron microscopy levels to reveal their location in pre- and postsynaptic elements. To date, very few studies have explored the functions of mGluRs in auditory processing at the systems and behavioral levels. Finally, alteration in mGluR morphology and physiology under abnormal hearing conditions is entirely unknown. Nonetheless, great efforts to learn about glutamate signaling in the auditory brainstem are ongoing. The young generation of auditory scientists, by standing on the shoulders of earlier giants such as Dr. Edwin Rubel, is surely going to make promising progress toward advancing our understanding of the whisper of the auditory system even in a noisy environment.

Compliance with Ethics Requirements Jason Tait Sanchez declares that he has no conflict of interest.

Yong Lu declares that he has no conflict of interest.

References

Anderson, C. T., Radford, R. J., Zastrow, M. L., Zhang, D. Y., et al. (2015). Modulation of extrasynaptic NMDA receptors by synaptic and tonic zinc. *Proceedings of the National Academy of Sciences of the USA*, 112(20), 2705–2714.

Barnes-Davies, M., & Forsythe, I. D. (1995). Pre- and postsynaptic glutamate receptors at a giant excitatory synapse in rat auditory brainstem slices. *The Journal of Physiology*, 488 (Pt 2), 387–406.

Bartlett, T. E., & Wang, Y. T. (2013). The intersections of NMDAR-dependent synaptic plasticity and cell survival. *Neuropharmacology*, 74, 59–68.

Berridge, M. J. (2014). Calcium regulation of neural rhythms, memory and Alzheimer's disease. *The Journal of Physiology*, 592 (Pt 2), 281–293.

Bilak, M. M., Bilak, S. R., & Morest, D. K. (1996). Differential expression of *N*-methyl-D-aspartate receptor in the cochlear nucleus of the mouse. *Neuroscience*, 75(4), 1075–1097.

Billups, B., Graham, B. P., Wong, A. Y., & Forsythe, I. D. (2005). Unmasking group III metabotropic glutamate autoreceptor function at excitatory synapses in the rat CNS. *The Journal of Physiology*, 565(Pt 3), 885–896.

Blackmer, T., Kuo, S. P., Bender, K. J., Apostolides, P. F., & Trussell, L. O. (2009). Dendritic calcium channels and their activation by synaptic signals in auditory coincidence detector neurons. *The Journal of Neurophysiology*, 102(2), 1218–1226.
Born, D. E., & Rubel, E. W. (1985). Afferent influences on brain stem auditory nuclei of the chicken: Neuron number and size following cochlea removal. *The Journal of Comparative Neurology*, 231(4), 435–445.
Born, D. E., & Rubel, E. W. (1988). Afferent influences on brain stem auditory nuclei of the chicken: Presynaptic action potentials regulate protein synthesis in nucleus magnocellularis neurons. *The Journal of Neuroscience*, 8(3), 901–919.
Brenowitz, S., & Trussell, L. O. (2001). Maturation of synaptic transmission at end-bulb synapses of the cochlear nucleus. *The Journal of Neuroscience*, 21(23), 9487–9498.
Brenowitz, S., David, J., & Trussell, L. (1998). Enhancement of synaptic efficacy by presynaptic GABA(B) receptors. *Neuron*, 20(1), 135–141.
Carzoli, K. L., & Hyson, R. L. (2011). In vivo analysis of the role of metabotropic glutamate receptors in the afferent regulation of chick cochlear nucleus neurons. *Hearing Research*, 272(1–2), 49–57.
Chalifoux, J. R., & Carter, A. G. (2011). GABAB receptor modulation of synaptic function. *Current Opinion in Neurobiology*, 21(2), 339–344.
Chanda, S., & Xu-Friedman, M. A. (2010). A low-affinity antagonist reveals saturation and desensitization in mature synapses in the auditory brain stem. *Journal of Neurophysiology*, 103(4), 1915–1926.
Chanda, S., & Xu-Friedman, M. A. (2011). Excitatory modulation in the cochlear nucleus through group I metabotropic glutamate receptor activation. *The Journal of Neuroscience*, 31(20), 7450–7455.
Code, R. A., & Carr, C. E. (1994). Choline acetyltransferase-immunoreactive cochlear efferent neurons in the chick auditory brainstem. *The Journal of Comparative Neurology*, 340(2), 161–173.
Couchman, K., Grothe, B., & Felmy, F. (2012). Functional localization of neurotransmitter receptors and synaptic inputs to mature neurons of the medial superior olive. *Journal of Neurophysiology*, 107(4), 1186–1198.
Cramer, K. S., Bermingham-McDonogh, O., Krull, C. E., & Rubel, E. W. (2004). EphA4 signaling promotes axon segregation in the developing auditory system. *Developmental Biology*, 269(1), 26–35.
Cramer, K. S., Cerretti, D. P., & Siddiqui, S. A. (2006). EphB2 regulates axonal growth at the midline in the developing auditory brainstem. *Developmental Biology*, 295(1), 76–89.
Denker, A., & Rizzoli, S. O. (2010). Synaptic vesicle pools: An update. *Frontiers in Synaptic Neuroscience*, 2, 135.
Dietz, B., Jovanovic, S., Wielsch, B., Nerlich, J., et al. (2012). Purinergic modulation of neuronal activity in developing auditory brainstem. *The Journal of Neuroscience*, 32(31), 10699–10712.
Dingledine, R., Borges, K., Bowie, D., & Traynelis, S. F. (1999). The glutamate receptor ion channels. *Pharmacology Reviews*, 51(1), 7–61.
Ebert, U., & Ostwald, J. (1995). GABA can improve acoustic contrast in the rat ventral cochlear nucleus. *Experimental Brain Research*, 104(2), 310–322.
Ene, F. A., Kullmann, P. H., Gillespie, D. C., & Kandler, K. (2003). Glutamatergic calcium responses in the developing lateral superior olive: Receptor types and their specific activation by synaptic activity patterns. *Journal of Neurophysiology*, 90(4), 2581–2591.
Ene, F. A., Kalmbach, A., & Kandler, K. (2007). Metabotropic glutamate receptors in the lateral superior olive activate TRP-like channels: Age- and experience-dependent regulation. *Journal of Neurophysiology*, 97(5), 3365–3375.
Farazifard, R., & Wu, S. H. (2010). Metabotropic glutamate receptors modulate glutamatergic and GABAergic synaptic transmission in the central nucleus of the inferior colliculus. *Brain Research*, 1325, 28–40.
Feldman, D. E., & Knudsen, E. I. (1994). NMDA and non-NMDA glutamate receptors in auditory transmission in the barn owl inferior colliculus. *The Journal of Neuroscience*, 14(10), 5939–5958.

Feldman, D. E., & Knudsen, E. I. (1997). An anatomical basis for visual calibration of the auditory space map in the barn owl's midbrain. *The Journal of Neuroscience*, 17(17), 6820–6837.
Feldman, D. E., & Knudsen, E. I. (1998). Pharmacological specialization of learned auditory responses in the inferior colliculus of the barn owl. *The Journal of Neuroscience*, 18(8), 3073–3087.
Fioravante, D., & Regehr, W. G. (2011). Short-term forms of presynaptic plasticity. *Current Opinion in Neurobiology*, 21(2), 269–274.
Flint, A. C., Maisch, U. S., Weishaupt, J. H., Kriegstein, A. R., & Monyer, H. (1997). NR2A subunit expression shortens NMDA receptor synaptic currents in developing neocortex. *The Journal of Neuroscience*, 17(7), 2469–2476.
Forsythe, I. D. (1994). Direct patch recording from identified presynaptic terminals mediating glutamatergic EPSCs in the rat CNS, in vitro. *The Journal of Physiology*, 479 (Pt 3), 381–387.
Fortune, E. S., & Rose, G. J. (2001). Short-term synaptic plasticity as a temporal filter. *Trends in Neuroscience*, 24(7), 381–385.
Franken, T. P., Roberts, M. T., Wei, L., Golding, N. L., & Joris, P. X. (2015). In vivo coincidence detection in mammalian sound localization generates phase delays. *Nature Neuroscience*, 18(3), 444–452.
Fujino, K., & Oertel, D. (2003). Bidirectional synaptic plasticity in the cerebellum-like mammalian dorsal cochlear nucleus. *Proceedings of the National Academy of Sciences of the USA*, 100(1), 265–270.
Gambrill, A. C., & Barria, A. (2011). NMDA receptor subunit composition controls synaptogenesis and synapse stabilization. *Proceedings of the National Academy of Sciences of the USA*, 108(14), 5855–5860.
Gillespie, D. C., Kim, G., & Kandler, K. (2005). Inhibitory synapses in the developing auditory system are glutamatergic. *Nature Neuroscience*, 8(3), 332–338.
Hall, B. J., & Ghosh, A. (2008). Regulation of AMPA receptor recruitment at developing synapses. *Trends in Neuroscience*, 31(2), 82–89.
Hall, B. J., Ripley, B., & Ghosh, A. (2007). NR2B signaling regulates the development of synaptic AMPA receptor current. *The Journal of Neuroscience*, 27(49), 13446–13456.
Harris, J. A., & Rubel, E. W. (2006). Afferent regulation of neuron number in the cochlear nucleus: Cellular and molecular analyses of a critical period. *Hearing Research*, 216–217, 127–137.
Harris, J. A., Iguchi, F., Seidl, A. H., Lurie, D. I., & Rubel, E. W. (2008). Afferent deprivation elicits a transcriptional response associated with neuronal survival after a critical period in the mouse cochlear nucleus. *The Journal of Neuroscience*, 28(43), 10990–11002.
Hassfurth, B., Grothe, B., & Koch, U. (2010). The mammalian interaural time difference detection circuit is differentially controlled by GABAB receptors during development. *The Journal of Neuroscience*, 30(29), 9715–9727.
He, S., Wang, Y. X., Petralia, R. S., & Brenowitz, S. D. (2014). Cholinergic modulation of large-conductance calcium-activated potassium channels regulates synaptic strength and spine calcium in cartwheel cells of the dorsal cochlear nucleus. *The Journal of Neuroscience*, 34(15), 5261–5272.
Henley, J. M., Barker, E. A., & Glebov, O. O. (2011). Routes, destinations and delays: Recent advances in AMPA receptor trafficking. *Trends in Neuroscience*, 34(5), 258–268.
Hoffpauir, B. K., Grimes, J. L., Mathers, P. H., & Spirou, G. A. (2006). Synaptogenesis of the calyx of Held: Rapid onset of function and one-to-one morphological innervation. *The Journal of Neuroscience*, 26(20), 5511–5523.
Hollmann, M., & Heinemann, S. (1994). Cloned glutamate receptors. *Annual Review of Neuroscience*, 17, 31–108.
Hyson, R. L. (1997). Transneuronal regulation of ribosomes after blockade of ionotropic excitatory amino acid receptors. *Brain Research*, 749(1), 61–70.
Hyson, R. L. (1998). Activation of metabotropic glutamate receptors is necessary for transneuronal regulation of ribosomes in chick auditory neurons. *Brain Research*, 809(2), 214–220.
Hyson, R. L., & Rubel, E. W. (1989). Transneuronal regulation of protein synthesis in the brainstem auditory system of the chick requires synaptic activation. *The Journal of Neuroscience*, 9(8), 2835–2845.

Irie, T., & Ohmori, H. (2008). Presynaptic GABA(B) receptors modulate synaptic facilitation and depression at distinct synapses in fusiform cells of mouse dorsal cochlear nucleus. *Biochemical and Biophysical Research Communications*, 367(2), 503–508.

Isaacson, J. S., & Walmsley, B. (1996). Amplitude and time course of spontaneous and evoked excitatory postsynaptic currents in bushy cells of the anteroventral cochlear nucleus. *Journal of Neurophysiology*, 76(3), 1566–1571.

Juiz, J. M., Albin, R. L., Helfert, R. H., & Altschuler, R. A. (1994). Distribution of GABAA and GABAB binding sites in the cochlear nucleus of the guinea pig. *Brain Research*, 639(2), 193–201.

Kalmbach, A., Kullmann, P. H., & Kandler, K. (2010). NMDAR-mediated calcium transients elicited by glutamate co-release at developing inhibitory synapses. *Frontiers in Synaptic Neuroscience*, 2, 27.

Kandler, K., & Friauf, E. (1995). Development of glycinergic and glutamatergic synaptic transmission in the auditory brainstem of perinatal rats. *The Journal of Neuroscience*, 15(10), 6890–6904.

Kandler, K., & Gillespie, D. C. (2005). Developmental refinement of inhibitory sound-localization circuits. *Trends in Neuroscience*, 28(6), 290–296.

Kantamneni, S. (2015). Cross-talk and regulation between glutamate and GABAB receptors. *Frontiers in Cellular Neuroscience*, 9, 135.

Kato, B. M., & Rubel, E. W. (1999). Glutamate regulates IP3-type and CICR stores in the avian cochlear nucleus. *Journal of Neurophysiology*, 81(4), 1587–1596.

Kato, B. M., Lachica, E. A., & Rubel, E. W. (1996). Glutamate modulates intracellular Ca^{2+} stores in brain stem auditory neurons. *Journal of Neurophysiology*, 76(1), 646–650.

Kim, S. J., Kim, Y. S., Yuan, J. P., Petralia, R. S., et al. (2003). Activation of the TRPC1 cation channel by metabotropic glutamate receptor mGluR1. *Nature*, 426(6964), 285–291.

Kopp-Scheinpflug, C., Steinert, J. R., & Forsythe, I. D. (2011). Modulation and control of synaptic transmission across the MNTB. *Hearing Research*, 279(1–2), 22–31.

Kotak, V. C., & Sanes, D. H. (1995). Synaptically evoked prolonged depolarizations in the developing auditory system. *Journal of Neurophysiology*, 74(4), 1611–1620.

Kotak, V. C., & Sanes, D. H. (1997). Deafferentation weakens excitatory synapses in the developing central auditory system. *European Journal of Neuroscience*, 9(11), 2340–2347.

Kuba, H., Koyano, K., & Ohmori, H. (2002). Synaptic depression improves coincidence detection in the nucleus laminaris in brainstem slices of the chick embryo. *European Journal of Neuroscience*, 15(6), 984–990.

Kuba, H., Oichi, Y., & Ohmori, H. (2010). Presynaptic activity regulates Na(+) channel distribution at the axon initial segment. *Nature*, 465(7301), 1075–1078.

Kushmerick, C., Price, G. D., Taschenberger, H., Puente, N., et al. (2004). Retroinhibition of presynaptic Ca^{2+} currents by endocannabinoids released via postsynaptic mGluR activation at a calyx synapse. *The Journal of Neuroscience*, 24(26), 5955–5965.

Lachica, E. A., Rubsamen, R., Zirpel, L., & Rubel, E. W. (1995). Glutamatergic inhibition of voltage-operated calcium channels in the avian cochlear nucleus. *The Journal of Neuroscience*, 15(3 Pt 1), 1724–1734.

Lachica, E. A., Kato, B. M., Lippe, W. R., & Rubel, E. W. (1998). Glutamatergic and GABAergic agonists increase $[Ca^{2+}]i$ in avian cochlear nucleus neurons. *Journal of Neurobiology*, 37(2), 321–337.

Levi-Montalcini, R. (1949). The development to the acoustico-vestibular centers in the chick embryo in the absence of the afferent root fibers and of descending fiber tracts. *The Journal of Comparative Neurology*, 91(2), 209–241.

Lu, T., & Trussell, L. O. (2007). Development and elimination of endbulb synapses in the chick cochlear nucleus. *The Journal of Neuroscience*, 27(4), 808–817.

Lu, Y. (2014). Metabotropic glutamate receptors in auditory processing. *Neuroscience*, 274, 429–445.

Lu, Y., & Rubel, E. W. (2005). Activation of metabotropic glutamate receptors inhibits high-voltage-gated calcium channel currents of chicken nucleus magnocellularis neurons. *Journal of Neurophysiology*, 93(3), 1418–1428.

Lu, Y., Harris, J. A., & Rubel, E. W. (2007). Development of spontaneous miniature EPSCs in mouse AVCN neurons during a critical period of afferent-dependent neuron survival. *Journal of Neurophysiology*, 97(1), 635–646.

Luscher, C., Nicoll, R. A., Malenka, R. C., & Muller, D. (2000). Synaptic plasticity and dynamic modulation of the postsynaptic membrane. *Nature Neuroscience*, 3(6), 545–550.

Ma, C. L., Kelly, J. B., & Wu, S. H. (2002). AMPA and NMDA receptors mediate synaptic excitation in the rat's inferior colliculus. *Hearing Research*, 168(1–2), 25–34.

Magnusson, A. K., Park, T. J., Pecka, M., Grothe, B., & Koch, U. (2008). Retrograde GABA signaling adjusts sound localization by balancing excitation and inhibition in the brainstem. *Neuron*, 59(1), 125–137.

Malenka, R. C. (1994). Synaptic plasticity in the hippocampus: LTP and LTD. *Cell*, 78(4), 535–538.

Martinez-Galan, J. R., Perez-Martinez, F. C., & Juiz, J. M. (2012). Signalling routes and developmental regulation of group I metabotropic glutamate receptors in rat auditory midbrain neurons. *Journal of Neuroscience Research*, 90(10), 1913–1923.

Mayer, F., Albrecht, O., Dondzillo, A., & Klug, A. (2014). Glycinergic inhibition to the medial nucleus of the trapezoid body shows prominent facilitation and can sustain high levels of ongoing activity. *Journal of Neurophysiology*, 112(11), 2901–2915.

Nicholas, A. H., & Hyson, R. L. (2004). Group I and II metabotropic glutamate receptors are necessary for the activity-dependent regulation of ribosomes in chick auditory neurons. *Brain Research*, 1014(1–2), 110–119.

Nicoletti, F., Wroblewski, J. T., Novelli, A., Alho, H., et al. (1986). The activation of inositol phospholipid metabolism as a signal-transducing system for excitatory amino acids in primary cultures of cerebellar granule cells. *The Journal of Neuroscience*, 6(7), 1905–1911.

Nicoletti, F., Bruno, V., Ngomba, R. T., Gradini, R., & Battaglia, G. (2015). Metabotropic glutamate receptors as drug targets: What's new? *Current Opinion in Pharmacology*, 20, 89–94.

Niedzielski, A. S., & Wenthold, R. J. (1995). Expression of AMPA, kainate, and NMDA receptor subunits in cochlear and vestibular ganglia. *The Journal of Neuroscience*, 15(3 Pt 2), 2338–2353.

Nishimaki, T., Jang, I. S., Ishibashi, H., Yamaguchi, J., & Nabekura, J. (2007). Reduction of metabotropic glutamate receptor-mediated heterosynaptic inhibition of developing MNTB-LSO inhibitory synapses. *European Journal of Neuroscience*, 26(2), 323–330.

Niswender, C. M., & Conn, P. J. (2010). Metabotropic glutamate receptors: Physiology, pharmacology, and disease. *Annual Review of Pharmacology and Toxicology*, 50, 295–322.

Okuda, H., Yamada, R., Kuba, H., & Ohmori, H. (2013). Activation of metabotropic glutamate receptors improves the accuracy of coincidence detection by presynaptic mechanisms in the nucleus laminaris of the chick. *The Journal of Physiology*, 591(Pt 1), 365–378.

Oleskevich, S., & Walmsley, B. (2002). Synaptic transmission in the auditory brainstem of normal and congenitally deaf mice. *The Journal of Physiology*, 540(Pt 2), 447–455.

Oleskevich, S., Youssoufian, M., & Walmsley, B. (2004). Presynaptic plasticity at two giant auditory synapses in normal and deaf mice. *The Journal of Physiology*, 560(Pt 3), 709–719.

Otis, T. S., & Trussell, L. O. (1996). Inhibition of transmitter release shortens the duration of the excitatory synaptic current at a calyceal synapse. *Journal of Neurophysiology*, 76(5), 3584–3588.

Otis, T. S., Raman, I. M., & Trussell, L. O. (1995). AMPA receptors with high Ca^{2+} permeability mediate synaptic transmission in the avian auditory pathway. *The Journal of Physiology*, 482 (Pt 2), 309–315.

Paoletti, P., Bellone, C., & Zhou, Q. (2013). NMDA receptor subunit diversity: Impact on receptor properties, synaptic plasticity and disease. *Nature Reviews Neuroscience*, 14(6), 383–400.

Parks, T. N. (2000). The AMPA receptors of auditory neurons. *Hearing Research*, 147(1–2), 77–91.

Petralia, R. S., Wang, Y. X., Mayat, E., & Wenthold, R. J. (1997). Glutamate receptor subunit 2-selective antibody shows a differential distribution of calcium-impermeable AMPA receptors among populations of neurons. *The Journal of Comparative Neurology*, 385(3), 456–476.

Petralia, R. S., Rubio, M. E., Wang, Y. X., & Wenthold, R. J. (2000). Differential distribution of glutamate receptors in the cochlear nuclei. *Hearing Research*, 147(1–2), 59–69.

Pliss, L., Yang, H., & Xu-Friedman, M. A. (2009). Context-dependent effects of NMDA receptors on precise timing information at the endbulb of Held in the cochlear nucleus. *Journal of Neurophysiology*, 102(5), 2627–2637.

Raman, I. M., Zhang, S., & Trussell, L. O. (1994). Pathway-specific variants of AMPA receptors and their contribution to neuronal signaling. *The Journal of Neuroscience*, 14(8), 4998–5010.

Ravindranathan, A., Parks, T. N., & Rao, M. S. (1997). New isoforms of the chick glutamate receptor subunit GluR4: Molecular cloning, regional expression and developmental analysis. *Brain Research Molecular Brain Research*, 50(1–2), 143–153.

Ravindranathan, A., Donevan, S. D., Sugden, S. G., Greig, A., et al. (2000). Contrasting molecular composition and channel properties of AMPA receptors on chick auditory and brainstem motor neurons. *The Journal of Physiology*, 523(Pt 3), 667–684.

Rondard, P., & Pin, J. P. (2015). Dynamics and modulation of metabotropic glutamate receptors. *Current Opinion in Pharmacology*, 20, 95–101.

Rubel, E. W., & Fritzsch, B. (2002). Auditory system development: Primary auditory neurons and their targets. *Annual Review of Neuroscience*, 25, 51–101.

Sahley, T. L., Hammonds, M. D., & Musiek, F. E. (2013). Endogenous dynorphins, glutamate and *N*-methyl-D-aspartate (NMDA) receptors may participate in a stress-mediated type-I auditory neural exacerbation of tinnitus. *Brain Research*, 1499, 80–108.

Sanchez, J. T., Gans, D., & Wenstrup, J. J. (2007). Contribution of NMDA and AMPA receptors to temporal patterning of auditory responses in the inferior colliculus. *The Journal of Neuroscience*, 27(8), 1954–1963.

Sanchez, J. T., Wang, Y., Rubel, E. W., & Barria, A. (2010). Development of glutamatergic synaptic transmission in binaural auditory neurons. *Journal of Neurophysiology*, 104(3), 1774–1789.

Sanchez, J. T., Seidl, A. H., Rubel, E. W., & Barria, A. (2012). Control of neuronal excitability by NMDA-type glutamate receptors in early developing binaural auditory neurons. *The Journal of Physiology*, 590(Pt 19), 4801–4818.

Sanchez, J. T., Ghelani, S., & Otto-Meyer, S. (2015). From development to disease: Diverse functions of NMDA-type glutamate receptors in the lower auditory pathway. *Neuroscience*, 285, 248–259.

Sanes, D. H., McGee, J., & Walsh, E. J. (1998). Metabotropic glutamate receptor activation modulates sound level processing in the cochlear nucleus. *Journal of Neurophysiology*, 80(1), 209–217.

Sanz-Clemente, A., Nicoll, R. A., & Roche, K. W. (2013). Diversity in NMDA receptor composition: Many regulators, many consequences. *Neuroscientist*, 19(1), 62–75.

Schneggenburger, R., Sakaba, T., & Neher, E. (2002). Vesicle pools and short-term synaptic depression: Lessons from a large synapse. *Trends in Neuroscience*, 25(4), 206–212.

Shore, S. E. (2011). Plasticity of somatosensory inputs to the cochlear nucleus: Implications for tinnitus. *Hearing Research*, 281(1–2), 38–46.

Sladeczek, F., Pin, J. P., Recasens, M., Bockaert, J., & Weiss, S. (1985). Glutamate stimulates inositol phosphate formation in striatal neurones. *Nature*, 317(6039), 717–719.

Stange, A., Myoga, M. H., Lingner, A., Ford, M. C., et al. (2013). Adaptation in sound localization: From GABA(B) receptor-mediated synaptic modulation to perception. *Nature Neuroscience*, 16(12), 1840–1847.

Steinert, J. R., Postlethwaite, M., Jordan, M. D., Chernova, T., et al. (2010). NMDAR-mediated EPSCs are maintained and accelerate in time course during maturation of mouse and rat auditory brainstem in vitro. *The Journal of Physiology*, 588(Pt 3), 447–463.

Stincic, T. L., & Hyson, R. L. (2011). The localization and physiological effects of cannabinoid receptor 1 in the brain stem auditory system of the chick. *Neuroscience*, 194, 150–159.

Sun, H., Ma, C. L., Kelly, J. B., & Wu, S. H. (2006). GABAB receptor-mediated presynaptic inhibition of glutamatergic transmission in the inferior colliculus. *Neuroscience Letters*, 399(1–2), 151–156.

Takahashi, T., Forsythe, I. D., Tsujimoto, T., Barnes-Davies, M., & Onodera, K. (1996). Presynaptic calcium current modulation by a metabotropic glutamate receptor. *Science*, 274(5287), 594–597.

Tang, Y. Z., & Carr, C. E. (2004). Development of NMDA R1 expression in chicken auditory brainstem. *Hearing Research*, 191(1–2), 79–89.

Tang, Y. Z., & Carr, C. E. (2007). Development of *N*-methyl-D-aspartate receptor subunits in avian auditory brainstem. *The Journal of Comparative Neurology*, 502(3), 400–413.

Tang, Z. Q., Liu, Y. W., Shi, W., Dinh, E. H., et al. (2013). Activation of synaptic group II metabotropic glutamate receptors induces long-term depression at GABAergic synapses in CNS neurons. *The Journal Neuroscience*, 33(40), 15964–15977.

Tomita, S., Chen, L., Kawasaki, Y., Petralia, R. S., et al. (2003). Functional studies and distribution define a family of transmembrane AMPA receptor regulatory proteins. *The Journal of Cell Biology*, 161(4), 805–816.

Traynelis, S. F., Wollmuth, L. P., McBain, C. J., Menniti, F. S., et al. (2010). Glutamate receptor ion channels: structure, regulation, and function. *Pharmacology Review*, 62(3), 405–496.

Trussell, L. (1998). Control of time course of glutamatergic synaptic currents. *Progress in Brain Research*, 116, 59–69.

Trussell, L. O. (1999). Synaptic mechanisms for coding timing in auditory neurons. *Annual Review of Physiology*, 61, 477–496.

Trussell, L. O. (2002). Transmission at the hair cell synapse. *Nature Neuroscience*, 5(2), 85–86.

Trussell, L. O., & Fischbach, G. D. (1989). Glutamate receptor desensitization and its role in synaptic transmission. *Neuron*, 3(2), 209–218.

Turecek, R., & Trussell, L. O. (2001). Presynaptic glycine receptors enhance transmitter release at a mammalian central synapse. *Nature*, 411(6837), 587–590.

Tzounopoulos, T. (2008). Mechanisms of synaptic plasticity in the dorsal cochlear nucleus: plasticity-induced changes that could underlie tinnitus. *American Journal of Audiology*, 17(2), S170–175.

Tzounopoulos, T., Kim, Y., Oertel, D., & Trussell, L. O. (2004). Cell-specific, spike timing-dependent plasticities in the dorsal cochlear nucleus. *Nature Neuroscience*, 7(7), 719–725.

Vale, C., & Sanes, D. H. (2002). The effect of bilateral deafness on excitatory and inhibitory synaptic strength in the inferior colliculus. *European Journal of Neuroscience*, 16(12), 2394–2404.

Voytenko, S. V., & Galazyuk, A. V. (2011). mGluRs modulate neuronal firing in the auditory midbrain. *Neuroscience Letters*, 492(3), 145–149.

Walmsley, B., Berntson, A., Leao, R. N., & Fyffe, R. E. (2006). Activity-dependent regulation of synaptic strength and neuronal excitability in central auditory pathways. *The Journal of Physiology*, 572(Pt 2), 313–321.

Wang, L. C., Tang, Z. Q., & Lu, Y. (2012). Synaptic activity-induced Ca(2+) signaling in avian cochlear nucleus magnocellularis neurons. *Neuroscience Reseatch*, 72(2), 129–139.

Wang, Y., Cunningham, D. E., Tempel, B. L., & Rubel, E. W. (2009). Compartment-specific regulation of plasma membrane calcium ATPase type 2 in the chick auditory brainstem. *The Journal of Comparative Neurology*, 514(6), 624–640.

Wenzel, A., Villa, M., Mohler, H., & Benke, D. (1996). Developmental and regional expression of NMDA receptor subtypes containing the NR2D subunit in rat brain. *Journal of Neurochemistry*, 66(3), 1240–1248.

Wong, A. Y., Billups, B., Johnston, J., Evans, R. J., & Forsythe, I. D. (2006). Endogenous activation of adenosine A1 receptors, but not P2X receptors, during high-frequency synaptic transmission at the calyx of Held. *Journal of Neurophysiology*, 95(6), 3336–3342.

Wu, S. H., & Fu, X. W. (1998). Glutamate receptors underlying excitatory synaptic transmission in the rat's lateral superior olive studied in vitro. *Hearing Research*, 122(1–2), 47–59.

Wu, S. H., Ma, C. L., Sivaramakrishnan, S., & Oliver, D. L. (2002). Synaptic modification in neurons of the central nucleus of the inferior colliculus. *Hearing Research*, 168(1–2), 43–54.

Xu-Friedman, M. A., & Regehr, W. G. (2004). Structural contributions to short-term synaptic plasticity. *Physiological Reviews*, 84(1), 69–85.

Yang, Y. M., Aitoubah, J., Lauer, A. M., Nuriya, M., et al. (2011). GluA4 is indispensable for driving fast neurotransmission across a high-fidelity central synapse. *The Journal of Physiology*, 589(Pt 17), 4209–4227.

Youssoufian, M., Oleskevich, S., & Walmsley, B. (2005). Development of a robust central auditory synapse in congenital deafness. *Journal of Neurophysiology*, 94(5), 3168–3180.

Zhang, H., & Kelly, J. B. (2001). AMPA and NMDA receptors regulate responses of neurons in the rat's inferior colliculus. *Journal of Neurophysiology*, 86(2), 871–880.

Zhang, S., & Trussell, L. O. (1994). Voltage clamp analysis of excitatory synaptic transmission in the avian nucleus magnocellularis. *The Journal of Physiology*, 480(Pt 1), 123–136.

Zirpel, L., & Parks, T. N. (2001). Zinc inhibition of group I mGluR-mediated calcium homeostasis in auditory neurons. *Journal of the Association of Research in Otolaryngology*, 2(2), 180–187.

Zirpel, L., & Rubel, E. W. (1996). Eighth nerve activity regulates intracellular calcium concentration of avian cochlear nucleus neurons via a metabotropic glutamate receptor. *Journal of Neurophysiology*, 76(6), 4127–4139.

Zirpel, L., Nathanson, N. M., Rubel, E. W., & Hyson, R. L. (1994). Glutamate-stimulated phosphatidylinositol metabolism in the avian cochlear nucleus. *Neuroscience Letters*, 168(1–2), 163–166.

Zirpel, L., Lachica, E. A., & Rubel, E. W. (1995). Activation of a metabotropic glutamate receptor increases intracellular calcium concentrations in neurons of the avian cochlear nucleus. *The Journal of Neuroscience*, 15(1 Pt 1), 214–222.

Zirpel, L., Lippe, W. R., & Rubel, E. W. (1998). Activity-dependent regulation of $[Ca^{2+}]i$ in avian cochlear nucleus neurons: Roles of protein kinases A and C and relation to cell death. *Journal of Neurophysiolology*, 79(5), 2288–2302.

Zirpel, L., Janowiak, M. A., Taylor, D. A., & Parks, T. N. (2000). Developmental changes in metabotropic glutamate receptor-mediated calcium homeostasis. *The Journal of Comparative Neurology*, 421(1), 95–106.

Zucker, R. S., & Regehr, W. G. (2002). Short-term synaptic plasticity. *Annual Review in Physiology*, 64, 355–405.

Chapter 5
Development and Function of Inhibitory Circuitry in the Avian Auditory Brainstem

R. Michael Burger

Abstract The avian auditory system has provided an excellent model circuit to explore important features of acoustic processing. For example, in the chicken (*Gallus gallus*) system, in 1975 Parks and Rubel first confirmed the coincidence detection based delay line model of sound localization in vertebrates that was originally proposed by Jeffress in 1948. Further, the system provides an unmatched experimental substrate to investigate cellular physiology and morphology in light of computational function. Many anatomical and physiologically specialized features of auditory neurons have been identified first in birds, and often complementary properties are observed in mammals. These discoveries have contributed substantially to our general understanding of processing of acoustic signals, and the function of inhibition specifically. However, many mechanistic features of inhibitory physiology in the avian system contrast sharply with those of mammals, while achieving similar computational outcomes. This chapter reviews the major progress made toward understanding inhibitory roles in auditory function with a focus on three areas: (1) development of inhibitory circuitry, (2) functional organization of the inhibitory network, and (3) synaptic physiology of inhibition in birds. Although many specific mechanisms of inhibition in birds differ from those of mammals, these circuits exhibit remarkable convergence when viewed from a functional perspective.

Keywords Binaural • Corelease • Depolarizing inhibition • GABA • Superior olivary nucleus

R.M. Burger (✉)
Department of Biological Sciences, Lehigh University,
111 Research Dr., Bethlehem, PA 18015, USA
e-mail: burger@lehigh.edu

K.S. Cramer et al. (eds.), *Auditory Development and Plasticity*, Springer Handbook of Auditory Research 64, DOI 10.1007/978-3-319-21530-3_5

5.1 Introduction: The Elegant Circuit

Studies of the auditory system have greatly contributed to the broader field of neuroscience by drawing a clear connection among the identity, physiology, and function of neurons composing auditory brainstem circuitry. This clarity derives both from the anatomical arrangement of groups of neurons that share functional homogeneity and from the functional context itself. This principle is most well exemplified by the auditory brainstem pathway of birds. Within this system, some of the most significant advances in cell and synaptic physiology derive from investigation of the neural circuitry devoted to processing acoustic cues that are important for sound localization. Sound localization in azimuth depends on neural computation of time or intensity difference cues at the two ears for low- or high-frequency signals, respectively. Birds are generally limited to low-frequency hearing (<5 kHz), compared to mammals, and have evolved an experimentally tractable circuit to compute interaural time differences (ITDs). One is hard pressed to identify any vertebrate neural circuit in which cellular physiology, morphology, and function have provided a more fruitfully comprehensive and mechanistic understanding.

Put simply, the auditory system extracts spatial information from temporal features of sound stimuli. ITDs are the time difference cues created at the two ears by sound emanating from different regions of space. ITDs vary systematically with a sound source's angle of incidence on the listener, and at a maximum are very small, in the range of tens to hundreds of microseconds, depending on head size. Given that the physiological range of ITDs is so small, processing ITD cues demands extreme temporal precision at the limits of what is theoretically possible by neurons. Several chapters in this volume describe the specializations in ion channels, receptors, and cell morphology that have been discovered in this circuit to enhance this precise computation. This chapter focuses on the development and physiology of inhibition in the avian auditory circuit, a component of the circuitry, which is now well illuminated by studies by the Rubel group and others.

Inhibitory circuitry in the avian auditory brainstem has been under intensive investigation in the last two decades, and its contributions to the broader ITD processing circuit are now well understood. It influences the computational processes of individual neurons, but also provides a broad network modulatory influence at the systems level. This chapter briefly describes the basic elements of this circuit, and then provides a detailed overview of the development and functional contributions made by inhibitory components of the circuit.

5.1.1 The Excitatory Pathways of the ITD Processing Circuit

The excitatory pathways in the temporal processing circuitry involved in ITD processing in birds and mammals have been discussed extensively in earlier works, but are briefly described here (Fig. 5.1) (Burger and Rubel 2008; Grothe et al. 2010).

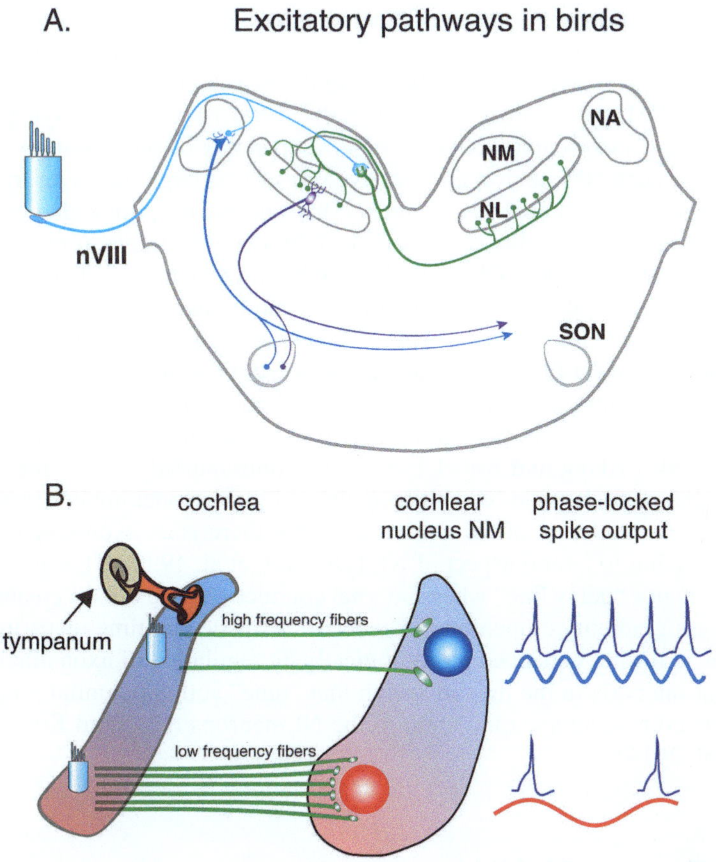

Fig. 5.1 (**A**) Excitatory circuitry in the avian brainstem auditory system. The auditory nerve enters the brain and bifurcates to innervate NA and NM. NA projects to the SON and higher order centers (blue pathway). NM is the first central synapse in the temporal processing pathway. NM projects bilaterally to NL. NL in turn also provides excitatory input to the SON. (**B**) Neurons in the temporal processing pathway respond to acoustic signals with phase-locked discharges originating in the frequency tuned auditory nerve fibers. Phase-locking is preserved or sharpened for ITD processing in NL. [**A** adapted from Burger et al. (2011), with permission]

The circuit leading to computation of ITDs begins with the auditory nerve fibers, where temporal features of the acoustic signal are encoded by phase-locked discharges. Each fiber is tuned to a specific frequency determined by its hair cell input on the cochlear basilar papilla. Phase locking converts temporal information in the acoustic signal into a reliable and precisely timed spike train (Fig. 5.1B). Phase-locked discharges originating from each ear are then compared by binaural neurons that compute ITDs.

The auditory fibers in the eighth cranial nerve (CN VIII) of the chicken enter the brain in the dorsal medulla and branches to innervate two nuclei (Fig. 5.1). The first, the nucleus angularis (NA), contains a diversity of neural types that represent various aspects of the acoustic signal, including sound intensity (Soares et al. 2002; Köppl and Carr 2003). The second target of auditory CN VIII axons, nucleus magnocellularis (NM), constitutes the first central synapse in the temporal processing pathway. In NM, the auditory nerve generally forms large endbulb synapses, providing a strong and reliable excitatory input. Depending on the frequency tuning of the neurons, each NM neuron may receive one or two large inputs or eight to ten smaller inputs from auditory nerve fibers. This synapse is optimized to ensure that phase locking is preserved or even sharpened through integrative summation at this synapse (Fukui et al. 2006; Oline and Burger 2014; Oline et al. 2016).

The key feature of ITD encoding arises in the bilateral projections of NM to nucleus laminaris (NL), the first site of binaural computation and the principal computing region for ITDs (Parks and Rubel 1975; Carr and Konishi 1990). Each NM neuron produces a bifurcating axon in which the ipsilateral branch loops back through NM and projects with equal length axons to the dorsal dendrites of the ipsilateral NL (Young and Rubel 1983). The contralaterally projecting NM collateral to the opposite side NL is distinguished by the branching of the NM axon along the ventral course of the lamina such that there is an axon length gradient from the medial to lateral aspect of NL (Overholt et al. 1992). This anatomically forms the major "delay line" where internal conduction time offsets created by the axon length gradient compensate for external acoustic cue time disparities. This delay line is further enhanced physiologically by regulation of axon diameter and internodal intervals in the myelin sheath that "tune" action potential propagation to provide coincident synaptic input to the NL neurons (Carr and Konishi 1990; Seidl et al. 2014).

5.1.2 Sources of Inhibition

In contrast to mammalian brainstem auditory circuitry, which comprises a myriad of functionally distinct clusters of inhibitory neurons, the inhibitory components of the avian circuit are primarily, but not exclusively, mediated by a single source, the superior olivary nucleus (SON). SON is a predominantly GABAergic nucleus that lies ventral to NM and NL (Carr et al. 1989; Lachica et al. 1994). The SON is driven by input from the ipsilateral nucleus angularis (NA) and NL, and feeds inhibition back on NA, NM, and NL with divergent unilateral projections. Single SON axons may innervate neurons in multiple target nuclei on a single side among NA, NM, and NL (Burger et al. 2005). A second output pathway from SON projects to its contralateral counterpart, forming a mutually inhibitory circuit between them. This reciprocally inhibitory pathway plays a key role in the circuit level function of the SON in mediating ITD selective responses in NL neurons. This is discussed in Sect. 5.4. A second source of inhibition in the circuit is a small cluster of GABAergic neurons situated between NM and NL that appear to provide a very specialized inhibitory input to low-frequency NL neurons (Sect. 5.3.3).

The architecture of the major excitatory and inhibitory projections, taken together, composes a relatively simple and experimentally tractable system to understand inhibition's role in auditory function (Fig. 5.2B). The principal inhibitory neurons are physically separated from their targets, allowing for independent manipulations of each SON. This arrangement has lent itself to major progress in

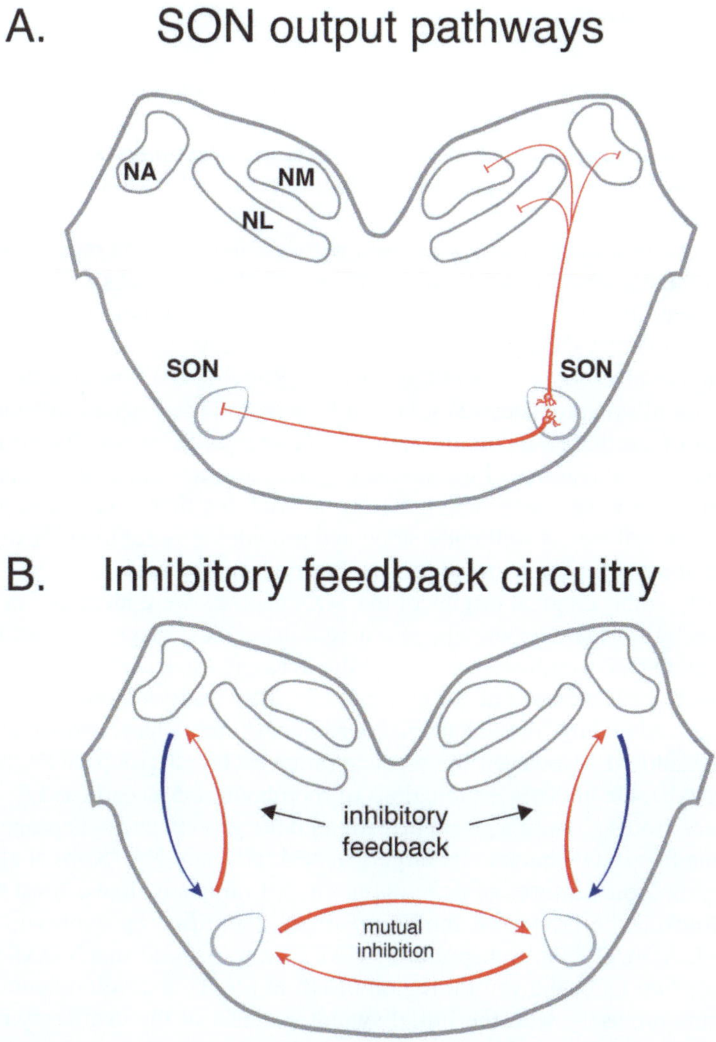

Fig. 5.2 Inhibitory pathways in the avian brainstem form negatively coupled feedback loops. (**A**) SON has two major output pathways: (1) a descending divergent pathway to NM, NA, and NL and (2) an ascending pathway to the contralateral SON. (**B**) When considered together with the excitatory drive derived from each ear, the SONs each form a negative feedback loop. A reciprocal feed-forward inhibition arrangement between the SONs will tend to balance levels of activity throughout the system bilaterally. [Adapted from Burger et al. (2011), with permission]

developmental, physiological, and computational neuroscience. The following sections describe the development, cell physiology, and finally systems level function of the inhibitory circuitry as it pertains to ITD computations in the avian system.

5.2 Embryonic Origins and Development of SON Innervation

A tremendous advantage of investigating neural function in birds is the remarkable access to the embryo at virtually any stage of development. As a result, the chick auditory system has no equal in terms of detailed descriptive and experimental development of the auditory pathway, including the development of the inhibitory components of the circuit.

The chick gestational period is 21 days, with hearing maturing rapidly compared to that in rodents, which generally do not hear until about 2 weeks after birth. Lippe (1994) showed that hearing onset occurs as early as embryonic day 14 (E14) in chickens, a time that shortly follows the initial excitatory innervation of NM by the auditory nerve (Jackson et al. 1982; Molea and Rubel 2003) 1 week before hatching. This initial synaptogenesis is followed by a period of synapse elimination and maturation of excitatory inputs through E17 (Rubel and Fritzsch 2002). By E18, most physiological features of the auditory system appear to be mature, and chicks begin to hear airborne sound within the egg around this time. The development of the excitatory pathway is well understood and provides a useful timeline to contextualize the more difficult to investigate inhibitory development.

The early embryological origins of the SON neurons were identified in innovative studies by Karina Cramer, who was able to label early stage, E2, precursors of NM, NL, and SON in embryos with lipophilic dyes and track their migration to their mature positions (Cramer et al. 2000) (Fig. 5.3). This study showed that SON neurons are derived mainly from the medial portion of rhombomere 5, which also gives rise to NM and NL. However, the precursors for each cell group, SON, NM, and NL, are localizable in distinct subregions of rhombomere 5 as early as E2.

In a later study, Cramer's group investigated the growth and synaptogenesis of projections from SON back to its targets in NM, NL, and NA (Korn et al. 2012). Using organotypic cultures of developing chick brainstem slices, Korn and colleagues tracked the growth of inhibitory fibers, identified by expression of the vesicular GABA/glycine transporter (VGAT). They showed that VGAT-positive SON fibers first enter the ventral region of NL at E9. This developmental period corresponds precisely with the initial synaptogenesis of the excitatory cochlear afferents in the NM occurring about 2 days before the first detectable evidence of hearing in the embryo (Jhaveri and Morest 1982; Rubel and Fritzsch 2002). GABAergic input to NM neurons is detected physiologically as early as E12 (Howard et al. 2007). VGAT-positive fibers gradually increase in density until at least E17, when they reach a stable staining intensity. Importantly, this study showed that astrocyte-secreted factors contribute to the maturation of these

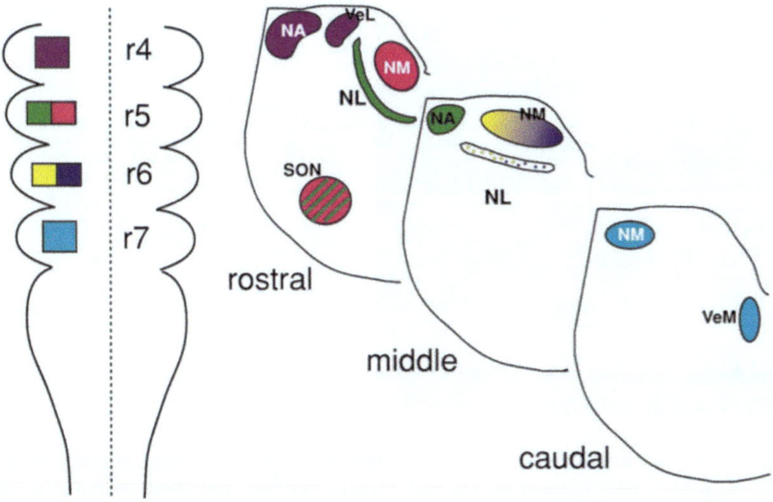

Fig. 5.3 The SON shares embryological origins with NM and NL. Dye injection in rhombomere 5 at E2 labels precursors of SON neurons. Injection mapping is indicated by colored squares. The same population of precursors gives rise to NM and NL neurons, particularly those that reside in the rostral portion of both nuclei in mature animals. [Reproduced from Cramer et al. (2000), with permission]

synapses during this period. Thus, based on the limited evidence available to date, the projections and maturation of the descending SON output pathway appear to mature in correspondence to the developmental timeline of the excitatory inputs, but with a delay of 1–2 days.

The central inhibitory components of the circuit had been difficult to investigate because the neurons originate within the auditory anlage with NM, NA, and NL neurons, then migrate to their positions in SON before projecting back to the cochlear nuclei and NL. Earlier embryological and developmental studies investigating the inhibitory innervation of NM and NL relied on immunohistochemical techniques to label synapse-forming fibers in NM, NA, and NL. These studies focused on identifying the neurotransmitter profile of inhibitory input and provided preliminary descriptions of inhibitory synaptogenesis and morphology.

Immunohistochemical studies, conducted primarily by Code and colleagues, showed that markers for GABAergic inhibitory input arise in chicken NM as early as E12. Strong GABAergic terminal labeling emerges by E15 and is characterized by small, diffuse, and pervasive boutons surrounding NM and NL somas (Fig. 5.4) (Code et al. 1989). This finding in chick complements the developmental studies by the Cramer group, and corresponds well to the description of GABAergic terminals in the NM of the barn owl (*Tyto alba*) (Fig. 5.4C) (Carr et al. 1989). The diffuse small bouton morphology suggests many available inhibitory synaptic release sites, a finding that is physiologically supported by graded GABAergic responses observed in physiological investigations of NM and NL (Yang et al. 1999; Lu and Trussell 2000).

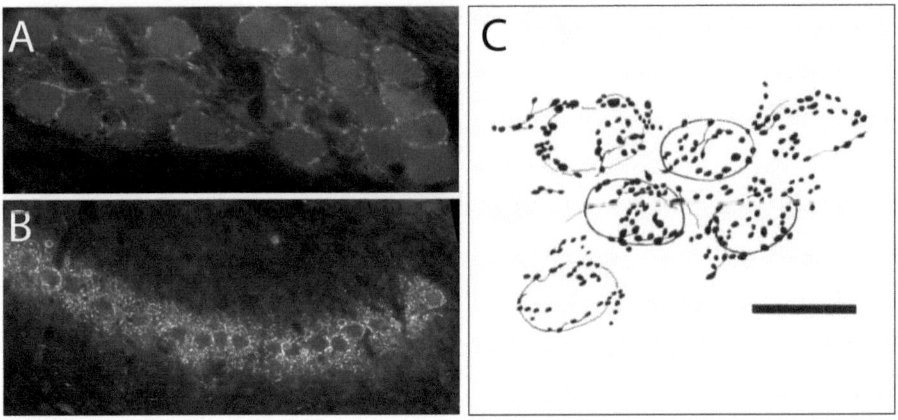

Fig. 5.4 GABAergic terminals in NM and NL form small diffuse boutons in chickens and barn owls. (**A**, **B**) Anti-GABA staining in NM and NL show diffuse small bouton-like terminals surrounding NM and NL somas. (**C**) A similar pattern was observed with anti-GAD staining in the barn owl as reconstructed in the camera lucida drawing. [**C** reproduced from Carr et al. (1989), with permission]

Carr et al. (1989) and Lachica et al. (1994) showed that the likely source of GABA was the SON. These studies confirmed the SON as the predominant inhibitory center in the avian auditory brainstem circuit, and strongly suggested that its output is primarily GABAergic.

These results were interesting in light of the fact that glycine is generally the dominant inhibitory neurotransmitter in the mature mammalian brainstem auditory system, but for many years has not been considered important in the avian system. Code and Rubel (1989) investigated markers for glycine, but found that the immunostaining was much weaker than GABA and thus unlikely to contribute substantially to the inhibition in the temporal processing nuclei. This finding was generally confirmed by later physiological studies that showed complete block of synaptically evoked inhibition at NM or NL by bicuculline, a $GABA_A$ receptor antagonist (Yang et al. 1999; Monsivais et al. 2000). Taken together, these physiological and anatomical findings led to an incorrect conclusion that glycinergic signaling was absent from the auditory brainstem in birds, in sharp contrast to NM and NL's analogs in the cochlear nucleus and medial superior olive of mammals. More recent work has shown that this conclusion is an oversimplification, and glycinergic transmission is indeed present, but only under certain conditions (Sect. 5.3.2).

5.3 Inhibitory Synaptic Physiology: Slow, Depolarizing, and Potent

A paradoxical and unusual property of the predominantly GABAergic influence on NM and NL is that it is depolarizing, but maintains a strong inhibitory effect. Depolarizing GABA and glycinergic input is a common feature of developing nervous systems in birds and mammals, where it contributes to synaptogenesis and synaptic maturation. The depolarizing nature of inhibition in the developing brain contributes to postsynaptic neuron survival by engaging Ca^{2+}-dependent mechanisms of plasticity (Ge et al. 2006). In most cases, the upregulation of chloride–potassium cotransporters at maturity shifts the reversal potential for inhibitory input into the hyperpolarizing range by reducing the internal chloride concentration of neurons (Represa and Ben-Ari 2005). In contrast, NM and NL appear to retain a depolarized Cl^- reversal potential throughout life.

5.3.1 GABAergic Input to NM and NL Is Depolarizing

Depolarizing inhibition was first observed, but misidentified as an excitatory input, in innovative in vitro whole brainstem preparations from the Rubel group (Hackett et al. 1982; Jackson et al. 1982). These studies featured intracellular recordings of NM neurons from acute whole brainstem preparations where stimulation of the auditory nerve root engaged the intact auditory system. Jackson et al. (1982) identified two "excitatory" synaptic inputs. First, they observed a very fast excitatory input that was marked by two to three discrete stepwise amplitudes depending on stimulation strength across multiple trials (Fig. 5.5A). Stepwise postsynaptic potentials (PSPs), as opposed to graded responses, are indicative of the presence of just a few large amplitude inputs. The "steps" in the PSP represent a summation of subset combinations or the full complement of inputs, such that each response amplitude corresponds to the recruitment of one, two, or up to a few individual input fibers. These inputs are now confirmed to provide fast phase-locked excitation to NM.

The second "excitatory input" was a very slow depolarizing input of unknown origin, and appeared over a finely graded range of amplitudes (Fig. 5.5A). In a later in vitro brain slice study of NL, Hyson et al. (1995) observed slow depolarizing responses to exogenously applied GABA in NL (Fig. 5.5B). It is now known that the input derived from the SON mediates depolarizing responses in both NM and NL via divergent inhibitory axons to both targets. Interestingly, over the period of early development preceding E18, when inhibitory synapses are being formed, the GABAergic input provides a net excitatory input to NM and NL neurons (Howard et al. 2007). This developmental sequence of excitatory followed by inhibitory postsynaptic responses is entirely consistent with the pattern of GABAergic development early in mammalian circuits (Kandler and Friauf 1995). It differs only in that both the early and late responses result from depolarizing Cl^- conductances.

Intensive investigation of GABA physiology in NM in the early 2000s by both the Rubel and Trussell groups yielded insights into the mechanistic underpinnings of the depolarizing inhibition. Two key features ensure that the depolarizing responses to GABA are inhibitory. First, the GABAergic depolarization is kinetically rather slow at onset, and generates a plateau depolarization at sustained input rates due to a buildup of asynchronous GABA release (Yang et al. 1999; Lu and Trussell 2000). The slow rate of depolarization subjects voltage-gated sodium channels to inactivation, thus raising the input threshold (Monsivais and Rubel 2001).

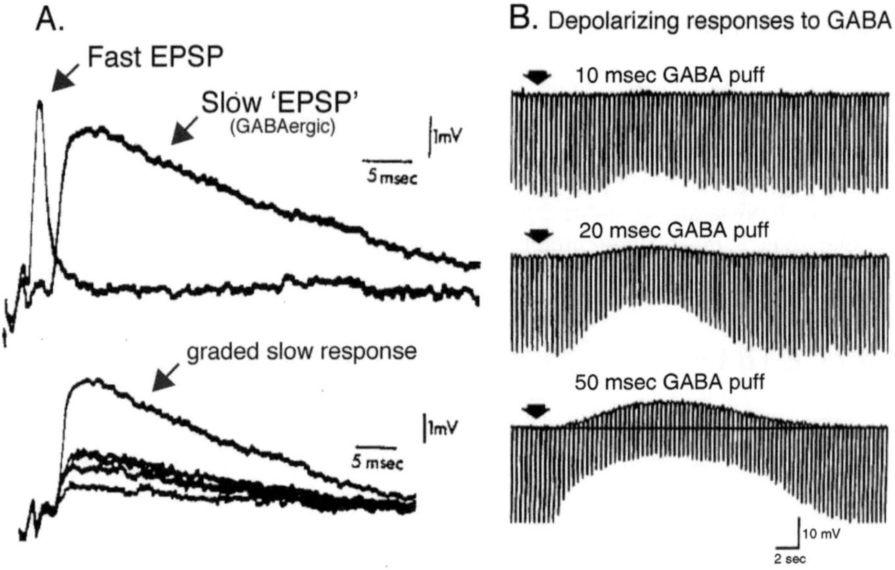

Fig. 5.5 GABAergic input is depolarizing in mature NM and NL neurons. (**A**) Fast excitatory and slow inhibitory depolarizing responses can be observed postsynaptically in NM neurons following electrical stimulation of synaptic inputs (*top panel*). "Slow" depolarizing responses yield graded increases in response amplitudes, as stimulation amplitude is increased (*bottom panel*). (**B**) Depolarizing responses are evoked from "puff" application of GABA directly onto NL neurons in vitro. The repetitive hyperpolarizing pulses were applied to assess input resistance. The input resistance declines during the GABA application. [**A** modified from Hackett et al. (1982). **B** reproduced from Hyson et al. (1995), with permission]

Second, the slow depolarization engages rapidly activating low threshold potassium currents at voltages just above rest. The potassium conductance dramatically decreases input resistance, and powerfully reduces membrane excitability. The combined effects of Na^+ channel inactivation and low-voltage K^+ channel activation generally produces a net inhibitory input that is more effective at spike suppression than a hyperpolarizing conductance of equal strength (Monsivais and Rubel 2001).

Contemporary with these studies, Lu and Trussell (2001) showed that under certain conditions in vitro, GABA could occasionally drive spiking in NM at stimulus onset before the plateau inhibition they described fully develops. Given that NM appeared specialized to detect temporally precise auditory nerve input, "false-positive" responses from a temporally imprecise inhibitory input seemed to be at odds with the circuit's function.

The ambiguity surrounding whether the GABAergic input is indeed inhibitory or excitatory was resolved by a later investigation of the physiology of GABA in NM. Howard et al. (2007) demonstrated that in the early post–hearing onset period (<E 18), GABAergic input is indeed excitatory, confirming the results of Lu and Trussell (2001), just as it is in young mammals. However, after E18, while the depolarizing response persists in birds, GABAergic input is almost always

strongly inhibitory, as described by Monsivais and Rubel (2001). The developmental "switch" depends on the upregulation of the low threshold potassium conductance, which matures at around E18. Following this stage, GABAergic depolarization and low-input resistance become physiologically coupled, and the net effect becomes inhibitory. Interestingly, this voltage gated coupling is so powerful that under a low level of inhibition, even large subthreshold excitatory PSPs from the auditory nerve can render NM neurons refractory, even in the total absence of spiking (Howard and Rubel 2010). The key feature of the refractory response is also the inhibition-coupled K^+ conductance, a strong mechanism that severely constrains the temporal summation available to NM neurons. As a result, phase-locking precision in the auditory nerve is preserved or improved through integration of excitation and inhibition in NM. The key question remaining unanswered by these studies then is what role depolarizing inhibition plays in shaping responses to sound in vivo.

Recordings of NM neurons in vivo are difficult owing to the strong acoustically driven neurophonic background. The neurophonic, patterned electrical background noise that is in phase with the activity of NM cells reduces the signal-to-noise ratio. Iwao Fukui succeeded in performing in vivo recordings from NM with high-impedance multibarrel electrodes, which allowed him to manipulate inhibition and directly observe the influence of GABA on NM responses (Fukui et al. 2010). Fukui showed that GABAergic input indeed contributed substantially to phase-locking precision of NM neurons. It also dampened the input–output function to prevent saturation at high intensities. Finally, the inhibition narrowed the excitatory tuning curve of NM neurons. Narrow tuning improves temporal processing by decreasing the bandwidth of frequencies that NM neurons will phase-lock to. Thus, Fukui et al. (2010) provided strong confirmation that GABAergic input to NM was indeed inhibitory in vivo in mature animals, and that it had important roles in contributing to temporal precision for ITD processing.

5.3.2 *An Unexpected Role for Glycine in the Avian Auditory Brainstem*

In vivo surveys of other auditory structures outside of NM and NL had been conducted in the barn owl, chicken, and a few other avian species over the years (Moiseff and Konishi 1983; Lachica et al. 1994), but until recently, there was no comprehensive study of the SON in any model. Acoustically driven responses from individual SON neurons, and the role of inhibition in shaping them, was investigated in a series of studies (Coleman et al. 2011; Tabor et al. 2012). These studies suggested that the inhibition received and provided by SON neurons is far more complex than previously thought in that although GABAergic signaling is clearly dominant, glycine is also likely to play a substantial role in avian hearing (Fischl et al. 2014; Fischl and Burger 2014).

Coleman et al. (2011) described sound-evoked responses in SON neurons to determine temporal precision, aural type, and input–output parameters, which had been previously only briefly described in two studies. SON neurons in general (1) are broadly tuned, (2) are predominantly monaural, and (3) generate either phasic or tonic responses to best frequency (BF) tones (Coleman et al. 2011; Tabor et al. 2012). However, these studies also showed that SON neurons are capable of phase locking below 2 kHz, albeit with a level of precision well below that observed in NM.

Finally, Coleman et al. (2011) demonstrated that for SON neurons, GABAergic inhibition had wide-ranging effects on frequency tuning, firing rate, and temporal precision in SON. These results are similar to the results of blocking GABAergic inhibition in NM as described in the Fukui et al. (2010) study. Surprisingly, however, similar outcomes were observed when glycinergic signaling was blocked, suggesting that glycine was more important than previously understood. Stimulation of the inhibitory fibers innervating SON neurons generated synaptic currents composed of both GABAergic and glycinergic components (Fig. 5.6A) (Coleman et al. 2011; Fischl et al. 2014). This was an intriguing result because the only documented inhibitory input to the SON was its contralateral SON counterpart. These results therefore suggested that the SON itself could be a source of glycinergic inhibition, despite the fact that no glycinergic inhibition had been detected in its other targets: NM, NL, or NA. However, this possibility was supported by immunohistochemical evidence of glycine receptor expression in these nuclei (Coleman et al. 2011), complementing the sparse glycine transmitter staining that had been reported by Code and Rubel (1989).

Physiological evidence of glycinergic signaling in NM, NA, and NL was established by Fischl et al. (2014), which demonstrated that sustained synaptic output was necessary to observe glycinergic transmission in these nuclei. Concurrent with this work in birds, Nerlich et al. (2014a) had recently shown that in the mammalian cochlear nucleus, GABA and glycine were coincidently released by synapses in analogous synapses to those being investigated in the chicken.

Nerlich's studies focused on the mammalian anteroventral cochlear nucleus, where bushy cells receive predominantly fast hyperpolarizing glycinergic input, in contrast to the slow depolarizing GABAergic input observed in NM. Nerlich and colleagues discovered that when synapses are activated at high rates, a GABAergic component of inhibition emerges and contributes to the summed inhibitory postsynaptic current (IPSC). The fraction of total inhibition attributable to GABA climbs steadily during the stimulus train. Fischl et al. (2014) adopted Nerlich's strategy in the avian system and observed perfectly complementary results (Fig. 5.6B). Inhibitory input to NM and NL neurons driven at high stimulation frequencies initially release only GABA, but gradually include a steadily rising glycinergic component. The glycinergic component remains small relative to the GABAergic fraction and does not exceed about 15% of the total current, but was shown to contribute substantially to the inhibition's ability to suppress spiking in NM (Fischl et al. 2014; Nerlich et al. 2014a, b).

This mixed, and putatively coreleased, inhibitory transmitter, observed in the cochlear nuclei of both birds and mammals, may eventually be regarded as a general

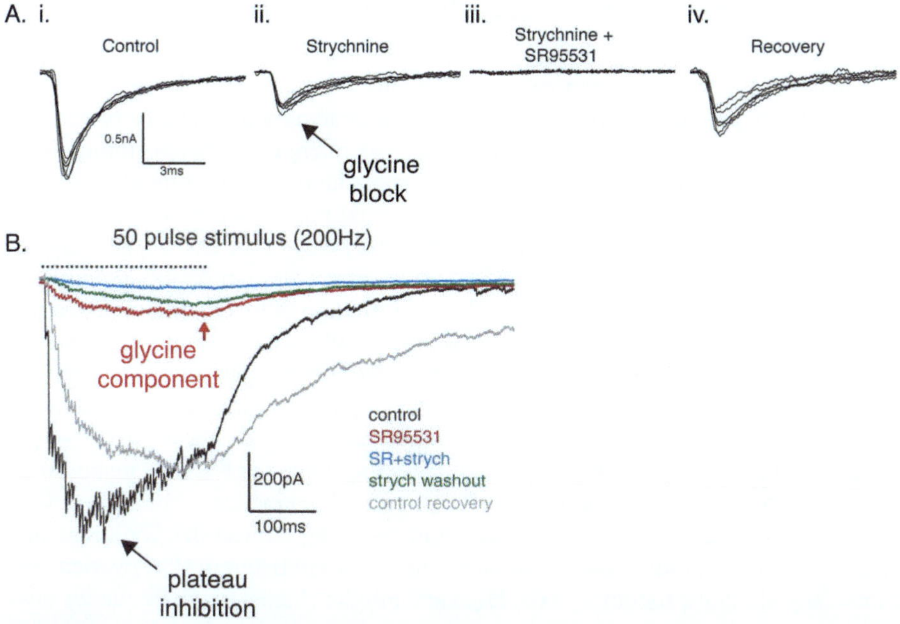

Fig. 5.6 Inhibition has glycinergic components in SON and NM. (**A**) Inhibitory postsynaptic current recorded in voltage clamp from an SON neuron. Blocking glycine receptors with strychnine caused substantial reduction in the synaptic current. Addition of SR95531 blocks GABA receptors and eliminates the IPSC. (**B**) In NM neurons, inhibition is dominated by GABA. However, IPSCs recorded during high-frequency stimulation show a glycinergic component that builds over the duration of the stimulus train (red trace), which is visible during GABA block. [**A** modified from Coleman et al. (2011). **B** modified from Fischl et al. (2014), with permission]

feature of inhibitory transmission. It is known that early in development, multitransmitter release is a common feature of inhibitory neurons (Gillespie et al. 2005), but generally over the course of maturation there is a switch toward a single detectable transmitter. In the auditory pathway this switch occurs soon after hearing onset (Nabekura et al. 2004; Gillespie et al. 2005). However, Trussell's group showed that in mature cochlear nucleus, experimentally manipulated depletion of glycine can result in an increased release of GABA (Apostolides and Trussell 2013). This presumably occurs because VGAT, the vesicular transporter for GABA and glycine, shows equal affinity for both transmitters (Wojcik et al. 2006). Thus, the working hypothesis that best explains the results in both the avian and mammalian cochlear nuclei is that under high, but physiologically relevant, stimulation rates such as those used by Nerlich and Fischl, depletion of the primary transmitter gives way to uptake and release of its complementary counterpart. Overall, this generalized cross-transmitter compensatory process may ensure that inhibition will remain competent to suppress spikes during ongoing intense stimulation. Further investigation will be required to firmly establish this model of corelease inhibition in birds and mammals.

5.3.3 A Role for Phase-Locked Inhibition in Birds?

The major distinction between roles for inhibitory circuitry in the mammalian or avian auditory pathways appears to be the disparate temporal dynamics of each. In mammals, hyperpolarizing and relatively fast glycinergic inhibition impinges on cochlear nucleus bushy cells. In mammalian binaural nuclei, principal neurons receive input from inhibitory neurons in the medial nucleus of the trapezoid body (MNTB), known phase-locking pathways that bear many of the classic hallmarks of temporal precision. In contrast, avian SON neurons are less temporally precise and provide a slow depolarizing inhibition. The contribution of this inhibition to temporal acuity depends on its coupling with voltage-gated channels rather than on temporally patterned activity. However, two recent studies suggest that there may be some exception to this arrangement in the low-frequency region of nucleus laminaris, where evidence for phase-coupled inhibition has emerged.

Tang and Lu (2012) described the first data suggesting that NL may make use of temporally patterned inhibition. They investigated the kinetics of inhibition with respect to the tonotopic axis in nucleus laminaris. They showed that for stimulated or spontaneous inhibition, the postsynaptic decay of GABAergic PSCs varied systematically along the tonotopic axis. High and middle characteristic frequency neurons showed two- to threefold slower synaptic current than low-frequency NL neurons. At higher stimulation rates, these kinetic differences became even more pronounced because IPSCs in middle- and high-frequency regions were more susceptible to asynchronous release and transmitter spillover, both effects extending PSC decay. These results clearly suggested that the presynaptic terminals of inhibitory neurons along NL's tonotopic axis vary in vesicle release physiology. No previous discovery had indicated that there were unique populations of SON neurons to mediate inhibitory effects specific to low-frequency NL. However, NL does have an alternative source of inhibition not shared by other nuclei in the brainstem.

von Bartheld and colleagues (1989) showed that there was a small population of diffuse GABAergic neurons residing in the internuclear neuropil between NM and NL (Fig. 5.7A, B). This small and difficult to investigate population of about 150 neurons was largely ignored for 20 years, until a recent study by Yamada and colleagues (2013). They showed that stimulation of NM evoked tightly coupled excitatory and inhibitory postsynaptic responses in NL neurons (Fig. 5.7C), putatively via these small interneurons. The IPSCs he observed were relatively rapid and GABAergic, consistent with the relatively fast inputs described in the Tang and Lu study. These data suggest that these intrinsic interneurons are part of a pathway originating in NM, which provides phase-coupled inhibition. Further, Yamada used computational modeling to demonstrate that this phase-coupled inhibition has the potential to strongly enhance NL neuron spike rate modulation with ITDs. This interneuron circuit in birds, if supported by further investigation, bears remarkable similarities with the feed-forward inhibitory pathway mediated by the MNTB in mammals, despite important differences in the interneuron location, synaptic convergence, and transmitter phenotypes.

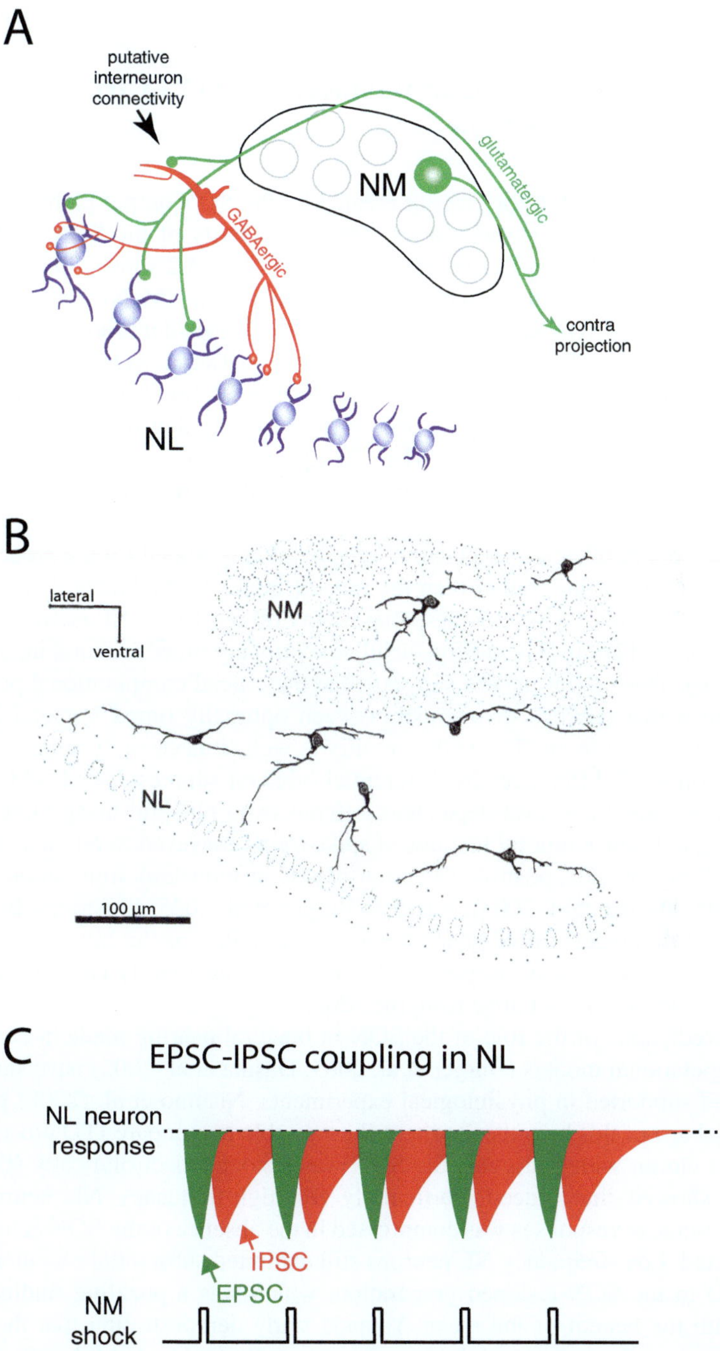

Fig. 5.7 Small inhibitory interneurons provide phase-coupled inhibition to low-frequency NL neurons. (**A**) Schematic drawing of putative interneuron circuitry. NM neurons provide putative excitatory collateral input to inhibitory interneurons (red) residing between NM and NL. (**B**) Camera lucida drawings of a sample of these GABAergic neurons first described in von Bartheld et al. (1989). (**C**) Excitatory–inhibitory coupled PSCs evoked in an NL neuron by Yamada et al. (2013) during stimulation of NM. Each excitatory PSC (green) is closely followed by a GABAergic PSC (red). [**B** reproduced from von Bartheld et al. (1989), with permission]

5.4 Binaural Processing in Balance: The SON Circuit from a Systems View

The function of the inhibitory circuit created by the SON output pathways is now well understood from a systems level. The connectivity of the SON is shown in Fig. 5.2, and described in Burger et al. (2005). As Fig. 5.2 shows, the SON is the principal provider of feedback inhibition to the ipsilateral NM, NL, and NA. A separate population of SON neurons also provides feed-forward inhibition to its contralateral counterpart. The reciprocal connectivity between SONs is the key to understanding the central importance of the SON in the broader circuit. The mutually inhibitory interaction between the SONs suggests that the system relies on negatively coupled feedback loops. The consequence of this arrangement is that any intensity-driven bilateral imbalances in activity would be mutually cancelled by the coupling.

The process of bilateral (or binaural) equalization is crucial for the processing of ITDs, where the timing, rather than the rate, of inputs to the binaural coincidence detectors in NL provide the crucial spatial cues (Colburn 1996). NL neurons respond well to monaural inputs, but synergistically to optimally timed binaural inputs (Peña et al. 1996). This output pattern suggests that the crucial computational process in NL requires that neurons distinguish between optimally timed binaural input or vigorous monaural input. The SON circuitry is well situated to dynamically factor out binaural level differences by differential bilateral suppression of NM spiking. This process equalizes level-dependent differences in binaural firing rates, ideally preserving only the temporal features of spike trains conveyed to NL and obviating the possibility of false-positive detection of strong monaural stimulation (Burger et al. 2005; Dasika et al. 2005); Fig. 5.8 (Burger et al. 2005) illustrates the model. This balancing process could generalize to any stimulus, as the bilateral inhibitory feedback loops will always oppose each other, and thus rapidly equalize the bilateral spike rates to NL that arise from the NMs.

The predictions of the role of the SON in binaural hearing made in conceptual and computational models (Burger et al. 2005; Dasika et al. 2005) have since been tested and supported in physiological experiments. Nishino et al. (2008) provided the crucial test of this hypothesis where she was able to compare ITD processing in the intact circuit with that when the SON was lesioned electrolytically (Fig. 5.9). Nishino showed that indeed, particularly for high-frequency NL neurons, the dynamic range of responses was compressed in the absence of the SON activity, just as predicted. Low-frequency NL neurons still exhibited substantial rate modulation with ITD in the SON-lesioned preparation, which was a puzzling finding at the time. With the benefit of the recent Yamada study demonstrating that these cells receive temporally patterned inhibition from von Bartheld's GABAergic neurons, perhaps the lack of change is not so surprising.

Fukui et al. (2010) confirmed the model at the level of NM showing that the SON input controls the firing rate as a function of intensity bilaterally in NM just as predicted by Burger et al. (2005). Together with the Nishino work described in the

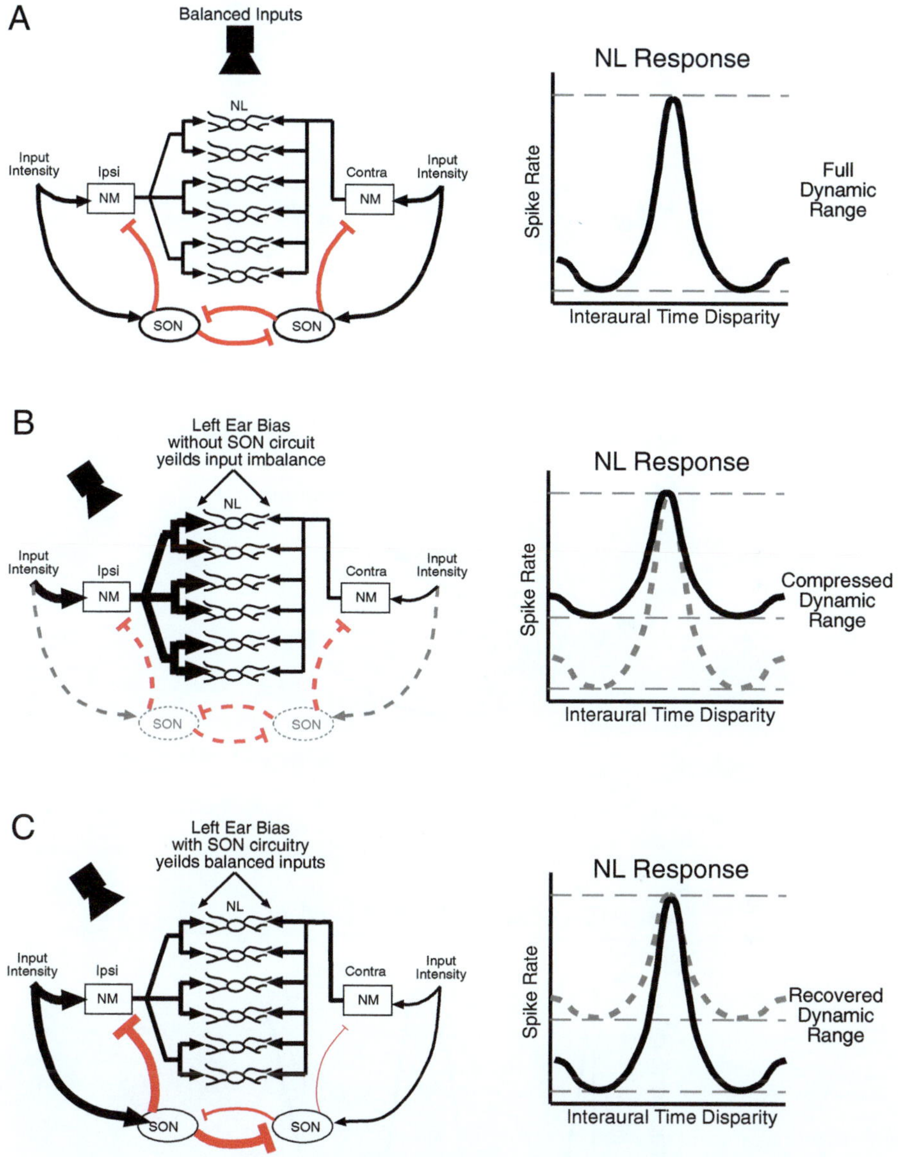

Fig. 5.8 Conceptual model of SON binaural function. (**A**) With sound arriving from the midline, the activity levels in NM and downstream regions will provide input of equal magnitude to NL neurons. The SONs will be equally driven, and inhibition (red lines) will be balanced in the system. (**B**) Lateralized sound will shift ITD, but also impose an intensity difference at the two ears. With the SON circuit disengaged (dashed lines), the excitatory circuitry through NM will be driven at higher firing rates ipsilateral to the sound source compared to the contralateral side, as indicated by line weight. This will cause a monaurally driven output in all NL neurons regardless of ITD. This monaural bias will compress the dynamic range available to encode ITDs. (**C**) With the speaker in the same position but with inhibition restored, the level difference will recruit stronger inhibition ipsilateral to the sound source and reduce the firing rate in the left NM. In addition, the ipsilateral SON will preferentially suppress the contralateral SON output, and *dis*inhibit the NM contralateral to the speaker. This process will bilaterally balance the input to NL neurons, and shift dynamically with speaker position. Removal of activity level differences allows the NL neurons to recover the full dynamic range to modulate output dependent primarily on temporal features of the stimulus. [Reproduced from Burger et al. (2005), with permission]

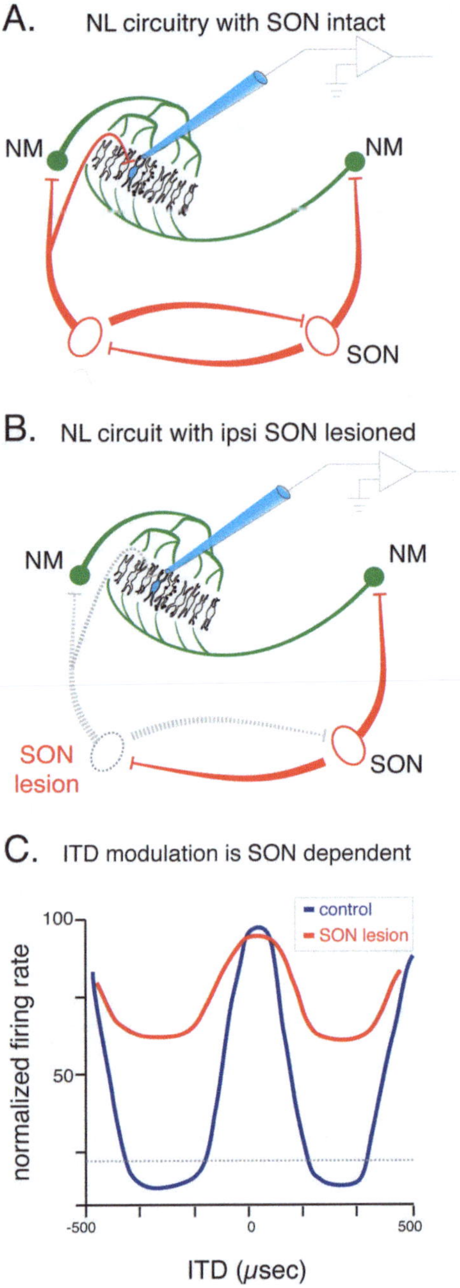

Fig. 5.9 Schematic summary of SON inactivation experiments supporting the binaural balance hypothesis. Nishino et al. (2008) showed that SON lesions reduced the ITD rate modulation of NL neurons in chick. This was especially true for NL neurons tuned to higher frequencies. (**A**, **B**) Schematic recording arrangements for (**A**) intact and (**B**) lesioned SON circuitry. (**C**) Schematic representation of results illustrated in Nishino et al. (2008), where NL response dynamic modulation is compressed in the absence of SON-derived inhibition. [Adapted from Burger et al. (2011), with permission]

preceding text, these studies show that SON indeed regulates activity evoked by either ear at the level of NM and NL. In doing so, the SONs and von Bartheld interneurons function together to have a major impact on rate modulation with ITD, the fundamental computational output of NL neurons.

5.5 Summary

In summary, the avian brainstem auditory system has been a source of continuous insight into neural development and function for decades. The bird auditory system presents a comprehensive and experimentally tractable brain circuit from which development, neural computation, and systems level function inform an integrated and mechanistic view of auditory processing. Many of the foundational discoveries in auditory neuroscience have been first described in birds, and these discoveries often have analogs or homologs in the mammalian system. The physiological properties of the inhibitory components of the circuit are in many ways distinct from those observed in mammals; however, when viewed from a functional standpoint the many parallels between birds and mammals are remarkable.

The developmental origins of the major auditory nuclei in chick have been fate mapped from the earliest stages following neural tube closure. Further, the migration, projections, and synaptogenesis of both the inhibitory and excitatory inputs to the cochlear nuclei and nucleus laminaris have been described in great detail. Physiologically, the depolarizing inhibition observed in birds is not unusual early in development, where its excitatory phenotype is likely to influence synaptogenesis. What is unusual is its persistence as a depolarizing but inhibitory input in the mature system. Its coupling to low-threshold potassium conductances generates a potent inhibition. This inhibition as manifest in NM shapes frequency tuning and phase-locking precision. The binaural neurons in NL appear to similarly benefit from inhibition for ITD tuning, with recent evidence suggesting that phase-coupled inhibition may enhance low-frequency ITD rate modulation. In the cochlear nucleus bushy cells and binaural MSO neurons of mammals, inhibition also controls low-threshold potassium channel activation, but with a hyperpolarizing input. Ultimately, the consequences of inhibition in mammalian circuitry are similar (Kopp-Scheinpflug et al. 2002; Roberts et al. 2013).

On a systems level, the bilaterally coupled circuitry of the SONs contributes to ITD computation in NL by balancing the bilateral firing rates of NM neurons dynamically. A similar binaural function has been observed in mammals, but is attributed to olivocochlear circuitry (Darrow et al. 2006).

Finally, recent work in both birds and mammals has shown that corelease of inhibitory transmitter persists into maturity in cochlear nuclei and elsewhere. This corelease is highly dependent on stimulus parameters and is most likely to be observed at high, but physiologically relevant, firing rates. It is possible that this property is a general feature of inhibitory neurons that utilize the same vesicular transporter for both of the major inhibitory transmitters, GABA and glycine.

Auditory research has benefitted tremendously from the vast volume of descriptive and experimental literature describing auditory development, physiology, and function in birds. These many lines of inquiry can often be traced back to the work of Ed Rubel and his colleagues, trainees, and collaborators, who have profoundly influenced the field for nearly five decades. Only a small portion of these achievements have been summarized here, but it is from this broad and deep foundation of knowledge that new discoveries and insights continue to emerge.

Acknowledgments Above all, I would like to thank my mentor and friend, Ed Rubel. Thank you for demonstrating high expectations, while never losing sight of science as a joyful pursuit. I would also like to thank the many colleagues who have contributed the work referenced in this chapter for continuing to inspire my own.

Compliance with Ethics Requirements R. Michael Burger declares that he has no conflict of interest.

References

Apostolides, P. F., & Trussell, L. O. (2013). Rapid, activity-independent turnover of vesicular transmitter content at a mixed glycine/GABA synapse. *The Journal of Neuroscience*, 33, 4768–4781.

Burger, R. M., & Rubel, E. W. (2008). Encoding of interaural timing for binaural hearing. In A. I. Baasbaum, A. Kaneko, G. M. Shepherd, G. Westheimer, et al. (Eds.), *The senses: A comprehensive reference*, Vol. 3: *Audition* (pp. 613–630). San Diego: Academic Press.

Burger, R. M., Cramer, K. S., Pfeiffer, J. D., & Rubel, E. W. (2005). The avian superior olivary nucleus provides divergent inhibitory input to parallel auditory pathways. *The Journal of Comparative Neurology*, 481, 6–18.

Burger, R. M., Fukui, I., Ohmori, H., & Rubel, E. W. (2011). Inhibition in the balance: Binaurally coupled inhibitory feedback in sound localization circuitry. *Journal of Neurophysiology*, 106, 4–14.

Carr, C. E., & Konishi, M. (1990). A circuit for detection of interaural time differences in the brain stem of the barn owl. *The Journal of Neuroscience*, 10, 3227–3246.

Carr, C. E., Fujita, I., & Konishi, M. (1989). Distribution of GABAergic neurons and terminals in the auditory system of the barn owl. *The Journal of Comparative Neurology*, 286,190–207.

Code, R. A., & Rubel, E. W. (1989). Glycine-immunoreactivity in the auditory brain stem of the chick. *Hearing Research*, 40, 67–172.

Code, R. A., Burd, G. D., & Rubel, E. W. (1989). Development of GABA immunoreactivity in brainstem auditory nuclei of the chick: Ontogeny of gradients in terminal staining. *The Journal of Comparative Neurology*, 284, 504–518.

Colburn, H. S. (1996). Computational models of binaural processing. In Hawkins, H. L., McMullen, T., Popper, A. N., & Fay, R. R. (Eds.), *Auditory computation* (pp. 332–400). New York: Springer-Verlag.

Coleman, W. L., Fischl, M. J., Weimann, S. R., & Burger, R. M. (2011). GABAergic and glycinergic inhibition modulate monaural auditory response properties in the avian superior olivary nucleus. *Journal of Neurophysiology*, 105, 2405–2420.

Cramer, K. S., Fraser, S. E., & Rubel, E. W. (2000). Embryonic origins of auditory brain-stem nuclei in the chick hindbrain. *Developmental Biology*, 224, 138–151.

Darrow, K. N., Maison, S. F., & Liberman, M. C. (2006). Cochlear efferent feedback balances interaural sensitivity. *Nature Neuroscience*, 9, 1474–1476.

Dasika, V. K., White, J. A., Carney, L. H., & Colburn, H. S. (2005). Effects of inhibitory feedback in a network model of avian brain stem. *Journal of Neurophysiology*, 94, 400–414.

Fischl, M. J., & Burger, R. M. (2014). Glycinergic transmission modulates GABAergic inhibition in the avian auditory pathway. *Frontiers in Neural Circuits*, doi:10.3389/fncir.2014.00019.

Fischl, M. J., Weimann, S. R., Kearse, M., & Burger, R. M. (2014). Slowly emerging glycinergic transmission enhances inhibition in the sound localization pathway of the avian auditory system. *Journal of Neurophysiology*, 111, 565–572.

Fukui, I., Sato, T., & Ohmori, H. (2006). Improvement of phase information at low sound frequency in nucleus magnocellularis of the chicken. *Journal of Neurophysiology*, 96, 633–641.

Fukui, I., Burger, R. M., Ohmori, H., & Rubel, E. W. (2010). GABAergic inhibition sharpens the frequency tuning and enhances phase locking in chicken nucleus magnocellularis neurons. *The Journal of Neuroscience*, 30, 12075–12083.

Ge, S., Goh, E. L., Sailor, K. A., Kitabatake, Y., et al. (2006). GABA regulates synaptic integration of newly generated neurons in the adult brain. *Nature*, 439(7076), 589–593.

Gillespie, D. C., Kim, G., & Kandler, K. (2005). Inhibitory synapses in the developing auditory system are glutamatergic. *Nature Neuroscience*, 8, 332–338.

Grothe, B., Pecka, M., & McAlpine, D. (2010). Mechanisms of sound localization in mammals. *Physiological Reviews*, 90, 983–1012.

Hackett, J. T., Jackson, H., & Rubel, E. W. (1982). Synaptic excitation of the second and third order auditory neurons in the avian brain stem. *Neuroscience*, 7, 1455–1469.

Howard, M. A., & Rubel, E. W. (2010). Dynamic spike thresholds during synaptic integration preserve and enhance temporal response properties in the avian cochlear nucleus. *The Journal of Neuroscience*, 30, 12063–12074.

Howard, M. A., Burger, R. M., & Rubel, E. W. (2007). A developmental switch from GABAergic excitation to inhibition controlled by K^+ conductances. *The Journal of Neuroscience*, 27, 2112–2123.

Hyson, R. L., Reyes, A. D., & Rubel, E. W. (1995). A depolarizing inhibitory response to GABA in brainstem auditory neurons of the chick. *Brain Research*, 677, 117–126.

Jackson, H., Hackett, J. T., & Rubel, E. W. (1982). Organization and development of brain stem auditory nuclei in the chick: Ontogeny of postsynaptic responses. *The Journal of Comparative Neurology*, 210, 80–86.

Jhaveri, S., & Morest, D. K. (1982). Sequential alterations of neuronal architecture in nucleus magnocellularis of the developing chicken: A Golgi study. *Neuroscience*, 7, 837–853.

Kandler, K., & Friauf, E. (1995). Development of glycinergic and glutamatergic synaptic transmission in the auditory brainstem of perinatal rats. *The Journal of Neuroscience*, 15, 6890–6904.

Köppl, C., & Carr, C. E. (2003). Computational diversity in the cochlear nucleus angularis of the barn owl. *Journal of Neurophysiology*, 89, 2313–2329.

Kopp-Scheinpflug, C., Dehmel, S., Dörrscheidt, G. J., & Rübsamen, R. (2002). Interaction of excitation and inhibition in anteroventral cochlear nucleus neurons that receive large endbulb synaptic endings. *The Journal of Neuroscience*, 22, 11004–11018.

Korn, M. J., Koppel, S. J., Li, L. H., Mehta, D., et al. (2012). Astrocyte-secreted factors modulate the developmental distribution of inhibitory synapses in nucleus laminaris of the avian auditory brainstem. *The Journal of Comparative Neurology*, 520, 1262–1277.

Lachica, E. A., Rübsamen, R., & Rubel, E. W. (1994). GABAergic terminals in nucleus magnocellularis and laminaris originate from the superior olivary nucleus. *The Journal of Comparative Neurology*, 348, 403–418.

Lippe, W. R. (1994) Rhythmic spontaneous activity in the developing avian auditory system. *The Journal of Neuroscience*, 14, 1486–1495.

Lu, T., & Trussell, L. O. (2000). Inhibitory transmission mediated by asynchronous transmitter release. *Neuron*, 26, 683–694.

Lu, T., & Trussell, L. O. (2001). Mixed excitatory and inhibitory GABA-mediated transmission in chick cochlear nucleus. *The Journal of Physiology* (London), 535, 125–131.

Moiseff, A., & Konishi, M. (1983). Binaural characteristics of units in the owl's brainstem auditory pathway: Precursors of restricted spatial receptive fields. *The Journal of Neuroscience*, 3, 2553–2562.

Molea, D., & Rubel, E. W. (2003). Timing and topography of nucleus magnocellularis innervation by the cochlear ganglion. *The Journal of Comparative Neurology*, 466, 557–591.

Monsivais, P., & Rubel, E. W. (2001). Accommodation enhances depolarizing inhibition in central neurons. *The Journal of Neuroscience*, 21, 7823–7830.

Monsivais, P., Yang, L., & Rubel, E. W. (2000). GABAergic inhibition in nucleus magnocellularis: Implications for phase locking in the avian auditory brainstem. *The Journal of Neuroscience*, 20, 2954–2963.

Nabekura, J., Katsurabayashi, S., Kakazu, Y., Shibata, S., et al. (2004). Developmental switch from GABA to glycine release in single central synaptic terminals. *Nature Neuroscience*, 7, 17–23.

Nerlich, J., Kuenzel, T., Keine, C., Korenic, A., et al. (2014a). Dynamic fidelity control to the central auditory system: Synergistic glycine/GABAergic inhibition in the cochlear nucleus. *The Journal of Neuroscience*, 34, 11604–11620.

Nerlich, J., Keine, C., Rübsamen, R., Burger, R. M., & Milenkovic, I. (2014b). Activity-dependent modulation of inhibitory synaptic kinetics in the cochlear nucleus. *Frontiers in Neural Circuits*, doi:10.3389/fncir.2014.00145.

Nishino, E., Yamada, R., Kuba, H., Hioki, H., et al. (2008). Sound-intensity-dependent compensation for the small interaural time difference cue for sound source localization. *The Journal of Neuroscience*, 28, 7153–7164.

Oline, S. N., & Burger, R. M. (2014). Short-term synaptic depression is topographically distributed in the cochlear nucleus of the chicken. *The Journal of Neuroscience*, 34, 1314–1324.

Oline, S. N., Ashida, G., & Burger, R. M. (2016). Tonotopic optimization for temporal processing in the cochlear nucleus. *The Journal of Neuroscience*, 36, 8500–8515.

Overholt, E. M., Rubel, E. W., & Hyson, R. L. (1992). A circuit for coding interaural time differences in the chick brainstem. *The Journal of Neuroscience*, 12, 1698–1708.

Parks, T. N., & Rubel, E. W. (1975). Organization and development of brain stem auditory nuclei of the chicken: Organization of projections from n. magnocellularis to n. laminaris. *The Journal of Comparative Neurology*, 164, 435–448.

Peña, J. L., Viete, S., Albeck, Y., & Konishi, M. (1996). Tolerance to sound intensity of binaural coincidence detection in the nucleus laminaris of the owl. *The Journal of Neuroscience*, 16, 7046–7054.

Represa, A., & Ben-Ari, Y. (2005). Trophic actions of GABA on neuronal development. *Trends in Neuroscience*, 6, 278–283.

Roberts, M. T., Seeman, S. C., & Golding, N. L. (2013). A mechanistic understanding of the role of feedforward inhibition in the mammalian sound localization circuitry. *Neuron*, 78, 923–935.

Rubel, E. W., & Fritzsch, B. (2002). Auditory system development: Primary auditory neurons and their targets. *Annual Review of Neuroscience*, 25, 51–101.

Seidl, A. H., Rubel, E. W., & Barría, A. (2014). Differential conduction velocity regulation in ipsilateral and contralateral collaterals innervating brainstem coincidence detector neurons. *The Journal of Neuroscience*, 34, 4914–4919.

Soares, D., Chitwood, R. A., Hyson, R. L., & Carr, C. E. (2002). Intrinsic neuronal properties of the chick nucleus angularis. *Journal of Neurophysiology*, 88, 152–162.

Tabor, K., Coleman, W. L., Rubel, E. W., & Burger, R. M. (2012). Tonotopic organization of the superior olivary nucleus in the chicken (*Gallus gallus*). *The Journal of Comparative Neurology*, 520, 1493–1508.

Tang, Z. Q., & Lu, Y. (2012). Two GABAA responses with distinct kinetics in a sound localization circuit. *The Journal of Physiology* (London), 590, 3787–3805.

von Bartheld, C. S., Code, R. A., & Rubel, E. W. (1989). GABAergic neurons in brainstem auditory nuclei of the chick: Distribution, morphology, and connectivity. *The Journal of Comparative Neurology*, 287, 470–483.

Wojcik, S. M., Katsurabayashi, S., Guillemin, I., Friauf, E., et al. (2006). A shared vesicular carrier allows synaptic corelease of GABA and glycine. *Neuron*, 50(4), 575–587.

Yamada, R., Okuda, H., Kuba, H., Nishino, E., et al. (2013). The cooperation of sustained and phasic inhibitions increases the contrast of ITD-tuning in low-frequency neurons of the chick nucleus laminaris. *The Journal of Neuroscience*, 33, 3927–3938.

Yang, L., Monsivais, P., & Rubel, E. W. (1999). The superior olivary nucleus and its influence on nucleus laminaris: A source of inhibitory feedback for coincidence detection in the avian auditory brainstem. *The Journal of Neuroscience*, 19, 2313–2325.

Young, S. R., & Rubel, E. W. (1983). Frequency-specific projections of individual neurons in chick brainstem auditory nuclei. *The Journal of Neuroscience*, 3, 1373–1378.

Chapter 6
Tuning Neuronal Potassium Channels to the Auditory Environment

Leonard K. Kaczmarek

Abstract Neurons in auditory brainstem nuclei have developed several molecular and cellular specializations to ensure that auditory information can be relayed and processed at high rates and with very high temporal accuracy. Among these is the expression of Kv3 family voltage-dependent potassium channels such as Kv3.1, which allow the neurons to fire many hundreds of times per second. The specific channel isoforms in auditory neurons change during development, allowing the activity and amount of Kv3.1 to be adjusted by changes in the amplitude and frequencies of sounds in the auditory environment. The molecular mechanisms that produce such adjustments include direct phosphorylation of the channels and rapid alterations in the rates at which the channels are synthesized. These mechanisms, in combination with changes in other types of channels, may maximize the accuracy of information transfer through brainstem nuclei in different auditory environments, and may contribute to the learning of auditory discrimination tasks.

Keywords Anteroventral cochlear nucleus • Calyx of Held • Ion channel • Kv3.1 • Kv3.3 • Medial nucleus of the trapezoid body • Protein kinase • Sound localization

6.1 Introduction

The mammalian auditory system has evolved to detect and discriminate among many different aspects of the auditory environment, such as the relative timing and pitch of sounds, as well as their spatial location. Both humans and other animals are able to improve their ability to discriminate such features of auditory stimuli when subjected to appropriate training paradigms (Mossbridge et al. 2006; Polley et al. 2006). For example, the superior ability of orchestra

L.K. Kaczmarek (✉)
Departments of Pharmacology and Cellular and Molecular Physiology, Yale University School of Medicine, 333 Cedar Street, New Haven, CT 06520, USA
e-mail: leonard.kaczmarek@yale.edu

K.S. Cramer et al. (eds.), *Auditory Development and Plasticity*, Springer Handbook of Auditory Research 64, DOI 10.1007/978-3-319-21530-3_6

conductors to detect small differences in the spatial location of sounds most likely results from years of training in this aspect of audition (Munte et al. 2001; Nager et al. 2003).

The way neurons process auditory information to detect subtle differences in the quality or direction of a sound differs from that used in other sensory pathways. For example, the detection of differences in parameters such as pitch and spatial location requires that neurons in auditory brainstem nuclei are capable of following sensory inputs with very high temporal precision (Spirou et al. 1990; Trussell 1999). Many neurons in auditory brainstem pathways are therefore capable of firing at very high rates (up to about 800 Hz). Moreover, many of these neurons lock their action potentials accurately to the phase of auditory stimuli, such as sine waves, at frequencies up to about 2000 Hz. For higher frequency sounds, the property of accurate phase locking allows neurons to lock their action potentials to the envelope of sounds that are amplitude modulated at these lower frequencies (Joris and Yin 1995; Joris 1996).

The ability of neurons to respond both rapidly and accurately to incoming stimuli requires that their intrinsic electrical properties are matched to the incoming patterns of stimulation. Perhaps more importantly, these intrinsic properties have to be adjusted in different auditory environments. This chapter discusses the regulation of one set of proteins, Kv3 family potassium channels, that are required for neurons to fire at high rates. Both during development and in the adult, these channels are regulated by short-term and long-term mechanisms, allowing auditory neurons to follow changes in the auditory environment and in patterns of afferent stimulation.

The work of Ed Rubel, to whom this volume is dedicated, as well the work of his many colleagues who have subsequently established their own laboratories, has provided numerous insights into basic mechanisms that regulate accuracy and timing of auditory neurons. The majority of these insights have come from experiments with the nervous system of chicks (Rubel et al. 1990). The exceptional ability of barn owls (*Tyto alba*) to localize sounds has also provided important insights into the role of Kv3 potassium channels in this process (Parameshwaran et al. 2001; Kullmann et al. 2013). Although this chapter primarily covers work on the mammalian auditory system, most mechanisms that regulate the excitability of neurons are conserved from invertebrates to mammals. This conservation is not limited to genes for channels such as Kv3 channels, but also extends to biochemical pathways such as those linked to the fragile X mental retardation protein (FMRP), which is now known to regulate the levels of Kv3 channels in neurons. Studies from the Rubel laboratory have used the more tractable cellular organization of auditory pathways in chick to provide insights into the mechanisms that regulate long-term changes in Kv3 channels (Lu et al. 2004), as well as the role that FMRP may play in shaping the properties of these neurons (Wang et al. 2014).

6.2 What Is So Special About Kv3 Potassium Channels?

There are four genes (*Kv3.1*, *Kv3.2*, *Kv3.3*, and *Kv3.4*, also known as *KCNC1*, *KCNC2*, *KCNC3*, and *KCNC4*) in the Kv3 potassium channel family. Of these, *Kv3.1* and *Kv3.3* are expressed at high levels in neurons of the auditory brainstem (Perney and Kaczmarek 1997; Li et al. 2001).

By alternative splicing of its RNA, the *Kv3.1* gene gives rise to two different isoforms, Kv3.1a and Kv3.1b. These channels differ in their sequence only in the length of the cytoplasmic C-terminus (Fig. 6.1A). As described in Sect. 6.5.2, the electrophysiological properties of these two variants are very similar and they differ primarily in their ability to be regulated by phosphorylation. The Kv3.1a variant is present early in embryonic development. In the majority of neuronal types, however, Kv3.1b increases expression during development and becomes the major isoform in adult neurons (Perney et al. 1992; Liu and Kaczmarek 1998a). In MNTB neurons, expression of the Kv3.1b isoform increases sharply between postnatal day P8 and P15, at the onset of hearing in rodents (Macica et al. 2003). (It is important to note that the Kv3.1b isoform was originally called Kv3.1α (Perney et al. 1992)).

Kv3.1 is a voltage-dependent potassium channel that belongs to the "delayed rectifier" class of channels (Yokoyama et al. 1989; Luneau et al. 1991). The major role of delayed rectifier channels, which was first defined in studies of the squid giant axon (Hodgkin and Huxley 1952), is to provide current for the repolarization of action potentials. The Kv3.1 channel has properties that differ from those of the majority of delayed rectifier channels, and is found at high levels in neurons that are capable of generating action potentials at very high rates (Perney et al. 1992; Rudy and McBain 2001). Kv3.1 currents activate and deactivate unusually rapidly with changes in voltage (Kanemasa et al. 1995; Rudy and McBain 2001). Moreover, in contrast to those of most other delayed rectifiers, Kv3.1 currents typically activate only when a cell is depolarized to potentials more positive than about −10 mV. Numerical simulations have demonstrated that these characteristics provide neurons with very narrow action potentials (Kanemasa et al. 1995). Perhaps more importantly, these action potentials are followed by only a very short relative refractory period. This property is to be expected because the relative refractory period results from the slow closing of potassium channels that were opened during the action potential. An important aspect of Kv3.1 channel function is that the channels generate a transient resurgent K^+ current during the falling phase of an action potential (Labro et al. 2015). This feature allows them to fully repolarize an action potential even though they close rapidly at negative potentials.

The overall biophysical properties of Kv3.3 closely resemble those of Kv3.1, and therefore, like Kv3.1, Kv3.3 promotes the ability of neurons in which it is expressed to follow high rates of stimulation. The major differences between Kv3.1 and Kv3.3 are in the way they inactivate with prolonged (nonphysiological) depolarization, the way they are regulated by protein kinases, and in the way their cytoplasmic domains interact with the cytoskeleton (Desai et al. 2008; Zhang et al. 2016).

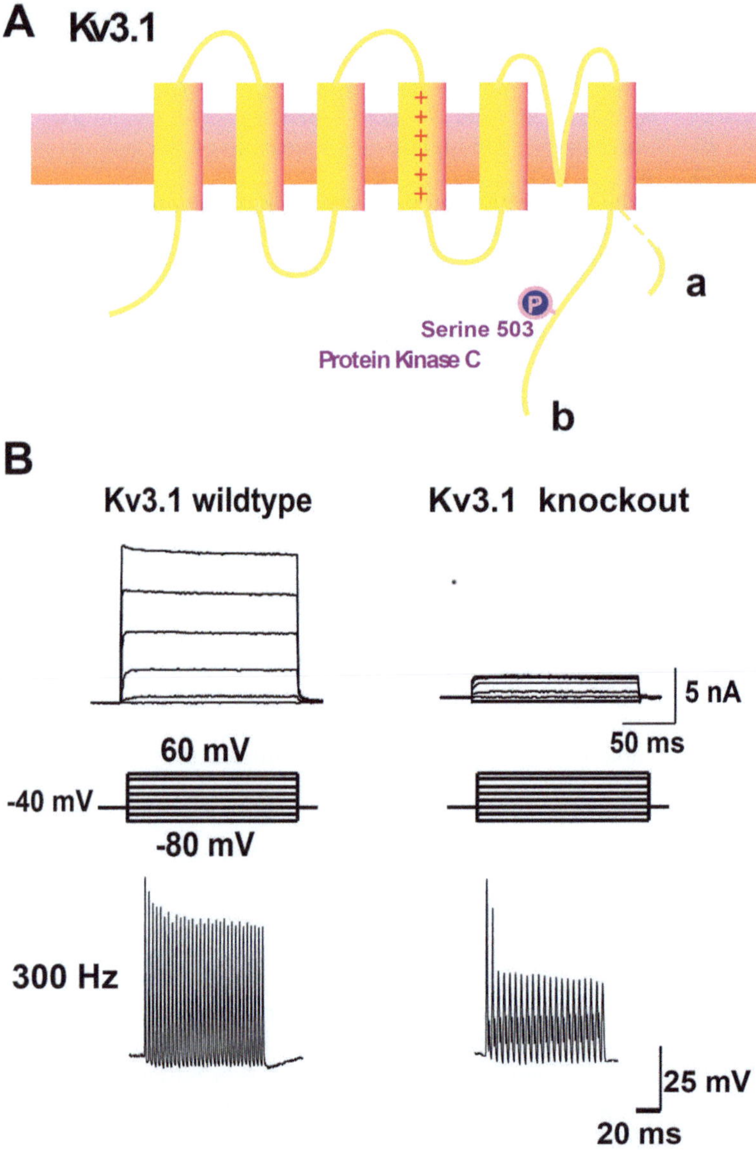

Fig. 6.1 The Kv3.1 voltage-dependent potassium channel. (**A**) Schematic diagram of transmembrane topology for the Kv3.1 potassium channel, indicating the intracellular site (serine 503) for phosphorylation by protein kinase C (Macica and Kaczmarek 2001; Macica et al. 2003). The Kv3.1a and Kv3.1b splice variants differ at the cytoplasmic C-terminal domain (Luneau et al. 1991). (**B**) The Kv3.1 channel is required for firing at high rates. The *upper left panel* shows voltage-clamp traces of Kv3.1 current in an MNTB neuron evoked by depolarization from −40 mV. This current is missing in neurons from animals in which the *Kv3.1* gene has been eliminated (*right panels*) (Macica et al. 2003). The current clamp traces at the bottom of the figure show that in *Kv3.1*$^{-/-}$ animals, medial nucleus of the trapezoid (MNTB) neurons are incapable of sustained action potential firing in response to stimulation at 300 Hz (Macica et al. 2003)

6.3 Brainstem Circuits Underlying Localization of Sounds Have High Levels of Kv3.1 and Kv3.3 Channels

6.3.1 Bushy Cells of the Anteroventral Cochlear Nucleus

Training that results in the improvement of performance in tasks related to the discrimination of pitch or timing has been shown to produce long-lasting changes in the organization of regions of the cerebral cortex (Polley et al. 2006). The basic neuronal circuitry that is required for the discrimination of these features, however, resides in the auditory brainstem. For example, differences in the spatial localization of sounds are detected by phase-locking neurons in the cochlear nuclei and the superior olivary complex (Fig. 6.2). Bushy cells in the anteroventral cochlear nucleus (AVCN) receive auditory information from hair cells in the ipsilateral cochlea via the spiral ganglion. In turn, these cells relay this information on the timing and intensity of sounds presented to the ipsilateral cochlea to the lateral superior olive (LSO) on the same side of the brain. Bushy cells also transmit the same information to the medial nucleus of the trapezoid body (MNTB) and to the medial superior olive (MSO) on the contralateral side. Finally, the LSO and the MSO, which receive inputs from both sides, make a comparison of the intensity or time of arrival of sounds arriving at the two ears. Both Kv3.1 and Kv3.3 are highly expressed in the bushy cells (Perney et al. 1992; Perney and Kaczmarek 1997).

The axons of bushy cells of the AVCN make an exceptionally large synaptic contact onto neurons of the MNTB (Fig. 6.2). This large presynaptic terminal,

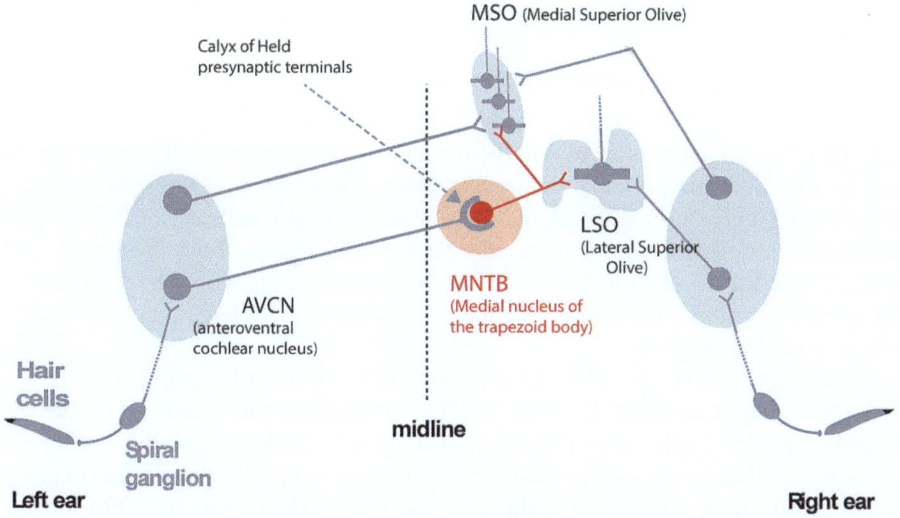

Fig. 6.2 Schematic diagram of connections between bushy cells in the anteroventral cochlear nucleus (AVCN), the medial nucleus of the trapezoid (MNTB), the lateral superior olive (LSO), and the medial superior olive (MSO)

termed the calyx of Held, provides strong glutamatergic excitatory input from the AVCN to the MNTB, and envelops the somata of the MNTB neurons (Morest 1968). Because this specialization allows patch-clamp electrodes to access the interior of a presynaptic terminal, it has been widely exploited to investigate general mechanisms of synaptic transmission in the nervous system (Schneggenburger and Forsythe 2006). Immunostaining of the calyx of Held terminals indicates that both Kv3.1b and Kv3.3 are present in the presynaptic plasma membrane (Song et al. 2005; Zhang et al. 2016). The voltage dependence, kinetics, and pharmacological properties of potassium currents in these terminals match those of both Kv3.1 and Kv3.3 (Wang and Kaczmarek 1998; Zhang et al. 2016). For example, like the Kv3 channels, the native current in the terminals is sensitive to very low levels of the potassium channel blocker tetraethyl ammonium (TEA) ions. The physiological significance of having both Kv3.1 and Kv3.3 subunits within these terminals is discussed in Sect. 6.5.

Experimental manipulations of the amplitude of Kv3 currents in the calyx of Held terminals produce major changes in both the amount of transmitter released from the terminals and the rate at which the readily releasable pool of transmitter recovers following repetitive stimulation (Wang and Kaczmarek 1998). This is because very small changes in the shape of the presynaptic action potential can produce significant alterations in the levels of intracellular Ca^{2+} ions, which trigger neurotransmitter release directly and also regulate replenishment of neurotransmitter-containing vesicles at release sites. Thus, in addition to being required for high-frequency firing, changes in Kv3.1 and Kv3.3 channel activity directly alter the strength of transmission of auditory information through the auditory brainstem.

6.3.2 *Principal Neurons of the MNTB*

The MNTB also plays a key role in the detection of differences in interaural intensity and timing. The principal neurons of this nucleus are glycinergic and inhibitory. They receive excitatory inputs from globular bushy cells of the AVCN and relay this information in the form of a strong inhibition to both the lateral and medial superior olives (Smith et al. 1998; Taschenberger and von Gersdorff 2000). This inhibitory input is required for the processing of both level and time differences (Brand et al. 2002).

Particularly high levels of Kv3.1 are found in all of the principal neurons of the MNTB (Wang et al. 1998; Li et al. 2001). Nevertheless, these neurons do not all express the same level of Kv3.1 channels. There exists a tonotopic gradient of Kv3.1 protein such that neurons in the medial (high-frequency) part of the nucleus express significantly more channel than those in the lateral (low-frequency) end of the nucleus (Li et al. 2001; Brew and Forsythe 2005). The significance of such gradients is discussed in Sect. 6.7.

Patch-clamp recordings from MNTB neurons have demonstrated the presence of a potassium current that precisely matches Kv3.1 in its kinetic and pharmacological characteristics (Wang et al. 1998). Moreover this component of potassium current is lacking in mice in which the *Kv3.1* gene has been deleted (Fig. 6.1B) (Macica et al. 2003). When MNTB neurons from *Kv3.1*$^{-/-}$ mice are stimulated with intracellular current pulses at frequencies less than about 200 Hz, their firing pattern does not differ substantially from that of wild-type cells. As the stimulus rate is increased beyond 200 Hz, however, they become incapable of generating action potentials to more than the first stimulus (Fig. 6.1B), confirming the requirement of Kv3.1 for high rates of action potential firing (Macica et al. 2003).

Messenger RNA for Kv3.3 channels is also present in MNTB neurons (Li et al. 2001), although genetic elimination of the Kv3.1 channel alone is sufficient to remove most of the Kv3-family–like current recorded at the somata of these neurons (Macica et al. 2003). Thus it is possible that Kv3.3 is expressed in the distal axon terminals of MNTB neurons, as in case for AVCN neurons. Genetic deletion of Kv3.1 does not, however, reduce total potassium current in MNTB neurons, suggesting a compensatory increase in other potassium currents, to which Kv3.3 channels may contribute (Macica et al. 2003).

6.3.3 Other Auditory Brainstem Nuclei

Although Kv3.1 is expressed at high levels in both the AVCN and MNTB, it is absent from the LSO and MSO, whose functions are to compare the firing patterns of inputs from AVCN and MNTB neurons activated by sounds coming from opposites sides of the head (Fig. 6.2). Instead, Kv3.3 is the major Kv3 family channel member in these nuclei (Li et al. 2001). The functional role and modulation of Kv3.3 in these nuclei has not yet been explored. Because of the similarity of Kv3.3 currents to those of Kv3.1, it might be expected that Kv3.3 contributes to the ability of cells in the LSO and MSO to fire at high rates, as found for Kv3.1 in the MNTB. However, because these two channels interact very differently with cytoskeletal proteins and proteins that regulate cell fate (Zhang et al. 2016), one can speculate that the presence of Kv3.3 channels in the LSO and MSO contributes to activity-dependent establishment of cell morphology and cell survival during development.

6.4 Too Much Kv3.1 Current Produces Errors in Timing

While having Kv3.1 current allows a cell to fire at high rates, too much Kv3.1 current may actually impair the transmission of auditory information. Figure 6.3A compares the firing pattern of an MNTB neuron in a brain slice to current pulses

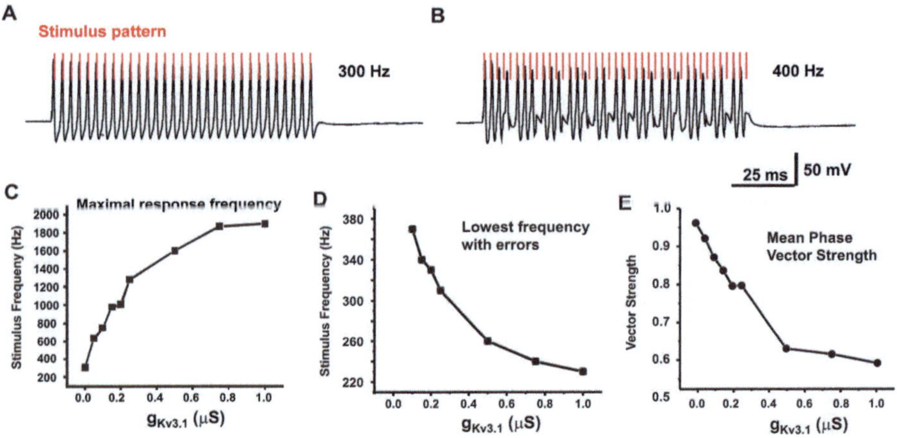

Fig. 6.3 Firing patterns of medial nucleus of the trapezoid (MNTB) neurons recorded in brainstem slices (Wang et al. 1998). (**A**) A neuron was stimulated at 300 Hz using intracellular current pulses applied through the patch pipette. The red bars above the action potentials show the stimulus pattern displaced in time such that the first stimulus is aligned to the peak of the first evoked action potential. (**B**) When the same neuron was stimulated at 400 Hz, phase locking to the stimulus pulses was lost. (**C–E**) Numerical simulations of the effects of increasing levels of Kv3.1 current on the maximal rates at which an MNTB neuron can generate sustained firing in response to repetitive stimulation (**C**), the lowest frequency at which irregular firing patterns occur (**D**), and the strength of phase vector averaged over stimulus frequencies from 200 Hz to the maximal response frequency (**E**) (Song et al. 2005; Yang et al. 2007)

applied at 300 and 400 Hz (Wang et al. 1998). At 300 Hz, this neuron fires an action potential in response to every stimulus, and each action potential is faithfully locked to each stimulus, providing an accurate representation of the stimulus pattern. In contrast, at 400 Hz, the neuron does not generate an action potential to every stimulus pulse (Fig. 6.3B). In response to the first four stimulus pulses, the latency from the stimulus pulse to the action potentials becomes progressively longer with each stimulus and the fifth stimulus fails to evoke an action potential. The subsequent stimuli evoke similar cycles of two or three evoked action potentials followed by a failure. Under these conditions, the firing pattern of the neurons is completely unrelated to the pattern of stimulation.

Such errors of timing occur because of a mismatch between the intrinsic electrical characteristic of a neuron and the pattern of stimulation to which it is exposed. The exact range of frequencies at which such "errors" of timing occur is slightly different for different neurons, and is very dependent on the level of Kv3.1 current. Neurons in Kv3.1 knockout animals are never capable of firing at rates that produce these timing errors (Macica et al. 2003). Numerical simulations of MNTB neurons show that as levels of Kv3.1 current increase, the ability of the neuron to follow high rates of stimulation increases (Fig. 6.3C). As Kv3.1 current is increased, however, such timing errors occur at progressively lower stimulus rates and the strength of the phase vector that represents the overall accuracy with which action potentials lock to the stimulus pulses progressively decreases (Fig. 6.3D, E) (Song et al. 2005). Note that the occurrence of these errors does not require that a cell fire at a high rate for a prolonged period of time. In the example of Fig. 6.3B, such errors are already detected with the first two or three action potentials at 400 Hz. This can be explained by the fact that increases in potassium currents, even those of Kv3.1, produce some enhancement of the relative refractory period, which, in turn, produces a progressive delay in the timing of successive action potentials evoked by a regular stimulus.

6.5 The Properties of Brainstem Neurons Are Plastic

A priori, one might expect that the very stereotyped processing of interaural intensity and timing differences at the brainstem level would require that the neurons in these circuits be relatively fixed in their electrical and synaptic properties. One of the surprises that emerged from studies of these neurons, however, is that their intrinsic electrical properties are plastic. Such plasticity allows the nervous system to adapt to various auditory environments and is likely to prevent errors in interpretation of auditory signals. It may also be that plasticity of brainstem neurons is also essential for the various types of improvement in auditory discrimination tasks that follow training.

This section focuses on three cellular mechanisms that regulate Kv3.1 potassium channels and that shape the intrinsic excitability of neurons in the MNTB over different time scales. These are (1) direct phosphorylation of Kv3.1 channel proteins, (2) regulation of the synthesis of Kv3.1 channels in response to auditory stimuli, and (3) transcriptional regulation of the gene encoding Kv3.1.

6.5.1 Casein Kinase 2 Adjusts the Voltage Dependence of Kv3.1 Current in MNTB Neurons

As is the case for many types of ion channels, the electrophysiological properties of the Kv3.1 channel are not fixed but can be altered by biochemical events within the cell. Indeed, some of the defining characteristics of Kv3.1 channels, such as the fact that they activate significantly only at very positive potentials (about −10 mV), depend on the channel protein being constitutively modified by the enzyme casein kinase 2 (Macica and Kaczmarek 2001). When MNTB neurons or cells transfected with Kv3.1 are treated with inhibitors of casein kinase 2, their voltage dependence is shifted to negative potentials and the currents inactivate more rapidly during sustained depolarization. Such changes in the characteristics of Kv3.1 current would be expected to have profound effects on the firing characteristics of MNTB neurons. How and when changes in the activity of casein kinase 2 occur under physiological conditions is, however, not yet understood.

6.5.2 Phosphorylation of Kv3.1 Channels Is Developmentally Regulated

A second enzyme that regulates Kv3.1 currents is protein kinase C. The actions of this enzyme on Kv3.1 current are selectively confined to the isoform of Kv3.1 that is present in mature neurons. By alternative splicing of its mRNA, the *Kv3.1* gene gives rise to two different isoforms, Kv3.1a and Kv3.1b. These channels differ in their sequence only in the length of the cytoplasmic C-terminus (Fig. 6.1A). The Kv3.1a variant is present early in embryonic development. In the majority of neuronal types, however, Kv3.1b increases during development and becomes the major isoform in adult neurons (Perney et al. 1992). In MNTB neurons, expression of the Kv3.1b isoform increases sharply between postnatal days P8 and P15, and the steep increase in Kv3.1b therefore correlates with the onset of hearing in mice and rats (Macica et al. 2003).

Under basal conditions, the currents produced by Kv3.1a and Kv3.1b appear identical (Macica et al. 2003). Both Kv3.1a and Kv3.1b proteins undergo phosphorylation following activation of protein kinase C. Only currents of the Kv3.1b isoform are, however, altered by the activation of this enzyme (Macica et al. 2003). This difference arises because the enzyme phosphorylates residue serine 503, which is present only in the longer cytoplasmic C-terminus of Kv3.1b (Fig. 6.1A). Phosphorylation at this site reduces the amplitude of Kv3.1b current by about 40% (Kanemasa et al. 1995; Macica et al. 2003).

In the absence of any external stimulation, the Kv3.1b channel in MNTB neurons, as well as in the presynaptic calyxes of Held, is basally phosphorylated (Song et al. 2005). This observation was established using an antibody that specifically recognizes Kv3.1b only when it is phosphorylated at serine 503. In brain slice preparations, stimulation of the presynaptic input from the AVCN at a high rate (600 Hz) for tens of seconds produces dephosphorylation of the channel at serine 503. Such stimulation activates protein phosphatases 1 or 2A, and the resultant dephosphorylation is blocked by inhibitors of these enzymes (Song et al. 2005). Recovery of the basal level of Kv3.1b phosphorylation in the brain slices occurs over 30–60 seconds and is mediated by several different isoforms of protein kinase C (Song and Kaczmarek 2006).

Stimulation-induced dephosphorylation of Kv3.1 produces an increase in Kv3.1b current recorded in MNTB neurons, and, as predicted from numerical simulations, this increased current allows the neurons to fire at higher rates than before the 600 Hz stimulus was applied (Song et al. 2005). As also predicted by simulations, "errors" of timing similar to those described in Fig. 6.2 occur at lower firing frequencies after dephosphorylation of Kv3.1b.

In intact animals, a change in the auditory environment also induces major changes in the phosphorylation state of Kv3.1b (Song et al. 2005). When rats are kept in a quiet environment such as a soundproof chamber, the phosphorylation state of Kv3.1b channels in the AVCN and MNTB is high, corresponding to a suppression of Kv3.1b current. A brief period (5 minutes) of exposure to a sound stimulus at a physiological sound level (70 dB, 600-Hz click trains,) causes a rapid decrease of approximately 90% in phosphorylation of the channel. If the sounds are applied to one ear only, by placing a plug in the contralateral ear, the dephosphorylation of Kv3.1b occurs selectively in the ipsilateral AVCN and contralateral MNTB (Fig. 6.4). In contrast, if the sound is applied binaurally, dephosphorylation occurs bilaterally.

These rapid changes in phosphorylation state in response to auditory stimulation are likely to adapt the intrinsic excitability of the neurons to an altered auditory environment. In quiet conditions, Kv3.1b currents are reduced through basal phosphorylation by protein kinase C. This allows the neurons to fire with maximal temporal accuracy when auditory stimuli evoke only low rates of action potential firing. When an animal enters a louder auditory environment, the dephosphorylation of serine 503 in Kv3.1b permits the neurons to generate action potentials at higher rates.

6.5.3 *Phosphorylation of Kv3.3 Channel Subunits*

Although much of the focus in this chapter has been on Kv3.1 channels, the gene for the closely related Kv3.3 channel is also expressed widely in the auditory brainstem, and is the major Kv3 family channel in nuclei such as the lateral superior olive (Li et al. 2001). Mutations in *KCNC3*, the human gene encoding Kv3.3, produce severe

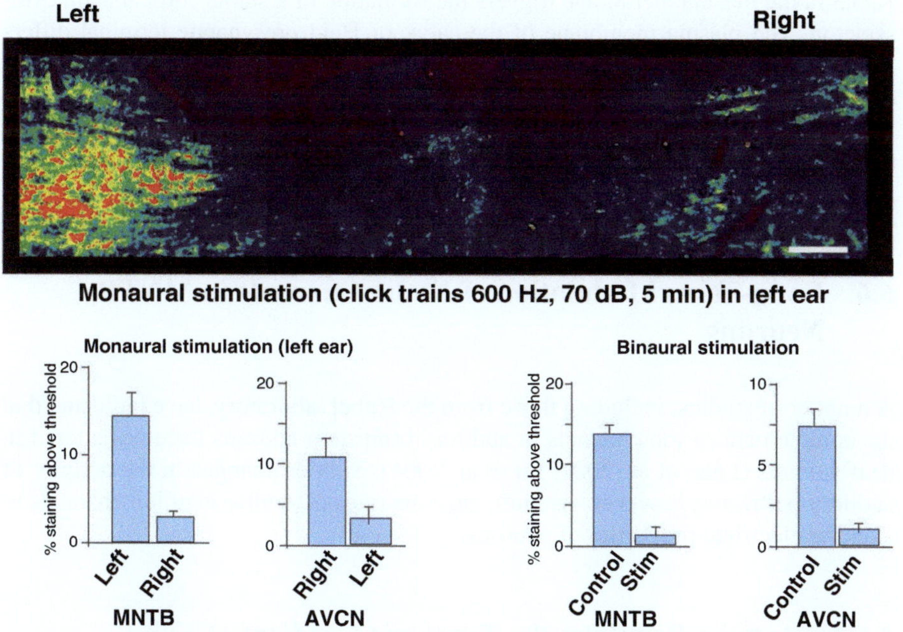

Fig. 6.4 Acoustic stimulation in vivo triggers the dephosphorylation of Kv 3.1b in MNTB neurons. Levels of phosphorylated Kv3.1b channel were detected using a serine 503 phospho-specific antibody (Song et al. 2005). The upper panel show high levels of phosphorylation (yellow-red) in the left MNTB of an animal with monaural plug in the right ear. A very much lower level of phosphorylation is detected in the contralateral MNTB, which received stimulation though a 70 dB 600-Hz click train for 5 minutes. The lower bar graphs show quantification of decreases in phosphorylation produced by monaural or binaural auditory stimulation (Song et al. 2005)

motor symptoms and also severely impair binaural processing needed for localization of auditory stimuli (Middlebrooks et al. 2013; Zhang and Kaczmarek 2016). Typically, control subjects can detect interaural timing differences of about 5–70 μs. In contrast, humans with disease-causing *Kv3.3* mutations cannot detect time differences of many hundreds of microseconds. A remarkable aspect of these findings is that the inability to process time or intensity differences occurs throughout the life of the subjects, even if they have not developed any other disease symptoms (Middlebrooks et al. 2013).

Like Kv3.1b, Kv3.3 is regulated by protein kinase C (Desai et al. 2008). In contrast to Kv3.1b, however, phosphorylation increases rather than decreases Kv3.3 currents, in major part by phosphorylation of two serine residues in the cytoplasmic N-terminus that regulate N-type inactivation of the channel (Desai et al. 2008; Zhang et al. 2016). Both Kv3.1 and Kv3.3 subunits are expressed in the presynaptic calyces of Held and they may combine there to form heteromeric channels. Sound-induced dephosphorylation of Kv3.1b occurs in these terminals, as well as in the postsynaptic somata (Song et al. 2005).

The significance of changes in the phosphorylation state of Kv3.1b/Kv3.3 heteromeric channels in the presynaptic terminals has yet to be determined. The long cytoplasmic C-terminus of Kv3.3, however, binds the cell survival protein Hax-1 and the Arp2/3 complex, which triggers actin nucleation (Zhang et al. 2016). Expression of Kv3.3 in the plasma membrane triggers the formation of a stable cortical actin cytoskeleton. The plasma membrane of the calyx of Held presynaptic terminal differs from that of the postsynaptic MNTB neuron by having exactly such a dense actin cytoskeleton under the membrane (Sakaba and Neher 2003). Thus it is possible that, in addition to altering the excitability of the terminal, sound-induced dephosphorylation of Kv3.1b/Kv3.3 channels contributes to modulation of the structure of the terminals during development.

6.6 Long-Term Modulation of Kv3.1 Currents in MNTB Neurons

A number of studies, including those from the Rubel laboratory, have indicated that the complement of ion channels in auditory brainstem neurons becomes altered in deaf animals (Leao et al. 2004; Lu et al. 2004). Subtler changes in the activity of auditory pathways, however, also influence the rates of synthesis of ion channels to alter the electrical properties of neurons.

6.6.1 Activity Regulates the Tonotopic Gradient of Kv3.1 Channels

A change in the phosphorylation state of Kv3.1b is the first and most rapid modification of these channels in response to everyday changes in sound levels of the auditory environment. With changes lasting more than a few minutes, however, neurons in the auditory brainstem begin to alter the rate at which these channels are synthesized (Strumbos et al. 2010a,b). To understand the significance of this change in protein synthesis, one first has to consider the normal distribution of Kv3.1 channels in auditory brainstem nuclei.

As described in Sect. 6.3.2, there is a tonotopic gradient of Kv3.1 channel protein in the MNTB, with medial (high-frequency) neurons expressing more Kv3.1 than lateral (low frequency) neurons (Li et al. 2001; Brew and Forsythe 2005). Early in development, such gradients may be established by guidance molecules such as ephrins (Cramer 2005). It is clear, however, that ongoing synaptic activity in auditory pathways contributes to the establishment and maintenance of such gradients. Evidence for this role for activity comes from a comparison of patterns of Kv3.1 staining in different strains of mice that undergo hearing loss at different ages (von Hehn et al. 2004). Normal tonotopic gradients of Kv3.1 expression are found in mice that have normal hearing, but as hearing loss begins to develop, these gradients are abolished.

When rats are exposed to auditory stimuli at levels comparable to those that produce dephosphorylation, but for a longer period (30 minutes), total levels of Kv3.1b increase within the MNTB, as detected by immunostaining (Strumbos et al. 2010a). The specific neurons in which levels are elevated depends, however, on the pattern and frequency of the sound stimulus. In experiments with rats, a low (4 kHz) or a high (32 kHz) pure tone carrier frequency was used to target lateral versus medial regions of the MNTB, respectively. At each carrier frequency, the temporal envelope of the stimulus was amplitude modulated at either a low (60 Hz) or a high rate (400 Hz). These two rates were used because the Kv3.1 channels are required for MNTB neurons to respond to stimulation at 400 Hz but not at 60 Hz (Macica et al. 2003). After 30 minutes of auditory stimulation with either the 4-kHz or 32-kHz carrier frequency, overall Kv3.1b immunoreactivity was enhanced by sounds modulated at 400 Hz compared to 60 Hz (Fig. 6.5A, B) (Strumbos et al. 2010a). The resultant tonotopic gradient of Kv3.1b was, however, different after exposure to the 4-kHz or 32-kHz carrier frequencies. Increases occurred preferentially in neurons of the medial (high-frequency) MNTB for the 32-kHz carried frequency and in the lateral MNTB for the 4 Hz carrier frequency (Fig. 6.5C).

6.6.2 Rapid Synthesis of Kv3.1 Channels in Response to Auditory Stimulation Is Regulated by the Fragile X Mental Retardation Protein

FMRP is an RNA-binding protein that can suppress the translation of mRNAs to which it is bound, allowing them to be translated only when a neuron is stimulated appropriately (Bassell and Warren 2008; Richter et al. 2015). Messenger RNA for Kv3.1 was one of the first mRNAs identified as a binding partner for FMRP (Darnell et al. 2001; Strumbos et al. 2010b). Loss of FMRP in humans results in fragile X syndrome, a condition associated with intellectual disability and extreme sensitivity

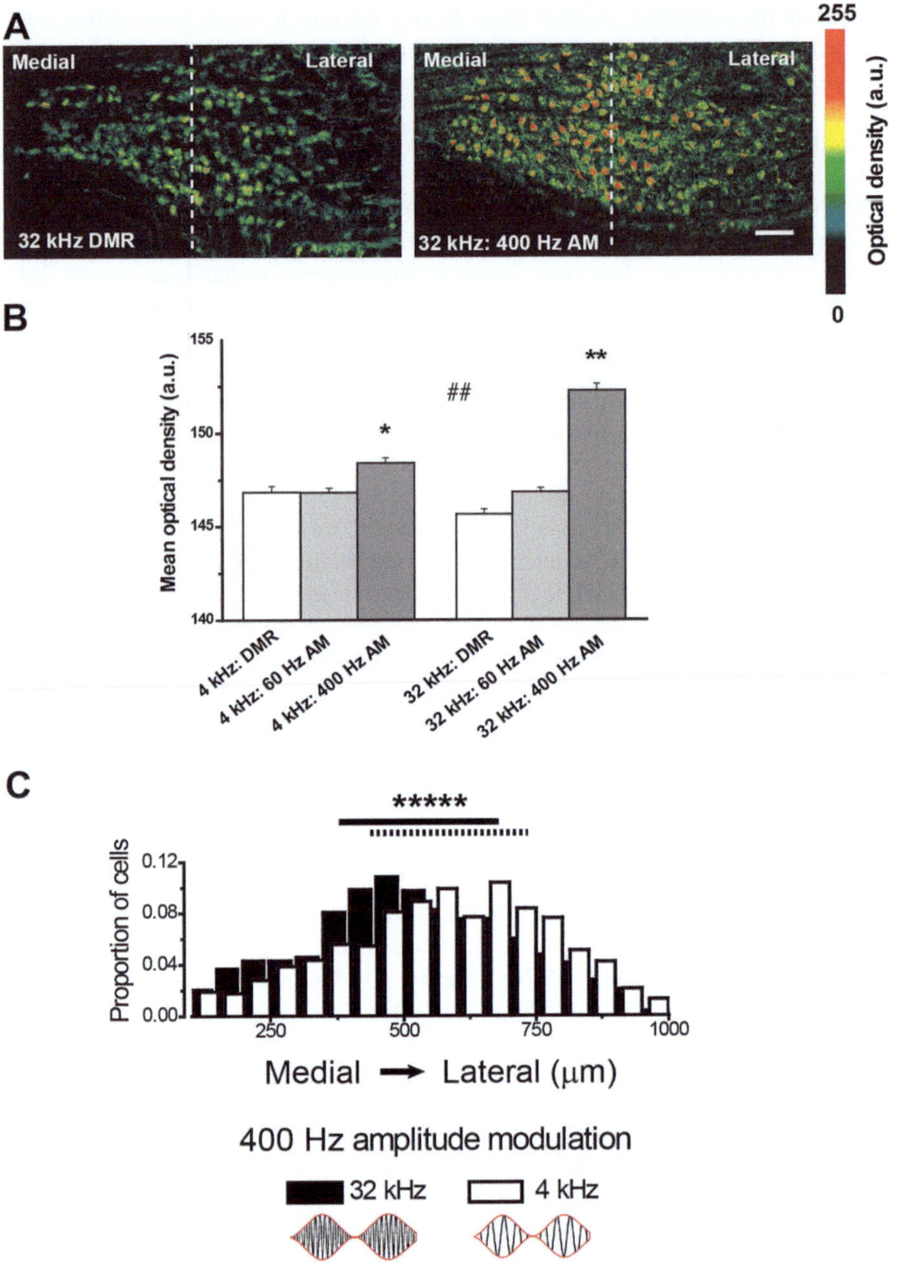

Fig. 6.5 Acoustic stimulation in vivo stimulates the net synthesis of new Kv3.1 channels in rat MNTB neurons. (**A**) Pseudocolor images showing intensity of Kv3.1b immunolabeling in medial and lateral section of medial nucleus of the trapezoid (MNTB) neurons in animals exposed to a control stimulus (DMR, for dynamic moving ripple, a more complex broad spectrum low-frequency amplitude-modulated stimulus) or to a 32-kHz pure tone carrier frequency amplitude

to auditory stimuli. The role of FMRP in regulating the rapid translation of Kv3.1 channels in response to auditory stimulation has been investigated by comparing the expression and regulation of the channels in wild-type mice with those in which the gene for FMPR has been deleted.

In wild-type mice, as in rats, a 30-minute exposure to a 32-kHz carrier frequency amplitude modulated at 400 Hz increases levels of Kv3.1b in both the AVCN and MNTB (Strumbos et al. 2010b). The situation is quite different in the FMRP knock-out animals. Even in the absence of stimulation, little or no gradient of Kv3.1b channels can be detected, either by immunostaining or by electrophysiological recordings of Kv3.1 currents in neurons along the tonotopic axis (Fig. 6.6A, B) (Strumbos et al. 2010b). Compared to wild-type animals, levels of Kv3.1 current are uniformly elevated across the medial to lateral aspect of the MNTB of animals lacking FMRP, consistent with the notion that FMRP normally acts to repress synthesis of Kv3.1 channels. Moreover, in contrast to the observation in wild-type animals, auditory stimulation fails to produce any further increase in levels of Kv3.1b (Fig. 6.6C) (Strumbos et al. 2010b).

The mechanism by which FMRP regulates protein translation is not yet fully understood. These findings suggest, however, that, in wild-type animals, synthesis of Kv3.1 channels is normally suppressed by the binding of Kv3.1 mRNA to FMRP along the tonotopic axis, with greatest suppression at the lateral (low-frequency) end of the MNTB. In response to auditory stimulation, biochemical mechanisms (most likely changes in phosphorylation state of FMRP) release mRNAs for translation, resulting in the rapid synthesis of new channels.

6.6.3 *Regulation of Kv3.1 Channels by Changes in Transcription*

Changes in Kv3.1 current produced by phosphorylation of the channel protein can occur within seconds of stimulation and are rapidly reversible. The changes in levels of Kv3.1 that can be attributed to an altered rate of translation of mRNA can be detected within minutes to tens of minutes of stimulation. Even longer-term changes in the expression of Kv3.1 may occur through changes in transcription, leading to the synthesis of new Kv3.1 mRNA.

As with both phosphorylation and translation, transcription of Kv3.1 is regulated by activity. Direct evidence for this regulation comes from investigations of the promoter for the *Kv3.1* gene (Gan and Kaczmarek 1998; Gan et al. 1999). There

modulated at 400 Hz for 30 minutes. (**B**) Bar graphs summarizing the effects of a 30-minute exposure to stimuli with different carrier frequencies and rates of amplitude modulation. (**C**) Probability histograms demonstrating that the tonotopic distribution of Kv3.1b along the medial-lateral axis of the MNTB is significantly different after exposure to the 32-kHz sound, modulated at 400 Hz, from that after exposure to the 4-kHz sound modulated at the same rate ($p < 4 \times 10^{-12}$). For details see Strumbos et al. (2010a)

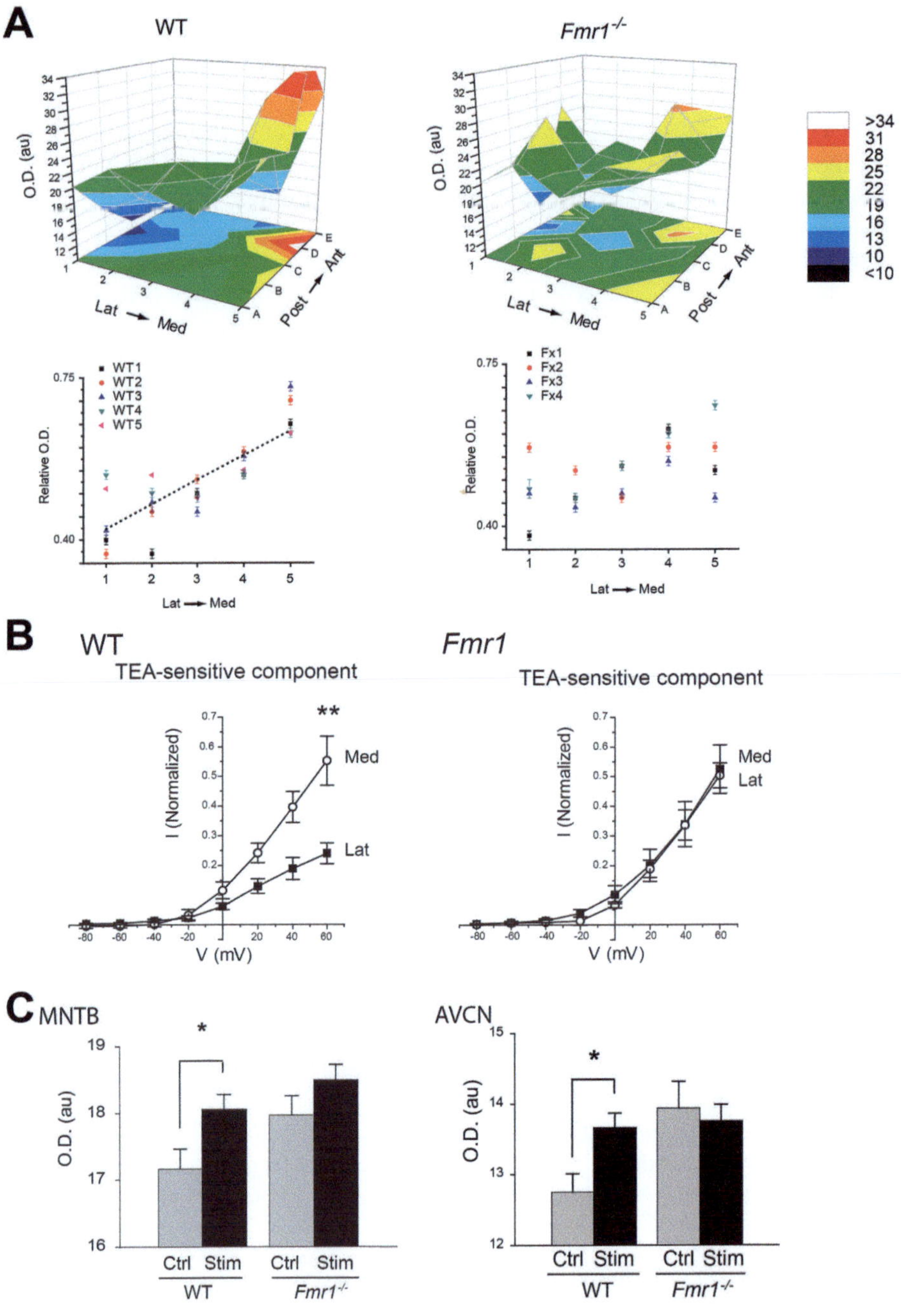

Fig. 6.6 Fragile X mental retardation protein (FMRP) is required for the tonotopic gradient of Kv3.1 and for auditory stimulation to stimulate channel synthesis. (**A**) Representative three-dimensional plots of mean Kv3.1b immunoreactivity (O.D.) across the lateral–medial and posterior–anterior axe of the MNTB in wild-type mice and in those lacking FMRP (*Fmr1*$^{-/-}$). Lower

exists a cyclic AMP/calcium response element (CRE) upstream of the start of transcription for Kv3.1 mRNA. This CRE binds the CRE-binding protein (CREB) transcription factor when CREB becomes phosphorylated following an elevation of cyclic AMP or calcium levels within a cell. Elimination of the CRE from the promoter produces a functional promoter that becomes insensitive to these cytoplasmic messengers (Gan et al. 1996). Because calcium levels are elevated during neuronal firing, the synthesis of Kv3.1 mRNA is likely to be enhanced by high-frequency stimulation of synaptic inputs to AVCN and MNTB neurons. A direct effect of elevated calcium on levels of Kv3.1 mRNA and currents has also been found for neurons of the inferior colliculus (Liu and Kaczmarek 1998b).

An increase in mRNA levels for the closely related Kv3.3 channel in response to depolarization have been reported in MNTB neurons maintained in an organotypic slice preparation, and this increase can be blocked by a CREB antagonist (Tong et al. 2010). We do not yet know how rapidly transcriptional mechanisms are likely to contribute to stimulation-induced changes in Kv3 channel levels or tonotopic gradients, nor how long they endure. Work with inferior colliculus neurons has, however, suggested that relatively large changes in Kv3.1 levels can occur over a few hours (Liu and Kaczmarek 1998b).

6.7 Changes in the Tonotopic Gradient of Kv3 Channels Contribute to the Processing of Auditory Information

What is the biological purpose of a tonotopic gradient of Kv3 channels and what is the benefit of modifying this gradient in response to auditory stimulation? Numerical simulations have established that increases in Kv3 current allow neurons to follow higher rates of stimulation, but can introduce errors in timing of individual action potentials (Macica et al. 2003; Song et al. 2005). The mere existence of a gradient, in which different neurons express different levels of Kv3.1, can largely eliminate such errors (Strumbos et al. 2010b). This is because such errors in individual neurons can be diluted out in the combined output of an entire ensemble of neurons. The lack of such a gradient may explain in part the deficits in auditory temporal discrimination in individuals with fragile X syndrome (Hall et al. 2009).

graphs show levels of Kv3.1b immunoreactivity as mean ± SEM for individual animals in five zones across the lateral–medial axis (Strumbos et al. 2010b). (**B**) Measurements of Kv3.1 current in neurons at the medial and lateral aspects of the MNTB show a clear difference in wild-type mice but no significant difference in FMRP knockout animals. Kv3.1 currents were quantified as the difference between total potassium current and that sensitive to 1 mM tetraethylammonium (TEA) ions, a potent Kv3.1 blocker (Strumbos et al. 2010b). (**C**) Bar graphs summarizing the effects of a 30-minute exposure to a 32-kHz sound stimulus modulated at 400 Hz on levels of Kv3.1 immunoreactivity in the medial nucleus of the trapezoid (MNTB) and anteroventral cochlear nucleus (AVCN) of wild-type and FMRP knockout animals (Strumbos et al. 2010b)

Computational studies have also demonstrated that, even with a preexisting tonotopic gradient, a further increase in accuracy of the output of an ensemble of neurons can usually be achieved by adjusting the level of Kv3 current in the neurons (Kaczmarek 2012). This adjustment can occur with any pattern of stimulation, including those containing multiple frequencies in the phase-locking range. Such optimization occurs when the levels of Kv3.1 conductance in a group of neighboring neurons is adjusted to those in the neurons that initially respond most accurately (Kaczmarek 2012). Thus, short-term and long-term changes in Kv3.1 current, either evoked by incoming stimuli or instructed by descending inputs from higher centers, may serve to adjust the overall fidelity of transmission through brainstem nuclei such as the MNTB.

6.8 Unresolved Questions About Channel Modulation in Auditory Neurons

6.8.1 Is Channel Modulation Local or Global?

The work described in the preceding text has demonstrated that Kv3.1b channels in the MNTB and the AVCN are modulated so as to adapt the excitability of these neurons to the auditory environment. There are still many remaining questions about how such modulation is controlled. It may be, for example, that all aspects of the channel modulation are controlled simply by the rate at which these neurons are driven by their inputs from the cochlea. For example, the phosphorylation of Kv3.1b in MNTB neurons could be determined solely by phosphatase/kinase pathways activated by glutamate release from the presynaptic calyces of Held. Nevertheless, experiments measuring changes in phosphorylation state in different auditory environments indicate that changes in phosphorylation occur in the presynaptic terminals as well as in the postsynaptic somata (Song et al. 2005). The mechanism by which phosphatase/kinase activity in the terminals is regulated by stimulation is not yet understood.

An additional uncertainty exists as to whether modulation occurs on a cell-by-cell basis, or whether the excitability of neurons in large regions of nuclei such as the MNTB is regulated coordinately. In in vivo experiments, the sound stimuli altered the phosphorylation state of channels in most or all of the neurons within the nucleus (Song et al. 2005). It is possible that auditory stimuli with more discrete frequency characteristics may alter phosphorylation only in a subset of neurons within a specific characteristic frequency range. Nevertheless, experiments using stimulation of the trapezoid body at 600 Hz also produced dephosphorylation of Kv3.1b in the majority of neurons (Song et al. 2005). It has been found that in such slice preparations only a minority of cells retain direct synaptic connection from trapezoid fibers in the midline of the slice, where the stimulating electrode is positioned (Billups et al. 2002). It remains a very intriguing possibility, therefore,

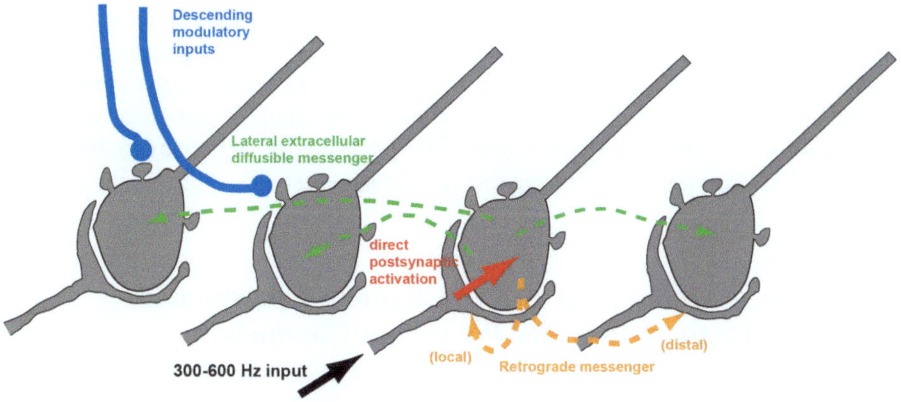

Fig. 6.7 Schematic diagram of pathways that could regulate the phosphorylation state and rates of synthesis of ion channels in medial nucleus of the trapezoid (MNTB) neurons following activation of synaptic inputs at a high rate

that stimulation of a subset of neurons at a high rate results in a coordinated change in the firing pattern of neighboring neurons that are not receiving such stimulation. This could occur, for example, if stimulation of one neuron resulted in the release of a diffusible extracellular messenger that acts in a retrograde fashion on nearby presynaptic terminals and laterally on neighboring somata, to alter the phosphorylation state of channels in these elements (Fig. 6.7). There is evidence that nitric oxide functions as such a messenger in MNTB neurons (Steinert et al. 2008). Stimulation of inputs from bushy cells for 2 minutes at a relatively low rate (100 Hz) triggers the production of NO and cyclic GMP. This occurs both in MNTB neurons receiving a direct presynaptic input as well as in adjacent neurons that are not stimulated synaptically. As a result, Kv3 currents are suppressed through a kinase/phosphatase pathway that has not been fully characterized. Such a mechanism could allow the excitability of the nucleus as a whole to be regulated by changes in the auditory environment.

6.8.2 *Can Accuracy of Transmission Be Altered by Pharmacological Agents?*

Many of the pharmacological agents that act on Kv3 family channels are relatively nonselective and block the channels (Grissmer et al. 1994). These agents include compounds as diverse as aminoglycosides and fluoxetine (Liu and Kaczmarek 2005; Sung et al. 2008). Recent studies have characterized novel imidazolidinedione derivatives that appear to act more selectively on Kv3 family channels and that increase current amplitude (Rosato-Siri et al. 2015; Taskin et al. 2015). The physiological actions of these compounds would not be expected to match exactly those of an increase in levels of Kv3.1 channels because they increase currents by shifting their voltage range of activation and inactivation to more negative potentials. Nevertheless, in brain slice recordings of mouse MNTB neurons, the compounds modulate firing rate at high rates of stimulation in ways that are consistent with their actions on Kv3.1 channels (Brown et al. 2016). Studies are underway to determine whether these compounds will have therapeutic benefits.

6.9 How Is Modulation of Kv3 Channels Coordinated with Changes in Other Potassium Channels?

Clearly, auditory brainstem neurons express channels other than Kv3.1 and Kv3.3. Only for Kv3.1, however, have studies been carried out on how phosphorylation and translation/transcription are altered in response to auditory stimulation. It is likely that stimulation-induced alterations in at least some of the other channels also contribute to changes in the way that auditory information is processed. Details of other channels in brainstem nuclei and their contributions to intrinsic excitability have been reviewed elsewhere (Johnston et al. 2010; Brown and Kaczmarek 2011). A brief account of these other channels is presented in the text that follows.

6.9.1 Kv1 Family Channels

While Kv3.1 channels determine the rate at which MNTB and AVCN neurons are able to generate sustained patterns of firing, a completely different set of potassium channels controls the properties of the membrane close to the threshold for firing. Among these is the Kv1 family of potassium channels (Johnston et al. 2010). Their activity is essential for maintaining temporal fidelity by ensuring that each incoming presynaptic action potential generates only a single postsynaptic action potential (Dodson et al. 2002). Kv1.1, Kv1.2, and Kv1.6 are localized along axons of MNTB neurons but are absent from the presynaptic calyx of Held terminal and from the postsynaptic cell bodies of MNTB neurons. Because they are located away from synaptic areas, they are unlikely to undergo rapid modulation by traditional receptor-linked mechanisms, but activation of protein kinases may regulate their synthesis and insertion into the plasma membrane (Winklhofer et al. 2003).

In contrast to the axonal location of the Kv1 family channels described in the preceding text, the Kv1.3 channel is localized predominantly to the presynaptic terminal itself, where it is found both in the plasma membrane and on intracellular vesicles (Gazula et al. 2010). Kv1.3 current is known to be regulated by phosphorylation of the channel protein by tyrosine kinases (Cook and Fadool 2002), but this regulation has not been tested in MNTB neurons. The finding that Kv1.3 is located on presynaptic vesicles raises the intriguing possibility that activity driven by sounds may trigger insertion of channels into the presynaptic membrane, resulting in activity-dependent alterations in presynaptic excitability.

6.9.2 Kv2.2 Channels in the Initial Segment

A major channel in the initial segment of MNTB neurons is the Kv2.2 voltage-dependent potassium channel (Johnston et al. 2008). Its major role appears to aid the recovery of sodium channel from inactivation by hyperpolarizing the membrane between action potentials during repetitive firing. In other neurons, phosphorylation of the closely related channel Kv2.1 rapidly alters its assembly into clusters in the membrane, as well as its biophysical properties (Misonou et al. 2004). It is not known, however, if anything similar occurs for Kv2.2 in auditory brainstem neurons.

6.9.3 Kv11 Channels

Kv11 channels are also termed ERG channels (named for *ether-a-go-go related gene*, the name of the corresponding channels in *Drosophila*). Immunolocalization and in situ hybridization studies indicate that Kv11.1 and Kv11.3 channel subunits are distributed in a punctate pattern on the plasma membrane of MNTB neuronal cell bodies (Hardman and Forsythe 2009). Although this distribution differs from that of the axonal Kv1 family channels, pharmacological experiments suggest that they play a similar role in regulating action potential threshold. Nothing is, however, known of their potential regulation by second messenger pathways in auditory neurons.

6.9.4 Sodium-Activated Potassium Channels

The sodium-activated potassium channels $K_{Na}1.1$ (Slack, Slo2.2) and $K_{Na}1.2$ (Slick, Slo2.1) are also expressed in MNTB neurons (Bhattacharjee et al. 2002, 2005). These channels may be activated by sodium entry through voltage-dependent sodium channels, through α-amino-3-hydroxy-5-methyl-4-isoxazolepropionic acid (AMPA) receptors in the postsynaptic membrane, or through other nonselective cation channels (Kaczmarek 2013). The greatest activation of such channels would therefore be expected during rapid firing induced by presynaptic stimulation.

In slices of auditory brainstem, pharmacological activation of these channels increases the temporal accuracy with which MNTB neurons lock their action potentials to stimulus pulses (Yang et al., 2007). The amplitude of both $K_{Na}1.1$ and $K_{Na}1.2$, as well as of heteromeric $K_{Na}1.1/K_{Na}1.2$ channels, is regulated by protein kinase C, which phosphorylates residue serine 407 in the cytoplasmic C-terminal domain of $K_{Na}1.1$ (Santi et al. 2006; Barcia et al. 2012). It is therefore probable that, as shown for Kv3.1, auditory stimulation may alter the phosphorylation state of these channels in MNTB neurons, and thereby adjust the temporal fidelity of transmission through the nucleus. Again, this hypothesis has not yet been tested directly.

FMRP binds $K_{Na}1.1$ mRNA, suggesting that, as described in Sect. 6.6.2 for Kv3.1, the synthesis of these channels will depend on activity evoked by the auditory environment. FMRP, however, also binds to the cytoplasmic C-terminus of Slack channels directly, increasing their open probability (Brown et al. 2010). As a result, the level of sodium-activated potassium current in MNTB neurons is reduced in FMRP knockout mice (Brown et al. 2010). The decrease in this sodium-dependent current, which activates near the resting potential, may contribute to the enhanced sensitivity of FMRP knockout animals to auditory stimuli.

A full understanding of how the intrinsic excitability of auditory brainstem neurons is altered by different patterns of auditory stimulation will require investigation of whether and how each of the channels covered in this section is modulated. In particular, how a neuron coordinates short-term and long-term levels in Kv3 proteins and currents with those of each of these other currents that regulate action potential threshold and other aspects of excitability remains a major question.

6.10 Do Changes in Excitability of Brainstem Neurons Contribute to Auditory Learning?

The changes in channel phosphorylation and synthesis described in this chapter may result directly from alterations in the patterns of sensory inputs, representing a "bottom up" form of regulation. It is possible, however, that control of the specific way that MNTB and AVCN neurons fire in response to changes in the auditory environment is also regulated by descending inputs from higher centers (Fig. 6.7). Such an input would also be capable of adjusting the electrical properties of either a subset of neurons or of the whole nucleus.

As described in Sect. 6.1, a high degree of accuracy of timing is required for tasks such as pitch discrimination and the ability to discern small changes in the spatial location of sounds. In no other place in the central auditory nervous system is such accuracy evident as in the brainstem neurons, whose function is to detect and process the small differences in the timing of incoming stimuli. Because auditory training results in improvements in the ability to detect small timing differences, it seems likely that the functioning of the AVCN and MNTB is altered during such learning. In particular, learning may result from changes in the ability of higher centers to "tune" the electrical properties of brainstem neurons to small differences in specific auditory stimuli. This tuning could occur either through changes in synaptic connections in higher centers, through changes in second messenger/protein kinase pathways in brainstem neurons, or through activity-driven changes in the expression of specific ion channel subunits within brainstem neurons.

6.11 Summary

Work covered in this chapter indicates that the intrinsic excitability of auditory brainstem neurons is remarkably plastic. Changes in the phosphorylation state of an ion channel can occur within seconds of a change in auditory environment, such as occurs, for example, in walking from a quite library to a noisy street or into a raucous party. These changes may adapt the evoked firing pattern of the neurons so as to maximize the transfer and decoding of information about the timing and direction of sounds. When exposure to a new sound environment persists for half an hour or more, translation and transcription of ion channel genes may provide longer-lasting changes in intrinsic excitability. Differences in the intrinsic excitability of brainstem neurons, and their ability rapidly to alter their excitability, may contribute to the superior ability of orchestra conductors to detect differences in the location of sound sources, or to the superior ability of musicians who play string instruments to detect differences in pitch.

Compliance with Ethics Requirements Leonard K. Kaczmarek has received research grants from Autifony Inc.

References

Barcia, G., Fleming, M. R., Deligniere, A., Gazula, V. R., et al. (2012). De novo gain-of-function KCNT1 channel mutations cause malignant migrating partial seizures of infancy. *Nature Genetics*, 44(11), 1255–1259.

Bassell, G. J., & Warren, S. T. (2008). Fragile X syndrome: Loss of local mRNA regulation alters synaptic development and function. *Neuron*, 60(2), 201–214.

Bhattacharjee, A., Gan, L., & Kaczmarek, L. K. (2002). Localization of the Slack potassium channel in the rat central nervous system. *The Journal of Comparative Neurology*, 454(3), 241–254.

Bhattacharjee, A., von Hehn, C. A., Mei, X., & Kaczmarek, L. K. (2005). Localization of the Na^+-activated K^+ channel Slick in the rat central nervous system. *The Journal of Comparative Neurology*, 484(1), 80–92.

Billups, B., Wong, A. Y., & Forsythe, I. D. (2002). Detecting synaptic connections in the medial nucleus of the trapezoid body using calcium imaging. *Pflugers Archiv – European Journal of Physiology*, 444(5), 663–669.

Brand, A., Behrend, O., Marquardt, T., McAlpine, D., & Grothe, B. (2002). Precise inhibition is essential for microsecond interaural time difference coding. *Nature*, 417(6888), 543–547.

Brew, H. M., & Forsythe, I. D. (2005). Systematic variation of potassium current amplitudes across the tonotopic axis of the rat medial nucleus of the trapezoid body. *Hearing Research*, 206(1–2), 116–132.

Brown, M. R., & Kaczmarek, L. K. (2011). Potassium channel modulation and auditory processing. *Hearing Research*, 279(1–2), 32–42.

Brown, M. R., Kronengold, J., Gazula, V. R., Chen, Y., et al. (2010). Fragile X mental retardation protein controls gating of the sodium-activated potassium channel Slack. *Nature Neuroscience*, 13(7), 819–821.

Brown, M. R., El-Hassar, L., Zhang, Y., Alvaro, G., et al. (2016). Physiological modulators of Kv3.1 channels adjust firing patterns of auditory brainstem neurons. *Journal of Neurophysiology*, jn 00174 02016.

Cook, K. K., & Fadool, D. A. (2002). Two adaptor proteins differentially modulate the phosphorylation and biophysics of Kv1.3 ion channel by SRC kinase. *The Journal of Biological Chemistry*, 277(15), 13268–13280.
Cramer, K. S. (2005). Eph proteins and the assembly of auditory circuits. *Hearing Research*, 206(1–2), 42–51.
Darnell, J. C., Jensen, K. B., Jin, P., Brown, V., et al. (2001). Fragile X mental retardation protein targets G quartet mRNAs important for neuronal function. *Cell*, 107(4), 489–499.
Desai, R., Kronengold, J., Mei, J., Forman, S. A., & Kaczmarek, L. K. (2008). Protein kinase C modulates inactivation of Kv3.3 channels. *The Journal of Biological Chemistry*, 283(32), 22283–22294.
Dodson, P. D., Barker, M. C., & Forsythe, I. D. (2002). Two heteromeric Kv1 potassium channels differentially regulate action potential firing. *The Journal of Neuroscience*, 22(16), 6953–6961.
Gan, L., & Kaczmarek, L. K. (1998). When, where, and how much? Expression of the Kv3.1 potassium channel in high-frequency firing neurons. *Journal of Neurobiology*, 37(1), 69–79.
Gan, L., Perney, T.M. & Kaczmarek, L.K. (1996). Cloning and characterization of the promoter for a potassium channel expressed in high frequency firing neurons, *The Journal of Biological Chemistry*, 271: 5859–5865.
Gan, L., Hahn, S. J., & Kaczmarek, L. K. (1999). Cell type-specific expression of the *Kv3.1* gene is mediated by a negative element in the 5′ untranslated region of the Kv3.1 promoter. *Journal of Neurochemistry*, 73(4), 1350–1362.
Gazula, V. R., Strumbos, J. G., Mei, X., Chen, H., et al. (2010). Localization of Kv1.3 channels in presynaptic terminals of brainstem auditory neurons. *The Journal of Comparative Neurology*, 518(16), 3205–3220.
Grissmer, S., Nguyen, A. N., Aiyar, J., Hanson, D. C., et al. (1994). Pharmacological characterization of five cloned voltage-gated K^+ channels, types Kv1.1, 1.2, 1.3, 1.5, and 3.1, stably expressed in mammalian cell lines. *Molecular Pharmacology*, 45(6), 1227–1234.
Hall, S. S., Walter, E., Sherman, E., Hoeft, F., & Reiss, A. L. (2009). The neural basis of auditory temporal discrimination in girls with fragile X syndrome. *Journal of Neurodevelopmental Disorders*, 1(1), 91–99.
Hardman, R. M., & Forsythe, I. D. (2009). Ether-a-go-go-related gene K^+ channels contribute to threshold excitability of mouse auditory brainstem neurons. *The Journal of Physiology*, 587(Pt 11), 2487–2497.
Hodgkin, A. L., & Huxley, A. F. (1952). A quantitative description of membrane current and its application to conduction and excitation in nerve. *The Journal of Physiology*, 117(Pt 4), 500–544.
Johnston, J., Griffin, S. J., Baker, C., Skrzypiec, A., et al. (2008). Initial segment Kv2.2 channels mediate a slow delayed rectifier and maintain high frequency action potential firing in medial nucleus of the trapezoid body neurons. *The Journal of Physiology*, 586(14), 3493–3509.
Johnston, J., Forsythe, I. D., & Kopp-Scheinpflug, C. (2010). Going native: Voltage-gated potassium channels controlling neuronal excitability. *The Journal of Physiology*, 588(Pt 17), 3187–3200.
Joris, P. X. (1996). Envelope coding in the lateral superior olive. II. Characteristic delays and comparison with responses in the medial superior olive. *Journal of Neurophysiology*, 76(4), 2137–2156.
Joris, P. X., & Yin, T. C. (1995). Envelope coding in the lateral superior olive. I. Sensitivity to interaural time differences.[erratum appears in *Journal of Neurophysiology* 1995 Jun;73(6):followi]. *Journal of Neurophysiology*, 73(3), 1043–1062.
Kaczmarek, L. K. (2012). Gradients and modulation of K(+) channels optimize temporal accuracy in networks of auditory neurons. *PLoS Computational Biology*, 8(3), e1002424.
Kaczmarek, L. K. (2013). Slack, Slick and sodium-activated potassium channels. *International Scholarly Research Notices Neuroscience*, 2013, 354262, 1–14.

Kanemasa, T., Gan, L., Perney, T. M., Wang, L. Y., & Kaczmarek, L. K. (1995). Electrophysiological and pharmacological characterization of a mammalian Shaw channel expressed in NIH 3T3 fibroblasts. *Journal of Neurophysiology*, 74(1), 207–217.

Kullmann, L., Schluter, T., Wagner, H., & Nothwang, H. G. (2013). Evolutionary conservation of Kv3.1 in the barn owl *Tyto alba*. *Brain, Behavior and Evolution*, 81(3), 187–193.

Labro, A. J., Priest, M. F., Lacroix, J. J., Snyders, D. J., & Bezanilla, F. (2015). Kv3.1 uses a timely resurgent $K(^{+})$ current to secure action potential repolarization. *Nature Communications*, 6, 10173.

Leao, R. N., Berntson, A., Forsythe, I. D., & Walmsley, B. (2004). Reduced low-voltage activated K^{+} conductances and enhanced central excitability in a congenitally deaf (*dn/dn*) mouse. *The Journal of Physiology*, 559(Pt 1), 25–33.

Li, W., Kaczmarek, L. K., & Perney, T. M. (2001). Localization of two high-threshold potassium channel subunits in the rat central auditory system. *The Journal of Comparative Neurology*, 437(2), 196–218.

Liu, S. J., & Kaczmarek, L. K. (1998a). The expression of two splice variants of the Kv3.1 potassium channel gene is regulated by different signaling pathways. *The Journal of Neuroscience*, 18(8), 2881–2890.

Liu, S. Q., & Kaczmarek, L. K. (1998b). Depolarization selectively increases the expression of the Kv3.1 potassium channel in developing inferior colliculus neurons. *The Journal of Neuroscience*, 18(21), 8758–8769.

Liu, S. Q., & Kaczmarek, L. K. (2005). Aminoglycosides block the Kv3.1 potassium channel and reduce the ability of inferior colliculus neurons to fire at high frequencies. *Journal of Neurobiology*, 62(4), 439–452.

Lu, Y., Monsivais, P., Tempel, B. L., & Rubel, E. W. (2004). Activity-dependent regulation of the potassium channel subunits Kv1.1 and Kv3.1. *The Journal of Comparative Neurology*, 470(1), 93–106.

Luneau, C. J., Williams, J. B., Marshall, J., Levitan, E. S., et al. (1991). Alternative splicing contributes to K^{+} channel diversity in the mammalian central nervous system. *Proceedings of the National Academy of Sciences of the USA*, 88(9), 3932–3936.

Macica, C. M., & Kaczmarek, L. K. (2001). Casein kinase 2 determines the voltage dependence of the Kv3.1 channel in auditory neurons and transfected cells. *The Journal of Neuroscience*, 21(4), 1160–1168.

Macica, C. M., von Hehn, C. A., Wang, L. Y., Ho, C. S., et al. (2003). Modulation of the kv3.1b potassium channel isoform adjusts the fidelity of the firing pattern of auditory neurons. *The Journal of Neuroscience*, 23(4), 1133–1141.

Middlebrooks, J. C., Nick, H. S., Subramony, S. H., Advincula, J., et al. (2013). Mutation in the kv3.3 voltage-gated potassium channel causing spinocerebellar ataxia 13 disrupts sound-localization mechanisms. *PloS ONE*, 8(10), e76749.

Misonou, H., Mohapatra, D. P., Park, E. W., Leung, V., et al. (2004). Regulation of ion channel localization and phosphorylation by neuronal activity. *Nature Neuroscience*, 7(7), 711–718.

Morest, D. K. (1968). The collateral system of the medial nucleus of the trapezoid body of the cat, its neuronal architecture and relation to the olivo-cochlear bundle. *Brain Research*, 9(2), 288–311.

Mossbridge, J. A., Fitzgerald, M. B., O'Connor, E. S., & Wright, B. A. (2006). Perceptual-learning evidence for separate processing of asynchrony and order tasks. *The Journal of Neuroscience*, 26(49), 12708–12716.

Munte, T. F., Kohlmetz, C., Nager, W., & Altenmuller, E. (2001). Neuroperception: Superior auditory spatial tuning in conductors. *Nature*, 409(6820), 580.

Nager, W., Kohlmetz, C., Altenmuller, E., Rodriguez-Fornells, A., & Munte, T. F. (2003). The fate of sounds in conductors' brains: an ERP study. *Cognitive Brain Research*, 17(1), 83–93.

Parameshwaran, S., Carr, C. E., & Perney, T. M. (2001). Expression of the Kv3.1 potassium channel in the avian auditory brainstem. *The Journal of Neuroscience*, 21(2), 485–494.

Perney, T. M., & Kaczmarek, L. K. (1997). Localization of a high threshold potassium channel in the rat cochlear nucleus. *The Journal of Comparative Neurology*, 386(2), 178–202.

Perney, T. M., Marshall, J., Martin, K. A., Hockfield, S., & Kaczmarek, L. K. (1992). Expression of the mRNAs for the Kv3.1 potassium channel gene in the adult and developing rat brain. *Journal of Neurophysiology*, 68(3), 756–766.

Polley, D. B., Steinberg, E. E., & Merzenich, M. M. (2006). Perceptual learning directs auditory cortical map reorganization through top-down influences. *The Journal of Neuroscience*, 26(18), 4970–4982.

Richter, J. D., Bassell, G. J., & Klann, E. (2015). Dysregulation and restoration of translational homeostasis in fragile X syndrome. *Nature Reviews Neuroscience*, 16(10), 595–605.

Rosato-Siri, M. D., Zambello, E., Mutinelli, C., Garbati, N., et al. (2015). A novel modulator of Kv3 potassium channels regulates the firing of parvalbumin-positive cortical interneurons. *The Journal of Pharmacology and Experimental Therapeutics*, 354(3), 251–260.

Rubel, E. W., Hyson, R. L., & Durham, D. (1990). Afferent regulation of neurons in the brain stem auditory system. *Journal of Neurobiology*, 21(1), 169–196.

Rudy, B., & McBain, C. J. (2001). Kv3 channels: Voltage-gated K^+ channels designed for high-frequency repetitive firing. *Trends in Neurosciences*, 24(9), 517–526.

Sakaba, T., & Neher, E. (2003). Involvement of actin polymerization in vesicle recruitment at the calyx of Held synapse. *The Journal of Neuroscience*, 23(3), 837–846.

Santi, C. M., Ferreira, G., Yang, B., Gazula, V. R., Butler, A., Wei, A., Kaczmarek, L. K., & Salkoff, L. (2006). Opposite regulation of Slick and Slack K^+ channels by neuromodulators. *The Journal of Neuroscience*, 26(19), 5059–5068.

Schneggenburger, R., & Forsythe, I. D. (2006). The calyx of Held. *Cell and Tissue Research*, 326(2), 311–337.

Smith, P. H., Joris, P. X., & Yin, T. C. (1998). Anatomy and physiology of principal cells of the medial nucleus of the trapezoid body (MNTB) of the cat. *Journal of Neurophysiology*, 79(6), 3127–3142.

Song, P., & Kaczmarek, L. K. (2006). Modulation of Kv3.1b potassium channel phosphorylation in auditory neurons by conventional and novel protein kinase C isozymes. *The Journal of Biological Chemistry*, 281(22), 15582–15591.

Song, P., Yang, Y., Barnes-Davies, M., Bhattacharjee, A., Hamann, M., et al. (2005). Acoustic environment determines phosphorylation state of the Kv3.1 potassium channel in auditory neurons. *Nature Neuroscience*, 8(10), 1335–1342.

Spirou, G. A., Brownell, W. E., & Zidanic, M. (1990). Recordings from cat trapezoid body and HRP labeling of globular bushy cell axons. *Journal of Neurophysiology*, 63(5), 1169–1190.

Steinert, J. R., Kopp-Scheinpflug, C., Baker, C., Challiss, R. A., et al. (2008). Nitric oxide is a volume transmitter regulating postsynaptic excitability at a glutamatergic synapse. *Neuron*, 60(4), 642–656.

Strumbos, J. G., Polley, D. B., & Kaczmarek, L. K. (2010a). Specific and rapid effects of acoustic stimulation on the tonotopic distribution of Kv3.1b potassium channels in the adult rat. *Neuroscience*, 167(3), 567–572.

Strumbos, J. G., Brown, M. R., Kronengold, J., Polley, D. B., & Kaczmarek, L. K. (2010b). Fragile X mental retardation protein is required for rapid experience-dependent regulation of the potassium channel Kv3.1b. *The Journal of Neuroscience*, 30(31), 10263–10271.

Sung, M. J., Ahn, H. S., Hahn, S. J., & Choi, B. H. (2008). Open channel block of Kv3.1 currents by fluoxetine. *Journal of Pharmacological Sciences*, 106(1), 38–45.

Taschenberger, H., & von Gersdorff, H. (2000). Fine-tuning an auditory synapse for speed and fidelity: Developmental changes in presynaptic waveform, EPSC kinetics, and synaptic plasticity. *The Journal of Neuroscience*, 20(24), 9162–9173.

Taskin, B., von Schoubye, N. L., Sheykhzade, M., Bastlund, J. F., et al. (2015). Biophysical characterization of KV3.1 potassium channel activating compounds. *European Journal of Pharmacology*, 758, 164–170.

Tong, H., Steinert, J. R., Robinson, S. W., Chernova, T., et al. (2010). Regulation of Kv channel expression and neuronal excitability in rat medial nucleus of the trapezoid body maintained in organotypic culture. *The Journal of Physiology*, 588(Pt 9), 1451–1468.

Trussell, L. O. (1999). Synaptic mechanisms for coding timing in auditory neurons. *Annual Review of Physiology*, 61, 477–496.

von Hehn, C. A., Bhattacharjee, A., & Kaczmarek, L. K. (2004). Loss of Kv3.1 tonotopicity and alterations in cAMP response element-binding protein signaling in central auditory neurons of hearing impaired mice. *The Journal of Neuroscience*, 24(8), 1936–1940.

Wang, L. Y., & Kaczmarek, L. K. (1998). High-frequency firing helps replenish the readily releasable pool of synaptic vesicles. *Nature*, 394(6691), 384–388.

Wang, L. Y., Gan, L., Forsythe, I. D., & Kaczmarek, L. K. (1998). Contribution of the Kv3.1 potassium channel to high-frequency firing in mouse auditory neurones. *The Journal of Physiology*, 509(Pt 1), 183–194.

Wang, Y., Sakano, H., Beebe, K., Brown, M. R., et al. (2014). Intense and specialized dendritic localization of the fragile X mental retardation protein in binaural brainstem neurons: A comparative study in the alligator, chicken, gerbil, and human. *The Journal of Comparative Neurology*, 522(9), 2107–2128.

Winklhofer, M., Matthias, K., Seifert, G., Stocker, M., et al. (2003). Analysis of phosphorylation-dependent modulation of Kv1.1 potassium channels. *Neuropharmacology*, 44(6), 829–842.

Yang, B., Desai, R., & Kaczmarek, L. K. (2007). Slack and Slick K(Na) channels regulate the accuracy of timing of auditory neurons. *The Journal of Neuroscience*, 27(10), 2617–2627.

Yokoyama, S., Imoto, K., Kawamura, T., Higashida, H., et al. (1989). Potassium channels from NG108-15 neuroblastoma-glioma hybrid cells: Primary structure and functional expression from cDNAs. *FEBS Letters*, 259(1), 37–42.

Zhang, Y., & Kaczmarek, L. K. (2016). Kv3.3 potassium channels and spinocerebellar ataxia. *The Journal of Physiology*, 594(16), 4677–4684.

Zhang, Y., Zhang, X. F., Fleming, M. R., Amiri, A., et al. (2016). Kv3.3 channels bind Hax-1 and Arp2/3 to assemble a stable local actin network that regulates channel gating. *Cell*, 165(2), 434–448.

Chapter 7
Ontogeny of Human Auditory System Function

Lynne A. Werner

Abstract When Edwin W Rubel published *Ontogeny of Auditory System Function* in 1984, little was known about human auditory development. This chapter summarizes the progress that has been made in that regard in the subsequent decades. It concludes that although the developmental course of many basic aspects of human auditory function has now been described, the morphological and physiological mechanisms responsible for that development are still largely unknown. Furthermore, it is not clear why some basic auditory abilities mature much later than others or why the developmental course of listening in complex sound environments extends into adolescence. Finally, the existence and timing of critical periods in human auditory development have only begun to be established. Although studies of development of the nonhuman auditory system may elucidate the mechanisms underlying human auditory development, there are difficulties inherent to the comparison of altricial and precocial species. Recent work combining behavioral and neural measures in both humans and nonhumans offers the promise of future progress in this field.

Keywords Afferent influences • Auditory scene analysis • Effects of experience • Frequency resolution • Hearing • Intensity • Loudness • Maturation • Psychoacoustics • Sensitive period • Sensitivity • Temporal processing

7.1 Introduction

In 1984, Edwin W Rubel published an *Annual Review* entitled "Ontogeny of Auditory System Function." This review was notable for two reasons. First, it attempted to integrate the limited data on human auditory development with those on auditory development in other species. Second, it emphasized the unanswered

L.A. Werner (✉)
Department of Speech and Hearing Sciences, University of Washington,
1417 N.E. 42nd Street, Seattle, WA 98105-6246, USA
e-mail: lawerner@uw.edu

K.S. Cramer et al. (eds.), *Auditory Development and Plasticity*, Springer Handbook of Auditory Research 64, DOI 10.1007/978-3-319-21530-3_7

questions in auditory development, pointing the way to potentially interesting research questions in that area.

In the decades since that review was published, tremendous progress has been made in the study of auditory system development, and many papers have been published on auditory function in human infants and children. The current chapter is intended as an update of Rubel (1984), but limited to studies of human auditory development. Like Rubel's review, the chapter focuses on the unanswered questions and suggests future directions for the field.

7.2 Development of Conductive Elements

The acoustics of the head, external ear, and middle ear set limits on all aspects of auditory processing. However, the conductive elements of the system have distinctive influences on two aspects of auditory function, absolute sensitivity, and spatial hearing. The resonance of the conductive apparatus is largely responsible for the frequency dependence of absolute thresholds. In addition, the head and pinna are critical to spatial hearing: The size of the head determines the range of interaural time differences (ITDs, the differences in arrival times of sound at the two ears) as well as the frequencies at which interaural level differences (ILDs, the differences in level of sound arriving at the two ears) are available to the binaural system; the resonance of the pinna shapes the spectrum of incoming sound in a directionally dependent fashion, providing the cues to sound source elevation and the resolution of front–back ambiguity.

7.2.1 External Ear

The head, pinna, and ear canal increase in size dramatically during infancy and childhood. The expectation, therefore, is that the external ear resonance frequencies will be higher in infants than in adults. Surprisingly few studies have quantified the physical characteristics of the head and external ear in human infants and children from an acoustic perspective and very few studies have quantified the acoustic consequences of the growth of the head and external ear. By acoustic estimates, ear canal length is approximately 15 mm at 1 month, 21 mm at 24 months, and 24 mm in adults (Keefe et al. 1994). Thus, the greatest change in the resonance of the external ear appears to occur during infancy. The diffuse field transfer function at 1 month of age shows a maximum gain of about 5 dB around 6000 Hz; by 24 months the peak gain has increased to more than 10 dB around 2400 Hz and is nearly adult-like (Keefe et al. 1994). Thus, the external ear limits hearing in natural environments during infancy. While the acoustic consequences of external ear growth depend on how sound is delivered to the ear, earphone measurements also indicate lower pressures generated in infant than in adult ears at frequencies above 2000 Hz, and only small differences between 24-month-olds and adults (Voss and Herrmann 2005). An exception of potential importance in clinical and laboratory studies is that insert earphones generate a 5-dB *greater* pressure in 1-month-olds than in adults across the frequency range. Thus, any comparison between developmental studies must consider any differences in the methods for delivering sound to the ear, how the sound level was calibrated, and the ages of the participants.

Clifton et al. (1988) measured interaural distance in a longitudinal sample of infants between birth and 22 weeks of age. Because of changes in head shape during infancy, interaural distance was found to increase more slowly than head circumference with age. However, the age differences in interaural distance were great enough to have substantial functional consequences for binaural hearing. The maximum ITD possible for a newborn's head would be only 400 μs compared to nearly 700 μs for an adult's head. By 22 weeks, the maximum ITD had increased to about 500 μs. Based on the measurements of Clifton et al., one would predict that newborns would not be able to generate substantial ILD below about 2500 Hz, compared to about 1500 Hz in adults. However, these predictions have not been tested empirically.

7.2.2 *Middle Ear*

Early studies of middle ear function in human infants used tympanometry, a clinical tool, to estimate admittance, the ease with which a medium transmits sound. As Rubel (1984) pointed out, such studies reported an increase in admittance over the first 3 months of life (Himelfarb et al. 1979). The problem with tympanometry is that it involves changing the air pressure in the ear canal, and because the stiffness of the ear canal increases during infancy, it is difficult to separate ear canal and middle ear effects. Keefe et al. (1993) measured ear canal impedance—the reciprocal of admittance—and energy reflectance—the proportion of sound energy that is not transmitted into the middle ear—at ambient air pressure in the ear canal of infants ranging from 1 month to 24 months of age. They concluded that the power transferred into the infant middle ear is less than that in the adult middle ear, largely because of the small infant middle ear cavity. At 3 months of age, about 8 dB less power is transferred into the middle ear at frequencies above 2000 Hz; the difference in power transfer is 5 dB or less above 4000 Hz at 24 months. A subsequent study of newborn infants reported changes in middle ear response between 33 and 46 weeks conceptional age (CA, gestational age + chronological age) that were modest compared to those observed later in infancy (Keefe et al. 2000).

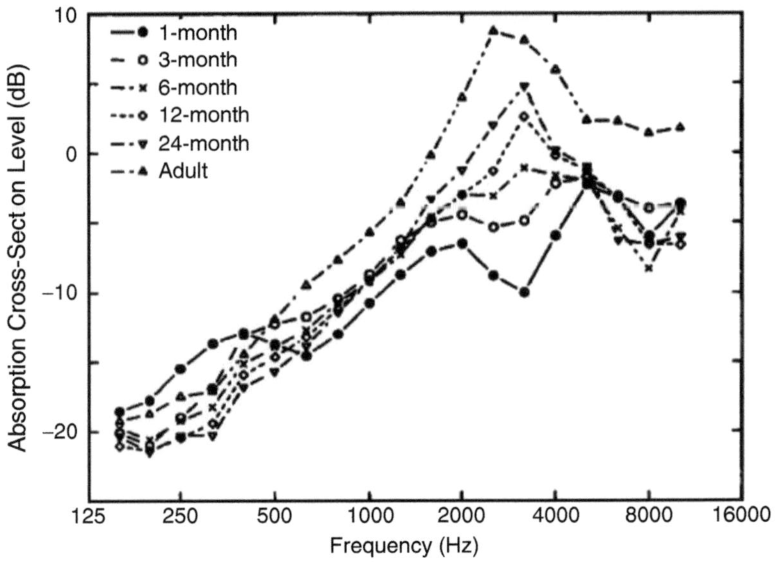

Fig. 7.1 Maturation of the conductive efficiency of the ear. Data are plotted in terms of sound absorption (cross-sectional level, 10 log A_D, in dB re: 1 cm^2) from the diffuse field to the ear canal as a function of frequency for six age groups. The greatest changes occur in the 2000- to 6000-Hz range during infancy, especially in the first 6 postnatal months; smaller changes occur at other frequencies. [From Keefe et al. (1994). Copyright © 1994 Acoustical Society of America]

7.2.3 Influences on Auditory Function

7.2.3.1 Absolute Sensitivity

Perhaps the most obvious effect of conductive immaturity on auditory function should be on absolute sensitivity: Less power transferred to the inner ear means that more power needs to be provided to elicit a response. Keefe et al. (1994) estimated the combined effect of external and middle ear immaturity on the absorption of sound into the ear (Fig. 7.1). Between about 2000 and 6000 Hz, the absorption level is about 15 dB lower in 1-month-olds than in adults. At lower and higher frequencies, the difference between 1-month-olds and adults is about 5 dB. By 24 months, however, the infant–adult difference in sound pressure at the tympanic membrane is no more than 5 dB at any frequency (Keefe et al. 1994).

While studies in infants and children uniformly indicate immature absolute sensitivity through the preschool period, age-related differences in behavioral absolute thresholds are generally greater than would be predicted from conductive immaturity alone. Keefe et al. (1994) estimated that the efficiency of conduction from the sound field into the middle ear improves by about 10 dB at 4000 Hz between 1 month and about 6 months of age, while behavioral thresholds are reported to decrease by 15–25 dB at 4000 Hz over the same age range (Trehub et al. 1980; Tharpe and Ashmead 2001). However, by the preschool period, the remaining small age-related differences in conductive efficiency probably account for the child–adult differences in behavioral absolute thresholds (Trehub et al. 1980).

7.2.3.2 Spatial Hearing

Spatial hearing improves over the course of infancy and early childhood, the period during which the greatest changes in head size occur. The threshold for discrimination of a change in spatial location decreases progressively between birth and 5 years of age, in both azimuth and elevation (Morrongiello 1988a; Litovsky 2012). Infants demonstrate at least a rough approximation of a map of auditory space (Clifton et al. 1991), and while the ability to pinpoint the location of a sound source is accurate in preschool children, their performance becomes less variable with age (Litovsky 2012). However, there is little evidence concerning the direct role of acoustic limitations in the development of spatial hearing. One exception is a report by Morrongiello (1987) that infants' minimum audible angle (MAA, the just discriminable difference in spatial location of two sounds) in elevation could be estimated only when the sound to be localized contained frequencies above 8 kHz, consistent with age-related changes in the resonant frequencies of the pinna. However, infants' MAA is poorer than would be expected based on acoustic limitations alone: For example, an adult's head creates a range of ITD about 40% greater than that created by a 6-month-old's, but the 6-month-old's MAA for a click is five to seven times greater than the adult's. Thus, if the limited ITD created by the infant head were the only limit on localization, the MAA would be much closer to that of an adult.

7.3 Differentiation of the Human Inner Ear

7.3.1 Morphology

Rubel (1984) pointed out that the onset of auditory function (i.e., the first cochlear response) is immediately preceded by the simultaneous and synchronous maturation of many mechanical and neural elements, including the properties of the basilar membrane, the differentiation of sensory hair cells of the cochlea, and the formation of synapses with the hair cells. He also described several gradients in the maturation of these elements: basal structures tend to mature before apical; inner hair cells tend to mature before outer; and afferent innervation tends to mature before efferent.

Lavigne-Rebillard and Pujol (1987, 1990) and their colleagues completed several scanning and transmission electron microscopic studies of this stage of cochlear development in human fetuses. Their results show that the events leading up to the onset of cochlear function (e.g., opening of the tunnel of Corti, release of the tectorial membrane, differentiation of inner hair cells, and afferent innervation) take place at around 18 weeks gestational age (GA). A lack of high-quality specimens at ages beyond 22 weeks GA made it difficult to accurately estimate the age at which the cochlea reached mature status. However, Lavigne-Rebillard and Pujol noted that cochlear development in humans follows the same course as that in nonhuman mammalian species, including the gradients along the length of the cochlea, between inner and outer hair cells, and between afferent and efferent innervation. More recently, Bibas et al. (2008) reported that several structural changes occur in the human cochlea after 25 weeks GA, including progressive opening of the tunnel of Corti toward the apex and thinning of the basilar membrane. Furthermore, Locher et al. (2015) argue that the characteristics of human stria vascularis at midgestation suggest that a few more weeks would be required to support an endocochlear potential adequate for cochlear response. It should be recognized, however, that all anatomical studies of human cochlear development use rather qualitative measures of maturity. As Abdala and Keefe (2012) note, a precise account of the developmental time course for the physical properties of the human basilar membrane and associated structures is not available. Thus, it is difficult to make precise predictions about the structural status of the cochlea at a particular point in gestation.

7.3.2 Physiology

7.3.2.1 Otoacoustic Emissions

Otoacoustic emissions (OAEs) have provided important insights into the development of the human cochlea for the past 30 years (e.g., Manley et al. 2017). Three types of OAE have been used in developmental studies: spontaneous otoacoustic emissions (SOAEs), which are recorded in the ear canal in absence of external stimulation; click-evoked otoacoustic emissions (CEOAEs), broadband responses elicited by a click; and distortion product otoacoustic emissions (DPOAEs), emissions at a distortion product frequency elicited by the presentation of two pure tones, or "primaries," with appropriate frequency and amplitude ratios. All three types of OAE are recorded in nearly all normal-hearing full-term newborns (Abdala and Keefe 2012). Difficulty recording reliable OAEs in infants of about 28 weeks CA has been reported, but OAEs are recorded in most infants by 30 weeks CA (Brienesse et al. 1998). Higher frequency SOAEs and CEOAEs tend to be recorded in younger infants, and all types of OAE are higher in amplitude in infants and preschool children than in adults (Prieve et al. 1997). These characteristics of OAE in young humans are largely accounted for by the acoustics of the external and middle ear (Abdala and Keefe 2012).

Increases in the levels of DPOAEs and of CEOAEs are consistently reported between 30 and 38 weeks CA, especially at frequencies between 2000 and 4000 Hz (Tognola et al. 2005; Abdala et al. 2008). The latency of CEOAEs decreases over the same age period and the spectrum of CEOAEs shifts toward higher frequencies (Ubbink et al. 2011). It is difficult to separate the contributions of middle ear and cochlear maturation to these age-related changes in OAEs in premature infants. Keefe et al. (2000) show little change in conductance measured in the ear canal between 33 weeks and 40 weeks CA. However, the efficiency of reverse transmission from the cochlea to the ear canal is also an important determinant of OAE characteristics; Keefe et al. (2003) reported that as much as 28% of the variance in OAE level could be accounted for by variations in ear canal measures of admittance. The unaccounted-for variance in OAE level could be a result of cochlear maturation, but as an indirect measure, ear-canal admittance may not capture some important aspects of forward or reverse acoustic transmission in the ear.

Abdala and her colleagues have carried out extensive studies of the development of DPOAEs in infants, including infants as young as 33 weeks CA, identifying several age-related changes in DPOAE during infancy and between infancy and adulthood. First, whereas adults' DPOAE amplitude is a nonlinear function of input level, growing nearly linearly at low levels and plateauing at midlevels, DPOAE amplitudes in newborns did not plateau or plateaued at higher levels than those in adults (Abdala and Chatterjee 2003). Second, tuning curves derived by suppressing DPOAE amplitude with tones of different frequencies were sharper in infants as old as 6 months postnatal age than in adults, but only for primary tone frequencies around 6000 Hz (Abdala 2003; Abdala et al. 2007). Third, DPOAE "fine structure," the quasiperiodic variation in DPOAE amplitude as a function of primary tone frequency, is more prevalent and higher in frequency in newborns than in adults. DPOAE fine structure is the result of the addition of two OAE components, a "distortion" component generated at the primary-frequency place on the basilar membrane and a "reflection" component linked to mechanical irregularities along the basilar membrane. The infant–adult difference in DPOAE fine structure is due to differences in only the reflection component, consistent with a greater number of mechanical irregularities on the newborn basilar membrane. Finally, the relationship between primary frequency and DPOAE distortion component phase is different in newborns than in adults for frequencies below about 1500 Hz, suggesting an age-related difference in the frequency–place map in the cochlear apex. The infant–adult differences in DPOAE growth functions and in suppression tuning curve width may be explained by the facts that the cochlea is a nonlinear system and that the input to the newborn cochlea at 6000 Hz may be 15 dB lower than that in adults (Abdala and Keefe 2012). However, the observed age-related differences in the reflection component of the DPOAE and in apical cochlear scaling suggest age-related changes in cochlear mechanics.

7.3.2.2 Evoked Neural Responses

The existence of cochlear responses has little functional significance if that response is not transmitted to the auditory nerve and beyond. Click-evoked wave I of the auditory brainstem response (ABR), reflecting the response of the auditory nerve, is first identified in most preterm infants at 28 weeks CA and in nearly all infants at 30 weeks CA (Rotteveel et al. 1987). This age is some weeks after morphological studies have estimated that the cochlea begins to respond to sound. However, repeatable responses that do not have the same amplitude peaks or peak latencies as the ABR have been reported in infants as young as 24–25 weeks CA (Amin et al. 1999). The delayed appearance of ABR wave I may indicate continued maturation of the hair cell–afferent nerve fiber synapse and factors such as myelination that can influence the synchrony in response across neurons that is required to elicit an evoked potential. Wave I latency is found to decrease with age between 25 and 40 weeks CA, but it appears to be adult-like shortly thereafter (Rotteveel et al. 1987).

Eggermont et al. (1996) addressed the nature of the age-related decrease in wave I latency using the derived response technique to determine the traveling wave delay between different locations along the basilar membrane. In that technique, the ABR to a click is recorded as the frequency of a high-pass masking noise is progressively decreased. Because only frequencies below those in the masker contribute to the masked ABR, the response latency will increase as the frequencies contributing to the response decrease (i.e., as the response originates more apically along the basilar membrane). It provides an estimate of cochlear travel time plus the time required to generate the neural response. Eggermont et al. compared the ABR wave I traveling wave delays to those obtained from DPOAE latencies in term newborns and in adults. While DPOAE latencies did not change with age, wave I traveling wave delays between the 11-kHz and 6-kHz regions were immature in term infants. Wave I traveling wave delays were mature across the board by 3–6 months of age (Eggermont et al. 1991b). This finding suggests that the immature wave I traveling wave delay reflects immaturity of hair cell–afferent nerve fiber synapse extending into the early months of infancy.

7.3.2.3 Behavioral Responses

Whereas nearly all data regarding cochlear and neural responses to sound that coincide with the final stages of inner ear differentiation have been obtained from premature infants, nearly all data regarding behavioral responses to sound during that period have been obtained from fetuses in utero. This research is difficult to carry out, and its results may be difficult to interpret for a variety of reasons (Gerhardt and Abrams 1996). Nonetheless, prenatal hearing remains an important topic of research, not only to refine models of human auditory development, but also to address the wisdom of current efforts by many expectant parents to promote their child's development by "optimizing" the fetal sound environment.

Changes in heart rate as well as motor responses have been observed in nearly all fetuses by 28 weeks of gestation (Birnholz and Benacerraf 1983), about the age when an identifiable ABR can be recorded in nearly all premature infants. In many of the early studies addressing this topic, the stimulus delivered to the fetus was produced by an artificial larynx applied to the maternal abdomen. Thus, the sound reaching the fetus was ill defined, and vibration may have been the effective stimulus.

Although not free of methodological concerns, a study by Hepper and Shahidullah (1994) provides a decent starting point for discussions of human auditory sensitivity

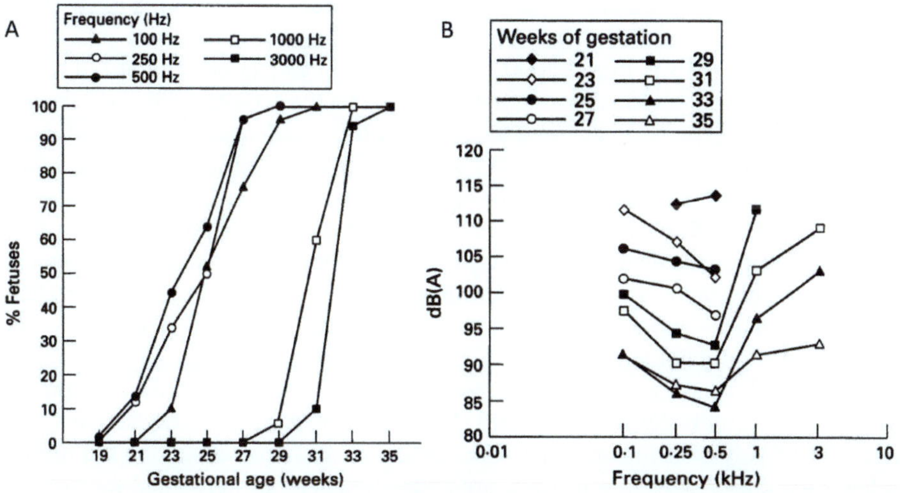

Fig. 7.2 Prenatal threshold development. (**A**) Proportion of fetuses who produced a motor response to a tone for five frequencies. (**B**) Sound pressure level required to produce a motor response as a function of frequency for eight age groups. Responses to higher frequencies appear later in development, but thresholds at higher frequencies develop rapidly once responses appear. [From Hepper and Shahidullah (1994). Copyright © 1994 BMJ Publishing Group Ltd. and the British Cardiovascular Society]

during the fetal period. These investigators delivered pure tones to fetuses through a loudspeaker placed on the maternal abdomen; the tones were calibrated in air near the speaker face. Fetal motor responses were recorded using ultrasound and scored from videotape offline. A total of 450 fetuses, ranging from 19 to 35 weeks GA, were observed in a cross-sectional design. The development of motor responses to sound was clearly dependent on tone frequency. At least 50% of fetuses responded to 100- to 500-Hz tones at 23 weeks GA, but it was not until 31 weeks GA that most fetuses responded to 1000 or 3000 Hz tones (Fig. 7.2A). Figure 7.2B shows estimated audibility curves for fetuses at different ages. At 21 weeks of gestation, responses were observed only at levels exceeding 110 dB SPL (A scale) and only at 250 or 500 Hz. The range of frequencies eliciting responses increased and the level at which responses were elicited decreased with increasing age. By 35 weeks GA, fetuses remained most sensitive to 500-Hz tones, responding to levels of about 85 dB SPL (A scale), while response levels to 100 and 3000 Hz were 5–10 dB higher.

The question here is how much of the fetus's sensitivity to sound can be attributed to cochlear factors. At 33 weeks GA, the fetus responds to a range of frequencies at levels measured at 85–105 dB SPL outside the maternal abdomen. Clearly, the conductive component is a major contributor to the fetal "threshold." Not only is sound attenuated by maternal tissue and amniotic fluid, but the route by which sound reaches the fetal inner ear is likely through bone conduction. Gerhardt and Abrams (1996) present an interesting analysis of the situation, using data from fetal sheep, which are like human fetuses in terms of prenatal acoustics and auditory function. Based on the cochlear microphonic of near-term fetal sheep in utero, these investigators estimate that while only 10 dB is lost in the conductive process at 125 Hz, about 45 dB is lost at 2000 Hz. If these figures can be generalized to the 33-week fetus—a questionable assumption given that the acoustics are likely changing as the fetus grows—that would suggest that the sound reaching the fetal inner ear at 2000 Hz is at least 60 dB SPL when the average fetus produces a motor response to sound. That an ABR can be elicited from a premature infant of 33 weeks CA by a 30-dB SPL click suggests that central nervous system immaturity is probably more important than cochlear factors, once conductive factors are accounted for. Presumably there is a point in development prior to 33 weeks when cochlear immaturity makes some contribution to auditory sensitivity, but when that point occurs is currently unknown.

Behavioral measures of absolute sensitivity show postnatal improvement in detection thresholds (Olsho et al. 1988). As noted in Sect. 7.2.3.1, conductive maturation is a likely contributor to the development of absolute sensitivity, but during infancy age differences in sensitivity are greater than can be accounted for by conductive immaturity. Early in infancy, maturation of the hair cell–afferent nerve fiber synapse may account for an additional portion of immature absolute thresholds. Between 3 and 6 months, the greatest change in absolute threshold is at frequencies above 4000 Hz (Olsho et al. 1988), the same age range and frequency range in which Eggermont et al. (1996) suggest that the hair cell–afferent nerve fiber synapse is maturing. By implication, even though OAE may indicate adult-like OHC function among premature infants, cochlear synaptic maturation, likely over a broader frequency range in younger premature infants and fetuses, may also limit sensitivity to sound in the prenatal period.

7.3.3 Development of the Place Principle

An important discovery in auditory development was reported by Rubel and Ryals (1983): The frequency-to-place map along the basilar membrane is not immutable, but shifts during development, such that the basal region of the basilar membrane responds to low- to mid-frequency sounds in younger animals. This phenomenon occurs in both birds and mammals, but appears to be restricted to higher frequency regions (Arjmand et al. 1988). In gerbils (*Meriones unguiculatus*), the shift in the cochlear frequency map occurs during the week following the earliest appearance of the cochlear microphonic (Harris and Dallos 1984). If it occurs in humans, the shift could well occur prenatally.

Mills and Rubel (1996) showed that the place code shift in gerbils was the result of age-related changes in the passive mechanics of the basilar membrane. Although there are not many anatomical studies of human auditory development in the last trimester of gestation, the data that are available suggest that while the basal structures of the cochlea at 20 weeks resemble those of other mammals at the onset of cochlear response, the tunnel of Corti continues to "grow" apically and the basilar membrane continues to thin well beyond 25 weeks of gestation. Presumably, such structural changes would be accompanied by changes in the passive response of the basilar membrane. Studies of OAE in premature infants generally report an increase in the frequency of SOAE and in the spectral content of CEOAE after 30 weeks CA (Brienesse et al. 1998; Ubbink et al. 2011) and there are reports of decreasing DPOAE latency in the 3- to 4-kHz range after 33 weeks CA (Quinonez and Crawford 1998). Both observations would also be consistent with a shift in the frequencies eliciting responses near the base of the cochlea in the last trimester of gestation. However, it should come as no surprise that the age-related changes in OAE could well be accounted for by maturation of the conductive apparatus (Eggermont et al. 1996). Thus, a convincing demonstration of a functional cochlear place code shift in humans has yet to be made.

7.4 Development of Neural Response Properties

The development of the human auditory nervous system is a prolonged process. Although click-evoked ABR can be recorded in premature infants, the components of the ABR that reflect the responses of structures central to the auditory nerve may take years to reach adult status. For example, Eggermont (1992) compared reported latencies of wave I and wave V of the ABR from several studies, finding that the difference between these latencies, the so-called I–V interval or central conduction time, did not become adult-like until 3 years or later. Maturation of cortical evoked potentials can extend into adolescence (Ponton et al. 2000). Although these studies provide a general time frame for auditory neural development, they provide little information about specific response properties of auditory neurons, making comparisons with nonhuman species difficult. Moreover, while authors of papers describing the development of evoked potentials often draw parallels between their results and those of behavioral studies of absolute sensitivity, hearing in noise, or speech perception, such comparisons are by their nature nonspecific. Of greater interest are studies that have attempted to relate neural responses to specific psychoacoustic abilities.

7.4.1 Frequency Resolution

Schneider et al. (1989) summarized a series of studies of detection of bandpass noises of different center frequencies masked by a broadband noise in individuals ranging from 6 months to 18 years of age, all tested with a similar psychoacoustic procedure. Their results showed that masked threshold decreases with age into the early school years, in parallel across frequencies (Fig. 7.3). This basic result was confirmed by several less comprehensive studies (Bargones et al. 1995). One explanation for this result is that frequency resolution, the ability to separate the

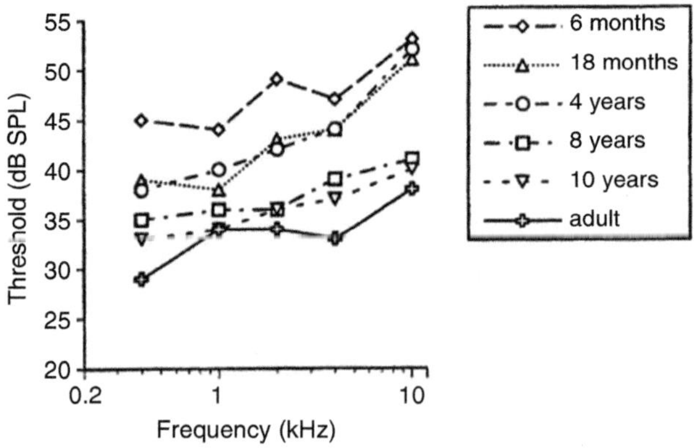

Fig. 7.3 Threshold for an octave band of noise masked by a broadband noise as a function of the center frequency of the noise band for six age groups. Masked thresholds decrease during infancy and childhood, reaching nearly adult-like values at 8–10 years. [Replotted from Schneider et al. (1989). Copyright © 1989 Acoustical Society of America]

components of a complex sound, develops over the course of infancy and into childhood. However, subsequent studies measuring frequency resolution more directly failed to identify any age-related changes after 6 months of age (Schneider et al. 1990; Hall and Grose 1991).

Not surprisingly, 3-month-old infants also have immature masked detection thresholds, but at this age frequency resolution does appear to contribute, at least at high frequencies. Spetner and Olsho (1990) demonstrated that 3-month-old infants are adult-like in frequency resolution at 500 and 1000 Hz, but that they have poorer resolution than adults at 8000 Hz. Folsom and Wynne (1987), and subsequently, Abdala and Folsom (1995), reported the same result based on masked ABR. Both ABR studies reported no difference between 6-month-olds and adults at any frequency, consistent with the behavioral studies. That the same result is observed in behavior and ABR at 3 months, while studies of OAE report mature frequency resolution among newborns (Abdala 2001), suggests a brainstem origin of early immature frequency resolution. While the synapse between hair cells and auditory nerve fibers responding to high frequencies has been reported to be immature early in infancy as well (Eggermont et al. 1996), how that immature synapse could contribute to immature frequency resolution is not clear.

7.4.2 Intensity Representation

7.4.2.1 Absolute Sensitivity

As noted in Sects. 7.2.3.1 and 7.3.2.3, absolute thresholds are reported to improve from the prenatal period through the school years (Lary et al. 1985; Schneider et al. 1986). Conductive maturation plays a major role in the development of absolute sensitivity (Okabe et al. 1988; Keefe et al. 1993), development of the organ of Corti plays some yet ill-defined role likely soon after the onset of hearing, and maturation of the hair cell–afferent nerve fiber synapse is involved into the early postnatal months (Eggermont et al. 1996). While "central" contributions are often cited as an additional source of threshold immaturity, the nature of such potential contributors has not been established. Brainstem maturation appears to be frequency specific, with the I–V interval to frequencies between 1000 and 8000 Hz developing faster and earlier than both higher and lower frequencies (Eggermont et al. 1991a). Two studies attempted to relate sensitivity measured by ABR to behavioral thresholds in a group of normal-hearing infants aged 3 and 6 months (Werner et al. 1993, 1994). The best ABR predictor of behavioral sensitivity was the I–V interval, but the relationship held only at 4000 and 8000 Hz, not 1000 Hz, and only for 3-month-old infants. Thus, synaptic efficiency at the level of the brainstem is another contributor to early immaturity of absolute sensitivity.

7.4.2.2 Intensity Discrimination

Intensity discrimination refers to the ability to detect a change or difference in intensity. It can be measured as the threshold for detecting an increase in the level of an ongoing sound, known as "increment detection" or a difference in intensity between two discrete sounds, generally referred to as gated intensity discrimination. Intensity discrimination has not been a topic of great interest among developmental researchers. Sinnott and Aslin (1985) reported that intensity difference thresholds of 7- to 9-month-old infants were about 6 dB for a 1000-Hz tone, under conditions in which adults could detect differences less than 2 dB. Increment detection thresholds have been reported to be immature in infants, particularly when the duration of the increment is short and when the noise band is centered at a low frequency, but adult-like in some conditions by 3 years (Berg and Boswell 1998, 2000). However, immature gated intensity discrimination is often reported in preschool or young school-age children (Buss et al. 2009). Buss et al. (2006) present evidence that children's relatively poor intensity discrimination is the result of relatively high levels of internal noise. Notably, Buss et al. showed that the internal noise does not result from lapses of attention, although they were unable to distinguish between variability in sensory coding, memory, and level-dependent inattentiveness. That 3-year-olds perform like adults in increment detection argues against a sensory coding explanation. Furthermore, Gorga et al. (1987, 1989) have shown that the effects of level on click-evoked ABR wave V latency in infants are like those seen in adults, suggesting comparable changes in brainstem neural response with changes in intensity. Additional psychophysical data will likely clarify the role of auditory memory and inattentiveness; this question might also be approached using evoked potentials to examine possible sensory contributions.

7.4.2.3 Loudness

Loudness refers to the perceived magnitude of sound. In adults, loudness is typically measured using scaling techniques such as magnitude estimation, which depends on the ability to generate numerical ratings of loudness, or cross-modality matching, which requires listeners to adjust another stimulus to a magnitude that matches the loudness of the sound. The variable of interest is generally the slope of the function relating intensity to loudness.

It is perhaps unsurprising that few studies have attempted to assess loudness growth in infants or toddlers. One exception is a study by Leibold and Werner (2002) that examined the relationship between response latency and sound level in infants and adults; the "response" for infants was a head turn in the direction of a mechanical toy that was paired with the tone. Response latency decreased with increasing level in both infants and adults, although the slope of the latency-level function was steeper in infants than in adults. Whether the difference in slope reflects a difference in loudness is not clear. In any case, magnitude estimation and cross-modality matching results indicate mature loudness growth by about 4 years of age (Collins and Gescheider 1989).

Given the inherent difficulty in measuring loudness in infant subjects, additional evoked potential studies would be useful in further investigation of potential age-related changes in loudness growth during infancy. Recent studies have reported that loudness growth is related to the latency-intensity function of the ABR in adult listeners (Silva and Epstein 2010), and one recent study reports steeper latency-level functions for speech- and tone-evoked cortical potentials for 4- to 12-month-old infants (Cone and Whitaker 2013).

7.4.3 Timbre and Spectral Shape Discrimination

Timbre refers to sound "quality;" it is that characteristic of sound that allows us to distinguish between different musical instruments playing the same note with the same sound pressure level. Timbre is largely determined by the shape of the amplitude spectrum of a sound and is thought to be important, for example, in the identification of vowels in speech. Timbre perception relies on at least two aspects of auditory coding: Frequency resolution determines the accuracy with which spectral shape is represented in the auditory pathway, but intensity resolution is also critical to the identification of changes in level across frequency. Because young infants discriminate between vowels (Marean et al. 1992), one might conclude that spectral shape discrimination is an early maturing ability. However, the results of studies directly addressing the development of spectral shape discrimination suggest a protracted period of development. For example, Clarkson et al. (1988) reported that 7-month-olds were relatively insensitive to changes in spectral shape; Allen and Wightman (1992) found that preschool children were also poor at spectral shape discrimination compared to adults. One difficulty in interpreting these results is that the overall level of the stimuli must vary from presentation to presentation to force listeners to rely on spectral shape to perform the task rather than on an intensity difference at some isolated spectral region. When level is varied in this way, it is possible that infants and children become confused about what to listen to.

Spectral ripple discrimination (SRD) is an alternative paradigm that has been used to assess spectral shape discrimination (Supin et al. 1998). In this paradigm listeners hear a noise with a sinusoidal spectral shape and are asked to identify a switch in the positions of the peaks and troughs in the spectrum. The frequency of the spectral envelope at which listeners can just detect the change in spectral shape is taken as a measure of spectral resolution. A recent study of infants by Horn et al. (2017) found that SRD was immature in infants. Peter et al. (2014) reported that 8- to 11-year-olds were immature at SRD, while 12- to 18-year-olds were adult-like. Given that "within-channel" intensity resolution is mature well before 12 years of age (see Sect. 7.4.2.2), an important question for future research is whether early immaturity of spectral shape processing results from immature across-frequency intensity resolution or from a more general inefficiency in task performance.

7.4.4 Temporal Processing

Rubel (1984) noted that among the neural response properties that had been studied in developing mammals, phase locking was likely to be one of the latest to reach mature status. The implication is that the course of development of auditory capacities that depend on phase locking—temporal resolution, pitch, and binaural hearing—would be prolonged.

7.4.4.1 Temporal Resolution

Temporal resolution is the ability to follow rapid changes in sound amplitude over time; it can be thought of as a measure of the accuracy of the representation of a sound's envelope. While the phase-locked response of auditory nerve fibers represents the envelope, as well as the fine structure of sound, the limits of temporal resolution in adults are controlled by other processes that operate on a longer time scale (Eddins and Green 1995). The psychophysical gold standard measure of temporal resolution is the temporal modulation transfer function (TMTF), which shows the modulation depth at which modulation is just detectable, as a function of modulation frequency. In adult listeners, the TMTF has the form of a low-pass filter function, with a cutoff frequency around 50–60 Hz and a high-frequency slope of about −3 dB/octave. The definitive study of children's TMTF was carried out by Hall and Grose (1994). Although children as old as 9–10 years required greater modulation depth to detect modulation than adults did, the TMTF of children as young as 4 years paralleled that of adults. This result leads to the conclusion that temporal resolution is mature by 4 years of age (Fig. 7.4).

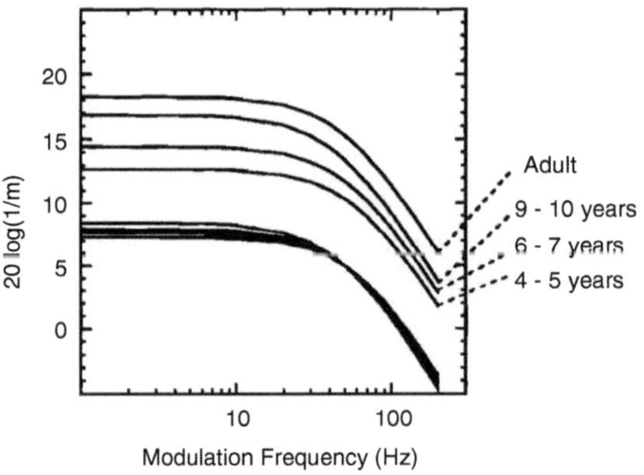

Fig. 7.4 Temporal modulation transfer functions (TMTFs, modulation depth at modulation detection threshold as a function of modulation frequency) for four age groups are shown at the top of the graph. At the bottom of the graph the TMTFs have all been shifted to the same range to illustrate the similarity in shape across age. The TMTF indicates that temporal resolution is mature by 4 years of age. [From Hall and Grose (1994). Copyright © 1994 Acoustical Society of America]

Because infants require much greater modulation depths to detect modulation than adults, it is difficult to establish the shape of the TMTF. However, in a recent study, Horn et al. (2016) obtained an abbreviated form of the TMTF to establish the cutoff frequency of 3-month-olds and adults, finding no difference between age groups. Furthermore, there is a variety of less direct measures that are consistent with the conclusion that temporal resolution is mature quite early. First, physiological measures of envelope following show adult-like responses at 1 month of age (Levi et al. 1993). Second, infants can use the envelope of speech to identify consonants by 3 months of age, and they are particularly dependent on higher frequency modulation to do so (Cabrera and Werner 2017. Finally, infants can take advantage of modulation in a masker to improve their detection and discrimination of target sounds (Werner 2013).

Other measures of temporal resolution provide a less consistent story about development. For example, the just detectable gap, or interruption, in a sound is much longer in infants than in adults, but reaches adult levels by 6 years (Werner et al. 1992). Both brainstem and cortical evoked potentials, on the other hand, suggest adult-like representation of gaps by 3 months of age (Trainor et al. 2001; Werner et al. 2001). Duration discrimination, in contrast, is poor in preschool children and approaches adult levels of performance only around 10 years (Elfenbein et al. 1993). Finally, newborns, 3-month-olds, and 10-year-olds are reported to be more susceptible than adults to nonsimultaneous masking effects, while 6- to 9-month-olds are reported to be adult-like (Lasky 1991; Buss et al. 1999; Werner, 1999). The consensus seems to be that the age-related differences in nonsimultaneous masking in older children result from maturation of processing efficiency rather than of temporal resolution (Hartley et al. 2000). The question of the discrepancy between the results of infant and child studies, however, remains to be addressed.

7.4.4.2 Pitch

Pitch refers to that aspect of sound that can be varied to create a melody (Plack 2014). Pitch perception depends on both frequency resolution and the coding of fine structure, as well as resolution of envelope frequencies in the 100–500 Hz range. Infants' preference for speech with highly modulated fundamental frequency (F0) contours (Fernald and Kuhl 1987), as well as their ability to discriminate between melodies (Trehub 2001), suggest an early sensitivity to pitch. However, the temporal code involved in pitch perception depends on phase locking at frequencies higher than those required to follow the envelope of sound. Thus, if phase locking is a late developing aspect of the auditory neural response, pitch may be a late developing aspect of auditory perception.

The results of studies of pure-tone frequency discrimination are consistent with that prediction. Frequency discrimination at frequencies under 2000 Hz is poor relative to adults at 3 months and improves little between 3 and 12 months (Olsho et al. 1987). Researchers consistently report that frequency discrimination continues to be poor in young children, particularly at low frequencies (Moore et al. 2008). In fact, in those studies the performance of 4- or 6-year-old children is poorer than that reported for infants.

In contrast, studies of the frequency following response, generated in the brainstem, consistently show that by 3 months, an accurate temporal representation of low-frequency tones is available to infants (Levi et al. 1995; Skoe et al. 2015). Another possible explanation is that while temporal information is available to them, infants and children have not yet learned to use that information efficiently. Buss et al. (2012) suggest that memory may be involved: Memory for the frequency of sound has been reported to achieve adult status only at 11 years of age (Keller and Cowan 1994).

The results of studies of complex pitch perception by infants paint a very different picture of infants' pitch-processing abilities. The pitch of a complex tone, typically consisting of a F0 component and a series of harmonics, corresponds to the pitch of the F0 component. Masking or filtering out the F0 component does not change the pitch of the complex, implying that the auditory system must integrate information across frequencies to derive pitch. While frequency resolution influences pitch perception, a temporal representation seems to be required for the perception of pitch (Plack 2014), and pitch is perceived when only temporal information is available (Burns and Viemeister 1981). Finally, studies of nonhuman primate (Wang and Walker 2012) and human (Plack et al. 2014) neural response have identified an area of auditory cortex that appears to respond based on complex pitch.

Despite a markedly immature auditory cortex (Eggermont and Moore 2012) and the apparent absence of a cortical evoked response to complex pitch change (He and Trainor 2009), 3-month-olds demonstrate surprisingly mature complex pitch processing in psychophysical studies. Not only do 3-month-olds respond to relatively large changes in (missing) fundamental frequency, but they do so when only temporal information is available (Lau and Werner 2012, 2014b). Moreover, they respond to pitch changes in a melodic context, when the melody is transposed to a different key from presentation to presentation (Lau et al. 2017). There are even conditions in which 3-month-olds are more likely than musically untrained adults to learn to respond to missing fundamental pitch changes and to ignore spectral changes (Lau and Werner 2014b). Preliminary results suggest that the just detectable change in complex pitch is already adult-like at 3 months (Lau and Werner 2014a).

A simple summary of the available data might be: Complex pitch perception matures early, but pure-tone frequency discrimination does not. At least two deficiencies in the available data make it difficult to accept that summary. First, neither pure-tone nor complex-tone discrimination has been assessed under the same stimulus conditions in infants and children. In fact, few data on the limits of children's complex pitch perception currently exist (but see Peter et al. 2014). Second, no attempts have been made to directly assess the impact of test method on pitch discrimination in infants and children. Additional data on the neural response to pitch changes would be valuable in trying to understand, first, why children have difficulty with some sorts of pitch processing, and second, how infants can achieve adult-like pitch processing with what is apparently a very immature auditory cortex.

7.4.4.3 Binaural Hearing

The MAA in azimuth improves from about 30° in newborns to about 5° at 18 months and is reported to be adult-like by 5 years of age (Fig. 7.5) (Litovsky 2012), and preschool age children have been reported to be accurate, but variable, in the ability to identify the absolute location of a sound source (Litovsky 2012). As summarized in Sect. 7.2.3.2, the accuracy of spatial hearing in infants is far poorer than would be predicted based on acoustics alone. Based on psychophysical studies of intensity coding (Sect. 7.4.2.2) and spectral shape discrimination (Sect. 7.5), immature monaural neural coding may also limit binaural processing during infancy and childhood. The extent to which immature binaural processing limits spatial hearing in young children is not clear: No psychophysical studies have established the limits of ILD discrimination in infants or children, and the literature on sensitivity to ITD in infants and children is sparse.

Several evoked potential studies report that a binaural interaction component is present in the ABR in newborns (Cone-Wesson et al. 1997; Furst et al. 2004) indicating that the circuitry involved in the calculation of interaural differences is in place at birth, and newborns demonstrate a mismatch response, the difference between the response evoked by a novel stimulus and that evoked by a repeated, "standard" stimulus to a relatively large change in ITD, ILD, or sound source position in sound field (Nemeth et al. 2015). Two psychophysical studies have assessed ITD discrimination in infants or children: While Ashmead et al. (1991) reported ITD discrimination thresholds on the order of 50–75 μs in 4- to 6-month-old infants, like ITD thresholds reported for naïve adults in other studies (Wright and Fitzgerald 2001), Van Deun et al. (2009) found that 4- to 9-year-old children, on average, had poorer ITD discrimination thresholds than adults tested in the same study. Thus, that

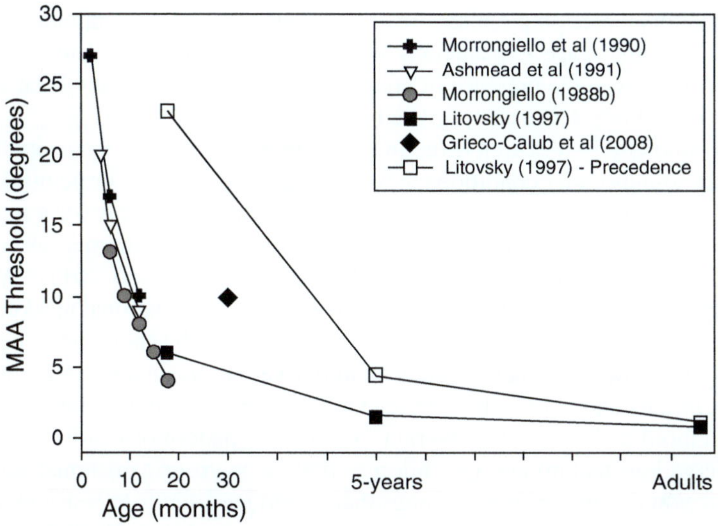

Fig. 7.5 Minimum audible angle (MAA) in azimuth as a function of age from six studies. Litovsky (1997) estimated the MAA under conditions that result in a precedence effect, or "echo suppression." [Reprinted from Litovsky (2012). Copyright © 2012 Springer. Data from the following sources: Morrongiello (1988b); Morrongiello et al. (1990), copyright © 1988, 1990 American Psychological Association; Ashmead et al. (1991) copyright © 1991 John Wiley & Sons; Litovsky (1997) copyright © 1997 Acoustical Society of America; Grieco-Calub et al. (2008) copyright © 2008 Wolters Kluwer Health, Inc]

infants and young children localize sounds is clear, but the cues they use to do so are not well understood.

7.4.5 Auditory Scene Analysis

Auditory scene analysis refers to the process by which a listener analyzes a complex sound mixture into its constituent components, groups together the components coming from each respective source, and follows the sound from a single source over time. Although Rubel (1984) did not mention this topic, it has garnered increasing interest. It involves the integration of multiple types of auditory information as well as top-down influences of memory and attention.

Grouping the components coming from each source is referred to here as simultaneous grouping, while following the sound from a single source over time is referred to as sequential grouping. Both types of grouping depend on the ability to identify commonalities across sound components, for example, in timing, harmonicity, and spatial location. Thus, auditory scene analysis requires the extraction of features such as temporal envelope, pitch, and spatial location, as well as accurate encoding of the spectral and temporal properties of sound. Because infants have generally been found to have mature auditory processing abilities (e.g., frequency resolution, temporal resolution, pitch) by 6 months of age, explanations for their immature ability to detect or discriminate target sounds in the presence of competing sounds have recently been sought in the realm of auditory scene analysis (Leibold 2012).

Infants and children exhibit difficulties in separating a target from simultaneous competing sounds under conditions in which adults exhibit little or no masking. For example, both infants and young children have higher thresholds for a tone in the presence of a masker far removed in frequency even when the masker spectrum does not vary (Werner and Bargones 1991; Leibold and Neff 2011). When the masker spectrum varies randomly (i.e., in the classic "informational masking" paradigm; Neff and Green 1987), infants' thresholds increase by about the same amount as adults' do. However, when the target and masker are made less similar to reduce masking, infants and young children tend to show less threshold improvement than adults (Leibold and Neff 2007; Werner 2013). This pattern of results is consistent with observations that infants and children continue to group a mistuned component with a harmonic complex at mistunings that would lead adults to hear the mistuned component as a separate sound (Alain et al. 2001; Folland et al. 2012). Thus, infants and children appear to be less efficient than adults at using the acoustic cues for simultaneous grouping, although only a few conditions have been studied.

The majority of studies of auditory scene analysis have examined sequential grouping or auditory stream formation. In all cases, sequential grouping processes have been reported to be qualitatively adult-like even during infancy (e.g., Smith and Trainor 2011). However, one of the major limitations of this literature is that the acoustic cues contributing to sequential grouping have not been varied independently, and only a few isolated acoustic cues have been considered (Fassbender 1993; McAdams and Bertoncini 1997). The few studies that have parametrically varied specific acoustic cues for sequential grouping in children report that children need much larger acoustic differences to form separate auditory streams (Sussman et al. 2001; Leibold and Bonino 2009). Young children appear to be unable to separate two speech streams based on overall intensity and to follow the less intense of two speech streams as adults do. Further, even teenagers show interference effects when the target and competing streams are presented in different ears, a condition in which adults show few interference effects (Wightman and Kistler 2005). Again, the available data suggest a protracted developmental course for sequential, as well as simultaneous, grouping, but considerably more research will be required to specify the conditions under which infants and children succeed and fail in auditory scene analysis. A glaring deficit in the literature is the failure to consider the effects of attention and top-down influences on auditory scene analysis in young listeners.

7.5 Afferent Influences on Auditory System Ontogeny

Some of the most interesting and potentially important questions in auditory development ask how experience influences the creation, maintenance, and refinement of central auditory neural pathways. In experiments with nonhuman animals, neural activity can be measured directly and controlled by manipulating the ear or the sound environment. Such studies have shown that a reduction or absence of input from the periphery can have profound effects on central circuitry and on the response properties of individual neurons (Moore 1990; Harris and Rubel 2006). Manipulation of the acoustic environment can change the tonotopic organization of auditory cortex (Han et al. 2007) and delay the development of binaural processing (Seidl and Grothe 2005). In humans, we depend on "natural experiments," such as hearing loss and exposure to native language speech, to determine whether such effects occur.

A central question related to the effects of experience on development is whether there is a sensitive period, a period of development during which the system is most susceptible to experiential effects and outside of which the effects of experience are difficult to reverse. In nonhuman mammals and birds the days following the onset of cochlear function have been identified as a sensitive period for many effects (Zhang et al. 2001), although the sensitive periods for other effects extend for weeks or months beyond that time (Keuroghlian and Knudsen 2007). The literature on human adult perception is replete with examples of the effects of early exposure to certain types of sounds. For example, adults readily perceive phonetic changes in their native language, but typically not in a nonnative language (Werker 1994). They are also more sensitive to melodic changes in music that conforms to the rules of their native culture (Schellenberg and Trehub 1999). Adults who experience early musical training or exposure to tonal languages may maintain the ability to identify pitch on an absolute basis, while others are sensitive only to relative pitch (Zatorre 2003). However, the prospective research that would allow researchers to establish the time course over which experiences are effective or to determine the existence and timing of sensitive periods has generally not been carried out. Two exceptions are described in this section.

7.5.1 Effects of Exposure to Speech

A well-established finding is that from early infancy, humans can discriminate phonetic contrasts that are not used in their native language and that are not discriminated by adult speakers of that native language (Saffran et al. 2006). Thus, young infants discriminate speech sounds that they have never heard used in speech. Around 10 months of age, however, whereas the preference for native language continues and sensitivity to native language speech contrasts improves, sensitivity to nonnative speech contrasts declines (Fig. 7.6) (Werker and Hensch 2015). In other words, perception narrows to the contrasts used in the speech to which the infant is exposed.

Recent research (summarized by Werker and Hensch 2015) has provided evidence that the formation of native speech categories along with the loss of nonnative speech categories displays the hallmarks of a sensitive period. The onset of this sensitive period, typically at about 30 weeks GA, appears to depend on neural

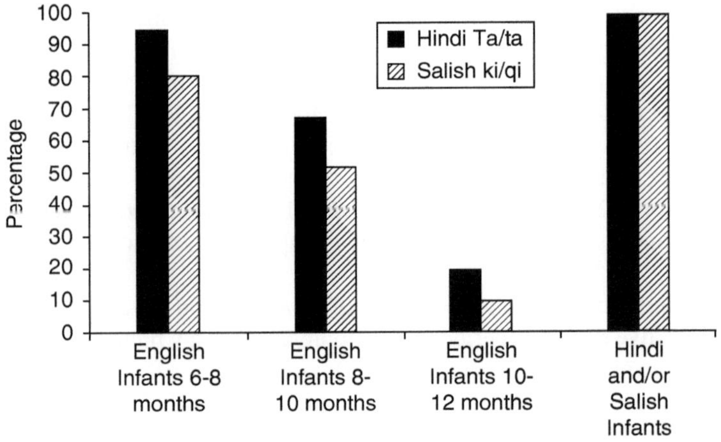

Fig. 7.6 Percentage of infants who discriminated a speech sound contrast from Hindi (black bars) or Salish (striped bars) languages in four groups. "English" infants were raised in monolingual English-speaking families and tested at one of three ages. "Hindi" and "Salish" infants were raised in families in which one of those languages was spoken and tested at 12 months. [From Werker and Tees (1984). Copyright © 1984 Elsevier]

maturity (Key et al. 2012), and can be accelerated pharmacologically (Weikum et al. 2012). The sensitive period is a period of plasticity during which experience shapes the auditory response to speech: the amplitude of evoked potentials in response to speech sound change increases (Key et al. 2012); behavioral responses indicate a recognition of native vowels and native-language speech generally (Mehler et al. 1988; Moon et al. 2015); and experimental manipulation of the distribution of speech sounds can readily influence the speech sound categories that infants form (Maye et al. 2002). The sensitive period typically ends at 10–12 months chronological age, and although the mechanisms responsible for the closure are not known, the total time that an infant is exposed to speech does not appear to be important (Pena et al. 2012). Finally, exposure to a second language under the right conditions can extend the duration of the sensitive period (Kuhl et al. 2003; Yeung and Werker 2009) and lack of sound stimulation may delay the sensitive period for some months (Houston et al. 2012a).

Thus, the biological systems that regulate early learning of speech sound categories are now beginning to be understood. Similar work will hopefully address the issues related to other sensitive periods in human auditory development. Another question of interest is how the abnormal early auditory experience of premature infants ultimately affects their auditory development. This question arises periodically as health professionals recognize the potential hazards of exposure to noise in the neonatal intensive care nursery (Lahav and Skoe 2014), but has not been addressed in any systematic way.

7.5.2 *Congenital Hearing Loss Treated with Cochlear Implants*

Cochlear implants (CIs) are devices that transduce sound into an electrical signal, encode its spectral and temporal characteristics, and directly stimulate auditory nerve fibers (Zeng and Fay 2004). When CIs first came into use, only individuals with profound impairment were implanted. The age at implantation varies across individuals, as does the period of deprivation prior to implantation, making it possible for researchers to attempt to establish sensitive periods in human auditory development.

Evoked potential studies of congenitally deaf CI listeners suggest that auditory nerve responses develop in the absence of stimulation during the first postnatal year, but that stimulation is required for the development of auditory brainstem responses (Gordon et al. 2006, 2010; Thai-Van et al. 2007). Even after long periods of deprivation, ABR latencies decrease with increasing "time-in-sound" along a trajectory like that observed in normal hearing infants (Ponton et al. 1996). Some cortical responses, in contrast, do not develop in individuals who experience a 2–3-year period of deprivation prior to 6–8 years of age (Eggermont and Moore 2012).

It is not surprising that most studies of CI users have focused on speech and language outcomes. Many congenitally deaf children who are implanted within the first 2–3 years of life develop speech and language skills within the normal range, but others do not (Nicholas and Geers 2007). Houston et al. (2012b) found that infants implanted prior to 12 months exhibited normal word learning abilities at 22–40 months, but that infants implanted after 12–15 months exhibited deficits in word learning. This effect is consistent with a sensitive period for speech sound category formation as discussed in Sect. 7.5.1.

Sensitive periods for the development of binaural hearing have been well documented in nonhuman species (King et al. 2001; Keuroghlian and Knudsen 2007). With the advent of bilateral CIs came an opportunity to determine whether a similar sensitive period exists in humans. Using an experimental processor to align bilateral CIs in time and tonotopy, Litovsky et al. (2010) compared ITD and ILD sensitivity in individuals implanted as adults who had experienced either no acoustic hearing, a short period of acoustic hearing, or a long period of acoustic hearing prior to becoming deaf. Interestingly, listeners in all groups maintained good ILD sensitivity, but those with no history of acoustic hearing were unable to discriminate between sounds based on ITD. A study of binaural hearing in children, using a similar design, but the subjects' own speech processors, suggests a sensitive period for the effects of experience on binaural hearing that ends some time prior to 3.5 years (Kilian et al. 2015). Gordon et al. (2014) found that children with no prior acoustic hearing experience with long-term bilateral CIs experience did demonstrate sensitivity, albeit poor sensitivity, to ITD, suggesting that with experience the auditory system may partially overcome the current limitations of bilateral CIs. In sum, it appears that early experience with bilateral inputs is a prerequisite to the development of ITD sensitivity, but a sensitive period has been only vaguely identified.

Studies of the development of auditory localization in barn owls (*Tyto alba*) with altered auditory and visual inputs indicate that with the right training, adult owls that experienced abnormal binaural inputs during a sensitive period may learn to localize normally (Knudsen 2004). It is possible that with improved CI engineering and the right training, CI users may also be able to take advantage of such neural plasticity even as adults.

7.6 Summary

Since 1984, the relative contributions of conductive, cochlear, and auditory neural coding factors to human auditory development have been elucidated for some aspects of auditory function. The development of absolute sensitivity and of frequency resolution in humans is fairly well understood. Far less is understood about the development of intensity resolution and the various sorts of temporal processing, and how spectral and temporal information is combined by immature humans to accomplish hearing in natural environments has only begun to be addressed. While afferent influences on the development of speech sound category formation and of binaural hearing are starting to be systematically described, similar effects on basic aspects of hearing, known to be affected by auditory experience in nonhumans, have not been examined at all. Finally, the cellular and molecular mechanisms responsible for human auditory development represent the great unknown. Thus, there is considerably more work to be done in the field of human auditory development.

Progress in this field depends on overcoming at least three obstacles. First, not many laboratories focus on human auditory development. For that reason, we are forced to draw broad conclusions about development based on one or two studies, often from a single laboratory. Expansion of the knowledge base in both neural and behavioral development should be a priority. Second, to identify the neural mechanisms involved in the development of auditory behavior, studies combining evoked potential, neural imaging, and psychophysical approaches are needed. The number of evoked potential (and increasingly, neural imaging) studies of infants and children is large. However, researchers conceive of neural measures as a substitute for behavioral measures, which at least by consensus among neuroscientists, are too difficult to obtain. Little progress will be made in this field until behavioral and neural measures are thought of as converging approaches to understanding auditory development. Researchers studying nonhuman auditory development have recently come around to this view (Sanes and Bao 2009). Third, with current technology the neural cellular and molecular mechanisms responsible for human auditory development cannot be addressed directly. At least for the moment, the only way to approach this topic is by studying nonhuman animals. Unfortunately, making comparisons between humans and nonhumans during development is currently very difficult. The time scales of development differ tremendously across species and attempts to convert the time scale from one species to that of another have not been informative. Further, researchers studying nonhuman auditory development tend to focus on ages close to the onset of cochlear function, a time point which is largely inaccessible in humans because it occurs prenatally, and to avoid studying nonhuman behavior, because it is too hard. Researchers studying human auditory development tend to focus on ages that are easiest to test (i.e., school-age children). Efforts are being made to enable better interspecies comparisons by studying older animals and including behavioral measures (Rosen et al. 2012). Such efforts must continue and be expanded if the ontogeny of auditory function is to be understood.

Acknowledgments Preparation of this chapter was supported by NIH, DC000396.

Compliance with Ethics Requirements Lynne Werner declares she has no conflict of interest.

References

Abdala, C. (2001). Maturation of the human cochlear amplifier: Distortion product otoacoustic emission suppression tuning curves recorded at low and high primary tone levels. *The Journal of the Acoustical Society of America*, 110, 1465–1476.

Abdala, C. (2003). A longitudinal study of distortion product otoacoustic emission ipsilateral suppression and input/output characteristics in human neonates. *The Journal of the Acoustical Society of America*, 114, 3239–3250.

Abdala, C., & Chatterjee, M. (2003). Maturation of cochlear nonlinearity as measured by distortion product otoacoustic emission suppression growth in humans. *The Journal of the Acoustical Society of America*, 114, 932–943.

Abdala, C., & Folsom, R. C. (1995). The development of frequency resolution in humans as revealed by the auditory brain-stem response recorded with notched-noise masking. *The Journal of the Acoustical Society of America*, 98, 921–930.

Abdala, C., & Keefe, D. H. (2012). Morphological and functional ear development. In L. A. Werner, A. N. Popper, & R. R. Fay (Eds.), *Human auditory development* (pp. 19–59). New York: Springer Science+Business Media.

Abdala, C., Keefe, D. H., & Oba, S. I. (2007). Distortion product otoacoustic emission suppression tuning and acoustic admittance in human infants: Birth through 6 months. *The Journal of the Acoustical Society of America*, 121, 3617–3627.

Abdala, C., Oba, S., & Ramanathan, R. (2008). Changes in the dp-gram during the preterm and early postnatal period. *Ear and Hearing*, 29, 512–523.

Alain, C., McDonald, K. L., Ostroff, J. M., & Schneider, B. (2001). Age-related changes in detecting a mistuned harmonic. *The Journal of the Acoustical Society of America*, 109, 2211–2216.

Allen, P., & Wightman, F. (1992). Spectral pattern discrimination by children. *Journal of Speech Language, and Hearing Research*, 35, 222–233.

Amin, S. B., Orlando, M. S., Dalzell, L. E., Merle, K. S., & Guillet, R. (1999). Morphological changes in serial auditory brainstem responses in 24 to 32 weeks' gestational age infants during the first week of life. *Ear and Hearing*, 20, 410–418.

Arjmand, E., Harris, D., & Dallos, P. (1988). Developmental changes in frequency mapping of the gerbil cochlea: Comparison of two cochlear locations. *Hearing Research*, 32, 93–96.

Ashmead, D. H., Davis, D., Whalen, T., & Odom, R. (1991). Sound localization and sensitivity to interaural time differences in human infants. *Child Development*, 62, 1211–1226.

Bargones, J. Y., Werner, L. A., & Marean, G. C. (1995). Infant psychometric functions for detection: Mechanisms of immature sensitivity. *The Journal of the Acoustical Society of America*, 98, 99–111.

Berg, K. M., & Boswell, A. E. (1998). Infants' detection of increments in low- and high-frequency noise. *Perception & Psychophysics*, 60, 1044–1051.

Berg, K. M., & Boswell, A. E. (2000). Noise increment detection in children 1 to 3 years of age. *Perception & Psychophysics*, 62, 868–873.

Bibas, A. G., Xenellis, J., Michaels, L., Anagnostopoulou, S., et al. (2008). Temporal bone study of development of the organ of corti: Correlation between auditory function and anatomical structure. *The Journal of Laryngology & Otology*, 122, 336–342.

Birnholz, J. C., & Benacerraf, B. R. (1983). The development of human fetal hearing. *Science*, 222, 516–518.

Brienesse, P., Maertzdorf, W. J., Anteunis, L. J. C., Manni, J. J., & Blanco, C. E. (1998). Click-evoked oto-acoustic emission measurement in preterm infants. *European Journal of Pediatrics*, 157, 999–1003.

Burns, E. B., & Viemeister, N. F. (1981). Played again SAM: Further observations on the pitch of amplitude-modulated noise. *The Journal of the Acoustical Society of America*, 70, 1655–1660.

Buss, E., Hall, J. W., Grose, J. H., & Dev, M. B. (1999). Development of adult-like performance in backward, simultaneous, and forward masking. *Journal of Speech, Language, and Hearing Research*, 42, 844–849.

Buss, E., Hall, J. W., & Grose, J. H. (2006). Development and the role of internal noise in detection and discrimination thresholds with narrow band stimuli. *The Journal of the Acoustical Society of America*, 120, 2777–2788.

Buss, E., Hall, J. W., & Grose, J. H. (2009). Psychometric functions for pure tone intensity discrimination: Slope differences in school-aged children and adults. *The Journal of the Acoustical Society of America*, 125, 1050–1058.

Buss, E., Hall, J. W., & Grose, J. H. (2012). Development of auditory coding as reflected in psychophysical performance. In L. A. Werner, A. N. Popper, & R. R. Fay (Eds.), *Human auditory development*. New York: Springer Science+Business Media.

Cabrera, L., & Werner, L. A. (2017). Infants' and adults' use of temporal cues in consonant discrimination. *Ear and Hearing* 38, 497–506.

Clarkson, M. G., Clifton, R. K., & Perris, E. E. (1988). Infant timbre perception: Discrimination of spectral envelopes. *Perception & Psychophysics*, 43, 15–20.

Clifton, R. K., Gwiazda, J., Bauer, J., Clarkson, M., & Held, R. (1988). Growth in head size during infancy: Implications for sound localization. *Developmental Psychology*, 24, 477–483.

Clifton, R. K., Perris, E. E., & Bullinger, A. (1991). Infants' perception of auditory space. *Developmental Psychology*, 27, 187–197.

Collins, A. A., & Gescheider, G. A. (1989). The measurement of loudness in individual children and adults by absolute magnitude estimation and cross-modality matching. *The Journal of the Acoustical Society of America*, 85, 2012–2021.

Cone, B., & Whitaker, R. (2013). Dynamics of infant cortical auditory evoked potentials (caeps) for tone and speech tokens. *International Journal of Pediatric Otorhinolaryngology*, 77, 1162–1173.

Cone-Wesson, B., Ma, E., & Fowler, C.G. (1997). Effect of stimulus level and frequency on ABR and MLR binaural interaction in human neonates. *Hearing Research*, 106, 163–178.

Eddins, D. A., & Green, D. M. (1995). Temporal integration and temporal resolution. In B. C. J. Moore (Ed.), *Hearing* (pp. 207–242). San Diego: Academic Press.

Eggermont, J. J. (1992). Development of auditory evoked-potentials. *Acta Oto-Laryngologica*, 112, 197–200.

Eggermont, J. J., & Moore, J. K. (2012). Morphological and functional development of the auditory nervous system. In L. A. Werner, A. N. Popper, & R. R. Fay (Eds.), *Human auditory development* (pp. 61–105). New York: Springer Science+Business Media.

Eggermont, J. J., Ponton, C. W., Coupland, S. G., & Winkelaar, R. (1991a). Frequency-dependent maturation of the cochlea and brain-stem evoked-potentials. *Acta Oto-Laryngologica*, 111, 220–224.

Eggermont, J. J., Ponton, C. W., Coupland, S. G., & Winkelaar, R. (1991b). Maturation of the traveling-wave delay in the human cochlea. *The Journal of the Acoustical Society of America*, 90, 288–298.

Eggermont, J. J., Brown, D. K., Ponton, C. W., & Kimberley, B. P. (1996). Comparison of distortion product otoacoustic emission (DPOAE) and auditory brainstem response (ABR) traveling wave delay measurements suggests frequency-specific synapse maturation. *Ear and Hearing*, 17, 386–394.

Elfenbein, J. L., Small, A. M., & Davis, M. (1993). Developmental patterns of duration discrimination. *Journal of Speech, Language, and Hearing Research*, 36, 842–849.

Fassbender, C. (1993). *Auditory grouping and segregation processes in infancy*. Norderstedt, Germany: Kaste Verlag.

Fernald, A., & Kuhl, P. (1987). Acoustic determinants of infant perception for motherese speech. *Infant Behavior and Development*, 10, 279–293.

Folland, N. A., Butler, B. E., Smith, N. A., & Trainor, L. J. (2012). Processing simultaneous auditory objects: Infants' ability to detect mistuning in harmonic complexes. *The Journal of the Acoustical Society of America*, 131, 993–997.

Folsom, R. C., & Wynne, M. K. (1987). Auditory brainstem responses from human adults and infants: Wave V tuning curves. *The Journal of the Acoustical Society of America*, 81, 412–417.

Furst, M., Bresloff, I., Levine, R. A., Merlob, P. L., & Attias, J. J. (2004). Interaural time coincidence detectors are present at birth: Evidence from binaural interaction. *Hearing Research*, 187, 63–72.

Gerhardt, K. J., & Abrams, R. M. (1996). Fetal hearing: Characterization of the stimulus and response. *Seminars in Perinatology*, 20, 11–20.

Gordon, K. A., Papsin, B. C., & Harrison, R. V. (2006). An evoked potential study of the developmental time course of the auditory nerve and brainstem in children using cochlear implants. *Audiology and Neuro-Otology*, 11, 7–23.

Gordon, K. A., Valero, J., Jewell, S. F., Ahn, J., & Papsin, B. C. (2010). Auditory development in the absence of hearing in infancy. *NeuroReport*, 21, 163–167.

Gordon, K. A., Deighton, M. R., Abbasalipour, P., & Papsin, B. C. (2014). Perception of binaural cues develops in children who are deaf through bilateral cochlear implantation. *PLoS ONE*, 9(12), e114841.

Gorga, M. P., Reiland, J. K., Beauchaine, K. A., Worthington, D. W., & Jesteadt, W. (1987). Auditory brainstem responses from graduates of an intensive care nursery: Normal patterns of response. *Journal of Speech, Language, and Hearing Research*, 30, 311–318.

Gorga, M. P., Kaminski, J. R., Beauchaine, K. L., Jesteadt, W., & Neely, S. T. (1989). Auditory brainstem responses from children three months to three years of age: Normal patterns of response II. *Journal of Speech, Language, and Hearing Research*, 32, 281–288.

Grieco-Calub, T. M., Litovsky, R. Y., & Werner, L. A. (2008). Using the observer-based psychophysical procedure to assess localization acuity in toddlers who use bilateral cochlear implants. *Otology and Neurotology*, 20(2), 235–239.

Hall, J. W., & Grose, J. H. (1991). Notched-noise measures of frequency selectivity in adults and children using fixed-masker-level and fixed-signal-level presentation. *Journal of Speech, Language, and Hearing Research*, 34, 651–660.

Hall, J. W., & Grose, J. H. (1994). Development of temporal resolution in children as measured by the temporal-modulation transfer-function. *The Journal of the Acoustical Society of America*, 96, 150–154.

Han, Y. K., Kover, H., Insanally, M. N., Semerdjian, J. H., & Bao, S. W. (2007). Early experience impairs perceptual discrimination. *Nature Neuroscience*, 10, 1191–1197.

Harris, D., & Dallos, P. (1984). Ontogenetic changes in frequency mapping in a mammalian ear. *Science*, 225, 741–743.

Harris, J. A., & Rubel, E. W. (2006). Afferent regulation of neuron number in the cochlear nucleus: Cellular and molecular analyses of a critical period. *Hearing Research*, 216, 127–137.

Hartley, D. E. H., Wright, B. A., Hogan, S. C., & Moore, D. R. (2000). Age-related improvements in auditory backward and simultaneous masking in 6- to 10-year-old children. *Journal of Speech, Language, and Hearing Research*, 43, 1402–1415.

He, C., & Trainor, L. J. (2009). Finding the pitch of the missing fundamental in infants. *The Journal of Neuroscience*, 29, 7718–7722.

Hepper, P. G., & Shahidullah, B. S. (1994). Development of fetal hearing. *Archives of Disease in Childhood*, 71, F81–F87.
Himelfarb, M. Z., Popelka, G. R., & Shanon, E. (1979). Tympanometry in normal neonates. *Journal of Speech, Language, and Hearing Research*, 22, 179–191.
Horn, D., Hutton-Gerhards, C., & Werner, L. (2016). Temporal modulation cutoff frequency in normal hearing infants. *Abstracts of the Midwinter Meeting of the Association for Research in Otolaryngology*, 39, 367.
Horn, D. L., Won, J. H., Rubinstein, J. T., & Werner, L. A. (2017). Spectral ripple discrimination in normal hearing infants. *The Journal of the Acoustical Society of America*, Indiana University 141, 613–623.
Houston, D. M., Beer, J., Bergeson, T. R., Chin, S. B., et al. (2012a). The ear is connected to the brain: Some new directions in the study of children with cochlear implants at Indiana University. *Journal of the American Academy of Audiology*, 23, 446–463.
Houston, D. M., Stewart, J., Moberly, A., Hollich, G., & Miyamoto, R. T. (2012b). Word learning in deaf children with cochlear implants: Effects of early auditory experience. *Developmental Science*, 15, 448–461.
Keefe, D. H., Bulen, J. C., Arehart, K. H., & Burns, E. M. (1993). Ear-canal impedance and reflection coefficient in human infants and adults. *The Journal of the Acoustical Society of America*, 94, 2617–2638.
Keefe, D. H., Burns, E. M., Bulen, J. C., & Campbell, S. L. (1994). Pressure transfer function from the diffuse field to the human infant ear canal. *The Journal of the Acoustical Society of America*, 95, 355–371.
Keefe, D. H., Folsom, R. C., Gorga, M. P., Vohr, B. R., et al. (2000). Identification of neonatal hearing impairment: Ear-canal measurements of acoustic admittance and reflectance in neonates. *Ear and Hearing*, 21, 443–461.
Keefe, D. H., Zhao, F., Neely, S. T., Gorga, M. P., & Vohr, B. R. (2003). Ear-canal acoustic admittance and reflectance effects in human neonates. I. Predictions of otoacoustic emission and auditory brainstem responses. *The Journal of the Acoustical Society of America*, 113, 389–406.
Keller, T. A., & Cowan, N. (1994). Developmental increase in the duration of memory for tone pitch. *Developmental Psychology*, 30, 855–863.
Keuroghlian, A. S., & Knudsen, E. I. (2007). Adaptive auditory plasticity in developing and adult animals. *Progress in Neurobiology*, 82, 109–121.
Key, A. P. F., Lambert, E. W., Aschner, J. L., & Maitre, N. L. (2012). Influence of gestational age and postnatal age on speech sound processing in NICU infants. *Psychophysiology*, 49, 720–731.
Kilian, C. F., Royle, N., Totten, C. L., Raine, C. H., & Lovett, R. E. S. (2015). The effect of early auditory experience on the spatial listening skills of children with bilateral cochlear implants. *International Journal of Pediatric Otorhinolaryngology*, 79, 2159–2165.
King, A. J., Kacelnik, O., Mrsic-Flogel, T. D., Schnupp, J. W. H., et al. (2001). How plastic is spatial hearing? *Audiology and Neuro-Otology*, 6, 182–186.
Knudsen, E. I. (2004). Sensitive periods in the development of the brain and behavior. *Journal of Cognitive Neuroscience*, 16, 1412–1425.
Kuhl, P. K., Tsao, F. M., & Liu, H. M. (2003). Foreign-language experience in infancy: Effects of short-term exposure and social interaction on phonetic learning. *Proceedings of the National Academy of Sciences of the USA*, 100, 9096–9101.
Lahav, A., & Skoe, E. (2014). An acoustic gap between the NICU and womb: A potential risk for compromised neuroplasticity of compromised neuroplasticity of the auditory system in preterm infants. *Frontiers in Neuroscience*, 8, 381.
Lary, S., Briassoulis, G., de Vries, L., Dubowitz, L. M. S., & Dubowitz, V. (1985). Hearing threshold in preterm and term infants by auditory brainstem response. *Journal of Pediatrics*, 107, 593–599.
Lasky, R. E. (1991). The effects of rate and forward masking on human adult and newborn auditory evoked response thresholds. *Developmental Psychobiology*, 24, 21–64.

Lau, B. K., & Werner, L. A. (2012). Perception of missing fundamental pitch by 3- and 4-month-old human infants. *The Journal of the Acoustical Society of America*, 132, 3874–3882.
Lau, B. K., & Werner, L. A. (2014a). Infant missing fundamental pitch sensitivity and melody discrimination. *Abstracts of the Midwinter Meeting of the Association for Research in Otolaryngology*, 37, 785.
Lau, B. K., & Werner, L. A. (2014b). Perception of the pitch of unresolved harmonics by 3- and 7-month-old human infants. *The Journal of the Acoustical Society of America*, 136, 760–767.
Lau, B. K., Lalonde, K., Oster, M. M., & Werner, L. A. (2017). Infant pitch perception: Missing fundamental melody discrimination. *The Journal of the Acoustical Society of America*, 141, 63–72.
Lavigne-Rebillard, M., & Pujol, R. (1987). Surface aspects of the developing human organ of corti. *Acta Oto-Laryngologica*, 436, 43–50.
Lavigne-Rebillard, M., & Pujol, R. (1990). Auditory hair cells in human fetuses: Synaptogenesis and ciliogenesis. *Journal of Electron Microsccopy Technique*, 15, 115–122.
Leibold, L. J. (2012). Development of auditory scene analysis and auditory attention. In L. A. Werner, A. N. Popper, & R. R. Fay (Eds.), *Human auditory development* (pp. 137–161). New York: Springer Science+Business Media.
Leibold, L. J., & Bonino, A. Y. (2009). Release from informational masking in children: Effect of multiple signal bursts. *The Journal of the Acoustical Society of America*, 125, 2200–2208.
Leibold, L. J., & Neff, D. L. (2007). Effects of masker-spectral variability and masker fringes in children and adults. *The Journal of the Acoustical Society of America*, 121, 3666–3676.
Leibold, L. J., & Neff, D. L. (2011). Masking by a remote-frequency noise band in children and adults. *Ear and Hearing*, 32, 663–666.
Leibold, L. J., & Werner, W. A. (2002). Relationship between intensity and reaction time in normal hearing infants and adults. *Ear and Hearing*, 23, 92–97.
Levi, E. C., Folsom, R. C., & Dobie, R. A. (1993). Amplitude-modulation following response (AMFR): Effects of modulation rate, carrier frequency, age, and state. *Hearing Research*, 68, 42–52.
Levi, E. C., Folsom, R. C., & Dobie, R. A. (1995). Coherence analysis of envelope-following responses (EFRs) and frequency-following responses (FFRs) in infants and adults. *Hearing Research*, 89, 21–27.
Litovsky, R. Y. (1997). Developmental changes in the precedence effect: Estimates of minimum audible angle. *The Journal of the Acoustical Society of America*, 102, 1739–1745.
Litovsky, R. (2012). Development of binaural and spatial hearing. In L. A. Werner, A. N. Popper, & R. R. Fay (Eds.), *Human auditory development* (pp. 163–195). New York: Springer Science+Business Media.
Litovsky, R. Y., Jones, G. L., Agrawal, S., & van Hoesel, R. (2010). Effect of age at onset of deafness on binaural sensitivity in electric hearing in humans. *The Journal of the Acoustical Society of America*, 127, 400–414.
Locher, H., de Groot, J., van Iperen, L., Huisman, M. A., et al. (2015). Development of the stria vascularis and potassium regulation in the human fetal cochlea: Insights into hereditary sensorineural hearing loss. *Developmental Neurobiology*, 75, 1219–1240.
Manley, G. A., Gummer, A. W., Fay, R. R., & Popper, A. N. (Eds.) (2017). *Understanding the cochlea*. New York: Springer Science+Business Media.
Marean, G. C., Werner, L. A., & Kuhl, P. K. (1992). Vowel categorization by very young infants. *Developmental Psychology*, 28, 396–405.
Maye, J., Werker, J. F., & Gerken, L. (2002). Infant sensitivity to distributional information can affect phonetic discrimination. *Cognition*, 82, B101–B111.
McAdams, S., & Bertoncini, J. (1997). Organization and discrimination of repeating sound sequences by newborn infants. *The Journal of the Acoustical Society of America*, 102(5 Pt 1), 2945–2953.
Mehler, J., Jusczyk, P. W., Lambertz, G., Halsted, N., et al. (1988). A precursor of language acquisition in young infants. *Cognition*, 29, 143–178.

Mills, D. M., & Rubel, E. W. (1996). Development of the cochlear amplifier. *The Journal of the Acoustical Society of America*, 100, 428–441.
Moon, C., Lagercrantz, H., & Kuhl, P. K. (2015). Language experienced in-utero affects vowel perception after birth: A two-country study. *Acta Pœdiatrica*, 102, 156–160.
Moore, D. R. (1990). Auditory brain-stem of the ferret: Early cessation of developmental sensitivity of neurons in the cochlear nucleus to removal of the cochlea. *The Journal of Comparative Neurology*, 302, 810–823.
Moore, D. R., Ferguson, M. A., Halliday, L. F., & Riley, A. (2008). Frequency discrimination in children: Perception, learning and attention. *Hearing Research*, 238, 147–154.
Morrongiello, B. A. (1987). Infants' localization of sounds in the median sagittal plane: Effects of signal frequency. *The Journal of the Acoustical Society of America*, 82, 900–905.
Morrongiello, B. A. (1988a). Infants' localization of sounds along two spatial dimensions: Horizontal and vertical axes. *Infant Behavior and Development*, 11, 127–143.
Morrongiello, B. A. (1988b). Infants' localization of sounds along the horizontal axis: Estimates of minimum audible angle. *Developmental Psychology*, 24, 8–13.
Morrongiello, B. A., Fenwick, K., & Chance, G. (1990). Sound localization acuity in very young infants: An observer-based testing procedure. *Developmental Psychology*, 26, 75–84.
Neff, D. L., & Green, D. M. (1987). Masking produced by spectral uncertainty with multicomponent maskers. *Perception and Psychophysics*, 41(5), 409–415.
Nemeth, R., Haden, G. P., Torok, M., & Winkler, I. (2015). Processing of horizontal sound localization cues in newborn infants. *Ear and Hearing*, 36, 550–556.
Nicholas, J. G., & Geers, A. E. (2007). Will they catch up? The role of age at cochlear implantation in the spoken language development of children with severe to profound hearing loss. *Journal of Speech, Language, and Hearing Research*, 50, 1048–1062.
Okabe, K. S., Tanaka, S., Hamada, H., Miura, T., & Funai, H. (1988). Acoustic impedance measured on normal ears of children. *Journal of the Acoustical Society of Japan*, 9, 287–294.
Olsho, L. W., Koch, E. G., & Halpin, C. F. (1987). Level and age effects in infant frequency discrimination. *The Journal of the Acoustical Society of America*, 82, 454–464.
Olsho, L. W., Koch, E. G., Carter, E. A., Halpin, C. F., & Spetner, N. B. (1988). Pure-tone sensitivity of human infants. *The Journal of the Acoustical Society of America*, 84, 1316–1324.
Pena, M., Werker, J. F., & Dehaene-Lambertz, G. (2012). Earlier speech exposure does not accelerate speech acquisition. *The Journal of Neuroscience*, 32, 11159–11163.
Peter, V., Wong, K., Narne, V. K., Sharma, M., et al. (2014). Assessing spectral and temporal processing in children and adults using temporal modulation transfer function (TMTF), iterated ripple noise (IRN) perception, and spectral ripple discrimination (SRD). *Journal of the American Academy of Audiology*, 25, 210–218.
Plack, C. J. (2014). *The sense of hearing*, 2nd ed. New York: Psychology Press.
Plack, C. J., Barker, D., & Hall, D. A. (2014). Pitch coding and pitch processing in the human brain. *Hearing Research*, 307, 53–64.
Ponton, C. W., Don, M., Eggermont, J. J., Waring, M. D., et al. (1996). Auditory system plasticity in children after long periods of complete deafness. *NeuroReport*, 8, 61–65.
Ponton, C. W., Eggermont, J. J., Kwong, B., & Don, M. (2000). Maturation of human central auditory system activity: Evidence from multi-channel evoked potentials. *Clinical Neurophysiology*, 111, 220–236.
Prieve, B. A., Fitzgerald, T. S., & Kemp, D. T. (1997). Basic characteristics of distortion product otoacoustice emissions in infants and children. *The Journal of the Acoustical Society of America*, 100, 3871–3880.
Quinonez, R. E., & Crawford, M. R. (1998). Longitudinal distortion product otoacoustic emission (dpe) latency changes in preterm neonates. *Acta Oto-Laryngologica*, 118, 26–31.
Rosen, M. J., Sarro, E. C., Kelly, J. B., & Sanes, D. H. (2012). Diminished behavioral and neural sensitivity to sound modulation is associated with moderate developmental hearing loss. *PLoS ONE*, 7, 13.
Rotteveel, J. J., de Graaf, R., Colon, E. J., Stegeman, D. F., & Visco, Y. M. (1987). The maturation of the central auditory conduction in preterm infants until three months post term. II. The auditory brainstem responses (ABRs). *Hearing Research*, 26, 21–35.

Rubel, E. W. (1984). Ontogeny of auditory system function. *Annual Review of Physiology*, 46, 213–229.

Rubel, E. W., & Ryals, B. M. (1983). Development of the place principle: Acoustic trauma. *Science*, 219, 512–514.

Saffran, J., Werker, J., & Werner, L. A. (2006). The infant's auditory world: Hearing, speech and the beginnings of language. In W. Damon, R. M. Lerner, D. Kuhn, & R. S. Siegler (Eds.), *Handbook of child psychology*, Vol. 2: *cognition, perception, and language*, 6th ed. (pp. 58–108). Hoboken, NJ: John Wiley & Sons.

Sanes, D. H., & Bao, S. W. (2009). Tuning up the developing auditory CNS. *Current Opinion in Neurobiology*, 19, 188–199.

Schellenberg, E. G., & Trehub, S. E. (1999). Culture-general and culture-specific factors in the discrimination of melodies. *Journal of Experimental Child Psychology*, 74, 107–127.

Schneider, B. A., Trehub, S. E., Morrongiello, B. A., & Thorpe, L. A. (1986). Auditory sensitivity in preschool children. *The Journal of the Acoustical Society of America*, 79, 447–452.

Schneider, B. A., Trehub, S. E., Morrongiello, B. A., & Thorpe, L. A. (1989). Developmental changes in masked thresholds. *The Journal of the Acoustical Society of America*, 86, 1733–1742.

Schneider, B. A., Morrongiello, B. A., & Trehub, S. E. (1990). The size of the critical band in infants, children, and adults. *Journal of Experimental Psychology [Human Perception and Performance]*, 16, 642–652.

Seidl, A. H., & Grothe, B. (2005). Development of sound localization mechanisms in the mongolian gerbil is shaped by early acoustic experience. *Journal of Neurophysiology*, 94, 1028–1036.

Silva, I., & Epstein, M. (2010). Estimating loudness growth from tone-burst evoked responses. *The Journal of the Acoustical Society of America*, 127, 3629–3642.

Sinnott, J. M., & Aslin, R. N. (1985). Frequency and intensity discrimination in human infants and adults. *The Journal of the Acoustical Society of America*, 78, 1986–1992.

Skoe, E., Krizman, J., Anderson, S., & Kraus, N. (2015). Stability and plasticity of auditory brainstem function across the lifespan. *Cerebral Cortex*, 25, 1415–1426.

Smith, N. A., & Trainor, L. J. (2011). Auditory stream segregation improves infants' selective attention to target tones amid distracters. *Infancy*, 16, 1–14.

Spetner, N. B., & Olsho, L. W. (1990). Auditory frequency resolution in human infancy. *Child Development*, 61, 632–652.

Supin, A. Y., Popov, V. V., Milekhina, O. N., & Tarakanov, M. B. (1998). Ripple density resolution for various rippled-noise patterns. *The Journal of the Acoustical Society of America*, 103, 2042–2050.

Sussman, E. S., Ceponiene, R., Shestakova, A., Naatanen, R., & Winkler, I. (2001). Auditory stream segregation processes operate similarly in school-aged children and adults. *Hearing Research*, 153, 108–114.

Thai-Van, H., Cozma, S., Boutitie, F., Disant, F., et al. (2007). The pattern of auditory brainstem response wave V maturation in cochlear-implanted children. *Clinical Neurophysiology*, 118, 676–689.

Tharpe, A. M., & Ashmead, D. H. (2001). A longitudinal investigation of infant auditory sensitivity. *American Journal of Audiology*, 10, 104–112.

Tognola, G., Parazzini, M., de Jager, P., Brienesse, P., et al. (2005). Cochlear maturation and otoacoustic emissions in preterm infants: A time-frequency approach. *Hearing Research*, 199, 71–80.

Trainor, L. J., Samuel, S. S., Desjardins, R. N., & Sonnadara, R. R. (2001). Measuring temporal resolution in infants using mismatch negativity. *NeuroReport*, 12, 2443–2448.

Trehub, S. E. (2001). Musical predispositions in infancy. In R. J. Zatorre & I. Peretz (Eds.), *The biological foundations of music. Annals of the New York Academy of Sciences*, Vol. 930, 1–16.

Trehub, S. E., Schneider, B. A., & Endman, M. (1980). Developmental changes in infants' sensitivity to octave-band noises. *Journal of Experimental Child Psychology*, 29, 282–293.

Ubbink, S. W. J., van Dijk, P., de Kleine, E., Brienesse, P., et al. (2011). Frequency shifts with age in click-evoked otoacoustic emissions of preterm infants. *The Journal of the Acoustical Society of America*, 129, 3788–3796.

Van Deun, L., van Wieringen, A., Van den Bogaert, T., Scherf, F., et al. (2009). Sound localization, sound lateralization, and binaural masking level differences in young children with normal hearing. *Ear and Hearing*, 30, 178–190.
Voss, S. E., & Herrmann, B. S. (2005). How does the sound pressure generated by circumaural, supra-aural, and insert earphones differ for adult and infant ears? *Ear and Hearing*, 26, 636–650.
Wang, X. Q., & Walker, K. M. M. (2012). Neural mechanisms for the abstraction and use of pitch information in auditory cortex. *The Journal of Neuroscience*, 32, 13339–13342.
Weikum, W. M., Oberlander, T. F., Hensch, T. K., & Werker, J. F. (2012). Prenatal exposure to antidepressants and depressed maternal mood alter trajectory of infant speech perception. *Proceedings of the National Academy of Sciences of the USA*, 109, 17221–17227.
Werker, J. F. (1994). Cross-language speech perception: Developmental change does not involve loss. In J. C. Goodman, & H. C. Nusbaum (Eds.), *The development of speech perception* (pp. 93–120). Cambridge, MA: MIT Press.
Werker, J. F., & Hensch, T. K. (2015). Critical periods in speech perception: New directions. *Annual Review of Psychology*, 66, 173–196.
Werker, J. F., & Tees, R. C. (1984). Cross-language speech perception: Evidence for perceptual reorganization during the first year of life. *Infant Behavior & Development*, 7, 49–63.
Werner, L. A. (1999). Forward masking among infant and adult listeners. *The Journal of the Acoustical Society of America*, 105, 2445–2453.
Werner, L. A. (2013). Infants' detection and discrimination of sounds in modulated maskers. *The Journal of the Acoustical Society of America*, 133, 4156–4167.
Werner, L. A., & Bargones, J. Y. (1991). Sources of auditory masking in infants: Distraction effects. *Perception & Psychophysics*, 50, 405–412.
Werner, L. A., Marean, G. C., Halpin, C. F., Spetner, N. B., & Gillenwater, J. M. (1992). Infant auditory temporal acuity: Gap detection. *Child Development*, 63, 260–272.
Werner, L. A., Folsom, R. C., & Mancl, L. R. (1993). The relationship between auditory brainstem response and behavioral thresholds in normal hearing infants and adults. *Hearing Research*, 68, 131–141.
Werner, L. A., Folsom, R. C., & Mancl, L. R. (1994). The relationship between auditory brainstem response latencies and behavioral thresholds in normal hearing infants and adults. *Hearing Research*, 77, 88–98.
Werner, L. A., Folsom, R. C., Mancl, L. R., & Syapin, C. (2001). Human auditory brainstem response to temporal gaps in noise. *Journal of Speech, Language, and Hearing Research*, 44, 737–750.
Wightman, F., & Kistler, D. J. (2005). Informational masking of speech in children: Effects of ipsilateral and contralateral distracters. *The Journal of the Acoustical Society of America*, 118, 3164–3176.
Wright, B. A., & Fitzgerald, M. B. (2001). Different patterns of human discrimination learning for two interaural cues to sound-source location. *Proceedings of the National Academy of Sciences of the USA*, 98, 12307–12312.
Yeung, H. H., & Werker, J. F. (2009). Learning words' sounds before learning how words sound: 9-month-olds use distinct objects as cues to categorize speech information. *Cognition*, 113, 234–243.
Zatorre, R. J. (2003). Absolute pitch: A model for understanding the influence of genes and development on neural and cognitive function. *Nature Neuroscience*, 6, 692–695.
Zeng, F. G., & Fay, R. R. (Eds.) (2004). *Cochlear implants: Auditory prostheses and electric hearing*. New York: Springer Science+Business Media.
Zhang, L. I., Bao, S. W., & Merzenich, M. M. (2001). Persistent and specific influences of early acoustic environments on primary auditory cortex. *Nature Neuroscience*, 4, 1123–1130.

Chapter 8
Early Experience and Auditory Development in Songbirds

Sarah M.N. Woolley

Abstract Vocal communication is critical for life in a wide range of vertebrate species. Mammals, birds, frogs, and fishes rely on auditory processing to perceive the vocal signals of others in the environment and gain social information such as the presence of potential mates or predators. Conspecific vocalizations convey information on sex, age, individual identity, and behavioral state. The importance of vocal communication for social behavior places auditory processing at the forefront of brain functions that directly impact fitness. Young humans and songbirds require experience of adult vocal communication to develop their own perceptual and vocal skills. Studies on songbird vocal development and auditory processing are revealing how early experience and developmental plasticity interact to specialize central auditory function for vocal communication. This chapter reviews research findings that shed light on the role of early song experience in shaping adult song perception and the auditory coding of songs.

Keywords Auditory cortex • Communication • Learning • Midbrain • Neural coding • Perception • Plasticity • Sensorimotor integration • Sensory • Social behavior • Vocal

8.1 Introduction

A broad range of animal species use hearing and voice to communicate social information that is essential for life. Animals in most vertebrate classes identify conspecifics and individuals, find and choose mates, and avoid danger by extracting information from vocal cues produced by others. The fitness advantages of auditory–vocal communication confer significant selective pressure on individuals to perceive and respond accurately to communication sounds (Bennur et al. 2013).

S.M.N. Woolley (✉)
Department of Psychology, Jerome L. Greene Science Center,
L3-031, 3227 Broadway, New York, NY 10027, USA
e-mail: sw2277@columbia.edu

K.S. Cramer et al. (eds.), *Auditory Development and Plasticity*, Springer Handbook of Auditory Research 64, DOI 10.1007/978-3-319-21530-3_8

Accurate perception of vocal communication signals depends on the auditory system's coding of vocal sounds that convey important social information. Thus, in vocal communicators, central auditory function should develop to include and possibly specialize for the neural representation of vocal communication sounds (Woolley and Moore 2011; Poremba et al. 2013).

In some birds and mammals, vocal communication skills must be learned. For these animals, auditory development may be shaped by sensory and social experience that is instructive for learning to perceive and produce vocal cues that convey information. Speech and language studies demonstrate the importance of early experience for human development of vocal communication skills. The age at which a child is first exposed to a specific spoken language is highly correlated with adult proficiency in that language (Werker and Tees 1999). Auditory perception is shaped for life by the native-language speech sounds infants hear in the first year of life, before speech develops (Kuhl 2010). Interestingly, social experience of speech sounds influences perceptual development more strongly than does strictly auditory experience of the same speech (Kuhl 2007). This observation suggests that social interaction gates auditory plasticity in young humans. The slow maturation of auditory perceptual skills relative to other modalities corresponds well with the importance of early experience in development of vocal communication skills (Saffran et al. 2006; Sanes and Woolley 2011). The impact of early auditory experience on adult speech skills demonstrates the brain's natural use of experience at specific life stages to develop complex abilities.

Understanding the neural mechanisms of vocal communication depends on animal models in which experience can be manipulated and the brain can be studied at high resolution. Animals that learn their vocalizations are particularly useful because manipulations of experience and neural activity permit the correlation of behavior and auditory processing. Whereas vocal communication is common in vertebrates, vocal learning is rare. Humans, three clades of birds (parrots, hummingbirds, and songbirds), and some cetaceans are established vocal learners (Brenowitz and Beecher 2005). Of nonhuman vocal learners, songbirds are the most studied and tractable for experimental research on auditory and vocal development. Songbirds such as the estrildid finches, a family that includes the zebra finch (*Taeniopygia guttata*), readily sing, choose mates, breed, rear and tutor young, and learn song in the laboratory, where behavior and neural activity can be precisely measured (Zann 1996; Woolley and Moore 2011). Decades of behavioral studies show that juvenile songbirds develop song behavior by hearing and interacting with singing adults and using auditory feedback to practice their own song production (Konishi 1964; Marler and Waser 1977). Recent studies of neural circuitry, neurochemistry, gene expression, and neurophysiology demonstrate important parallels between avian pallium and mammalian neocortex (Karten 2013; Calabrese and Woolley 2015). The detailed knowledge about song development and evolutionary conservation of auditory circuitry make the songbird a robust system for working out how experience and central auditory plasticity interact during the development of vocal communication skills (Sanes and Woolley 2011). Here, research findings on the influence of early experience on song perception and the auditory coding of communication vocalizations in songbirds are described. Learning to sing also relies on auditory processing and perception; thus, the role of auditory experience in song motor development is also considered.

8.2 Song Behavior

Bird songs are hierarchically organized sequences of complex sounds. The smallest acoustic units are "notes." Notes are grouped into "syllables." A series of syllables that is produced in a predictable sequence is a song "phrase" or "motif." Song differs by species in the acoustic features of notes and syllables and temporal patterns of phrases and motifs (Beecher and Brenowitz 2005; Brenowitz and Beecher 2005). Figure 8.1 shows spectrograms of zebra finch, Bengalese finch (*Lonchura striata domesticata*), and European starling (*Sturnus vulgaris*) song, illustrating the acoustic features and temporal patterns in the unique songs of individuals of the same species and in the songs of different species. Distinct sequences of syllables or phrases that one bird consistently produces are song "types." Birds of some species sing one song type while others sing many song types (Beecher and Brenowitz 2005; Brenowitz and Beecher 2005). Adults use the unique acoustics of conspecific song to identify potential mates and competitors and juveniles use the same cues to identify appropriate tutors. Because song is learned, individuals' songs differ in the acoustic features of notes, syllables, and phrases. Both sexes use the unique songs of known birds to maintain social relationships such as mates and territory neighbors (Catchpole and Slater 2008).

In most species, adult males sing to court females and to engage in aggressive encounters with other males. Adult females use male song as a mate choice cue (Beecher and Brenowitz 2005; Catchpole and Slater 2008). The amount of singing and the acoustic features of songs convey fitness information (Beecher and Brenowitz 2005; Richner 2016) and influence female attraction to males (Riebel 2009; Hauber et al. 2010). Juveniles of both sexes imprint on the songs of adults that rear them (Zann 1996; Catchpole and Slater 2008). Because zebra finches and European starlings are highly social birds that sing and display behavioral preferences for song features in the laboratory, they are most commonly studied in research on auditory processing and perception in songbirds.

8.3 Basic Hearing

Songbird species have similar but not identical U-shaped audibility curves, with maximal sensitivities between 2 and 5 kHz (Okanoya and Dooling 1987; Gleich and Langemann 2011). Low-frequency audibility parallels that of many mammals but the high-frequency range of hearing is limited compared to that in mammals. Most birds, including songbirds, hear only frequencies up to about 6 kHz (Dooling 1982; Okanoya and Dooling 1987), while most mammals hear far higher frequencies

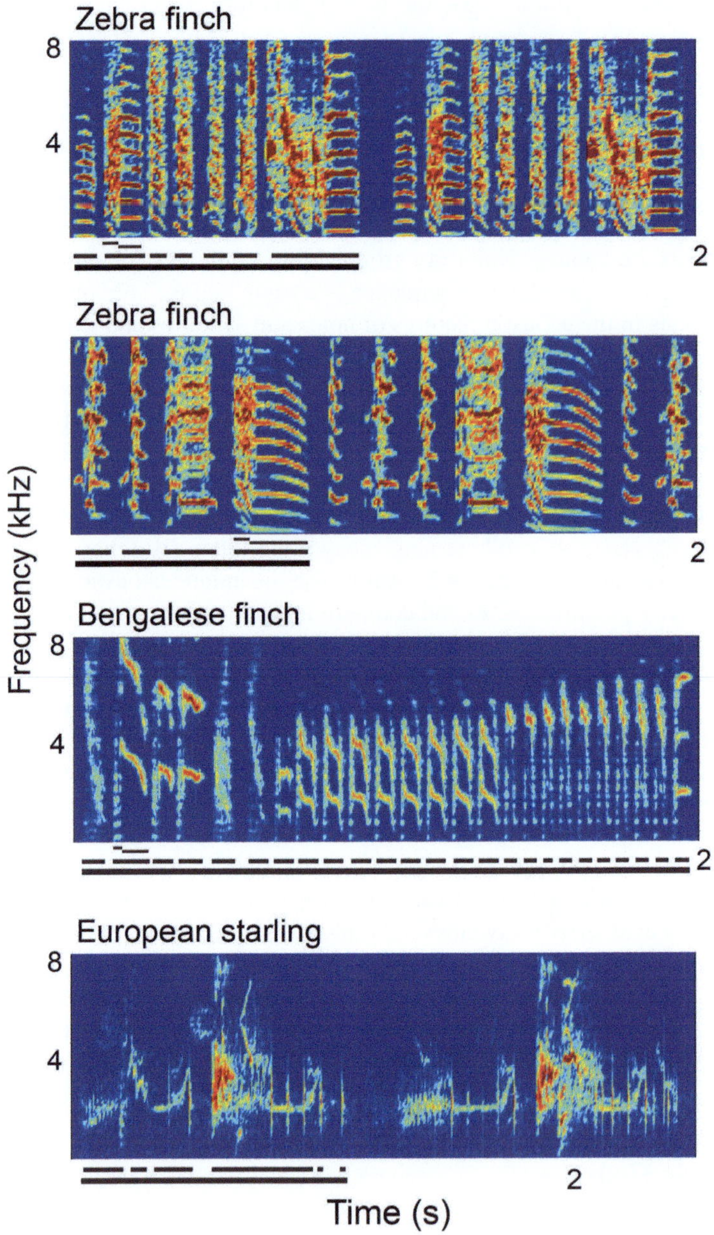

Fig. 8.1 Spectrograms show the acoustic features of song notes and syllables and the temporal patterns of syllables in the songs of different conspecific males and of three species. Color indicates intensity. Blue is low and red is high. Below each spectrogram, lines label notes (*top*), syllables (*middle*), and a motif (*bottom*)

(Heffner and Heffner 2007; Heffner and Heffner 2008). The avian high-frequency limit of hearing corresponds to a shorter basilar papilla length compared to the organ of Corti in mammals of similar size (Gleich and Langemann 2011; Woolley and Portfors 2013). Maximal sensitivity to frequencies between 2 and 5 kHz may be due to the physical properties of the sensory hair cells that encode those frequencies. Hair cell damage and regeneration studies indicate that frequencies above 2 kHz are encoded by the hair cells along the basal half of the songbird cochlea (Woolley and Rubel 1999; Woolley et al. 2001). This region of the avian epithelium has a high relative density of short hair cells, which have larger stereocilia bundles and presumably larger numbers of sensory transduction channels than do tall hair cells (Köppl 2011).

Basic hearing onset in songbirds is insufficiently studied, but it is hypothesized that, unlike precocial birds such as ducks and chickens, juvenile songbirds hear very little until after hatching. Zebra finches develop adult-like auditory brainstem response thresholds at 20 days of age (Amin et al. 2007). The hypothesis that hearing onset occurs after hatching is supported by studies indicating that the effects of auditory experience on behavioral development begin after hatching (Immelmann 1969; Eales 1985).

8.4 Early Song Experience and Behavior

8.4.1 Song Production

Normal song learning requires experience of others' adult songs and the juvenile's own vocalizations. Young males learn their songs through social and auditory exposure to singing adults and vocal practice. Birds that are raised in isolation or deafened as juveniles develop highly abnormal songs that lack the acoustic structure of normal adult song (Konishi 1964, 2004). In most species, song learning occurs before sexual maturity in a two-stage process of auditory memorization followed by auditory–vocal integration. The instructive role of auditory experience changes over the course of song learning. During the memorization stage, a juvenile hears and memorizes the song of an adult "tutor" (Doupe and Kuhl 1999). Experience with a tutor before the juvenile begins to practice singing is sufficient for an accurate copy to be produced later, in some cases months after the tutoring (Catchpole and Slater 2008). This delay indicates that juveniles remember tutors' songs until the age at which they practice and perfect their own songs. At sexual maturity (~100 days of age), male zebra finches normally produce copies of the tutor songs they experienced between 20 and 60 days of age. However, if song experience is withheld until 60 days, the period of song memorization can be extended (Eales 1985). Thus, the developmental period during which tutor song exposure clearly shapes adult song production depends on both age and experience. Social interactions with tutors also affect the likelihood and accuracy of song copying (Doupe and Kuhl 1999; Catchpole and Slater 2008).

During the practice stage of song learning, juveniles build vocal motor skills and progress through reliable phases of song development. Initially, young birds produce quiet, highly variable vocalizations that are called subsong. This stage of behavioral development is similar to the babbling stage in human speech development. Later, vocalizations become louder and more structured. Singing behavior during this "plastic song" phase begins to be rhythmic but is far more variable than adult singing. With practice, birds perfect their songs into sequences of acoustically well-defined notes, syllables, and phrases, forming motifs and song types. At sexual maturity, song behavior stabilizes and birds produce stereotyped songs that no longer change over time. Some improvisation occurs such that no two birds' songs are exactly alike (Zann 1996; Doupe and Kuhl 1999). During practice, progress toward normal adult song behavior depends critically on auditory feedback, which birds use to monitor and sculpt their motor performance. Deafening during this stage prevents normal song development, whereas social isolation does not (Konishi 2004). The maintenance of normal adult song also depends on auditory feedback, but to a lesser extent than song development (Woolley 2008). Thus, auditory experience is necessary for normal song production at all life stages but transitions from a reliance on hearing other birds sing to an age-dependent reliance on hearing self-generated singing.

8.4.2 Song Perception

As in humans, the perception of vocal communication sounds in songbirds is shaped by early auditory experience of vocalizations (Doupe and Kuhl 1999). This is the case for both sexes, even in species in which females do not sing. One clear way in which developmental exposure to song influences behavior is the memorization of individual songs heard early in life, including, but not restricted to, songs that are copied. Both males that learn to sing and females that do not sing remember their fathers' songs as adults (Miller 1979). As discussed in Sect. 8.4.1, young juveniles memorize the songs of adult tutors, which serve as templates for the development of song behavior in males and song perception in both sexes. Song experience during this developmental stage organizes long-term perception; male and female adults remember and show behavioral preferences for songs they experience most closely as juveniles (Miller 1979; Riebel 2009). Miller (1979) first used spatial proximity assays to demonstrate that adult zebra finches have perceptual memories of their fathers' songs and are attracted to those songs over other conspecific songs. In support of the conclusion that young juveniles accurately memorize tutor songs, song discrimination abilities appear to be well developed in juveniles that are learning to sing (Braaten et al. 2006). While the tutor song memory is important for song development in birds that learn to sing, its significance for nonsinging females is less clear. Females may sexually imprint on their fathers' songs to guide their attraction to particular song features or regional dialects during mate choice (Riebel 2009).

Juvenile experience of adult song influences general perceptual preferences for the acoustic features of songs. Birds of both sexes develop preferences for hearing conspecific song over heterospecific song based, at least in part, on early experience. Adults that were raised hearing conspecific song are more attracted to conspecific males and/or songs than to heterospecific males and songs (Cate and Mug 1984; Clayton 1988). This preference may begin with juvenile perceptual biases for the acoustic features of conspecific song but can subsequently wax or wane based on a young bird's experience (Marler 2004). Early experience may shape behavioral song preferences by reinforcing initial biases or by counteracting them, as in the case of cross-fostering (see Sect. 8.5.1.2). Birds raised without exposure to song show behavioral preferences for conspecific song over vastly different heterospecific song such as European starling or canary song (Braaten and Reynolds 1999; Lauay et al. 2004). However, some studies have reported the absence of behavioral preferences for conspecific song in isolation-reared females (Riebel 2009). More studies are needed to determine whether birds of all ages and both sexes show consistent preferences for the songs of their own species and the role of genetics in species-specific song preferences.

Cross-fostering studies suggest that species-specific song preferences are influenced by juvenile song experience. The cross-fostering approach gives all juveniles developmental experience of song and social interactions with adults of both sexes. But the acoustics of songs that are heard, memorized, and copied differ from those for juveniles that are reared by adults of their own species. This manipulation impacts behavioral song preferences in both males and females. Birds that are raised by adults of another species fail to show consistent preferences for conspecific songs as adults, and show stronger preferences for heterospecific songs than are observed in birds tutored by conspecific adults (Riebel 2009; Hauber et al. 2010).

The effects of early experience manipulations such as rearing birds in isolation from adult song and cross-tutoring on basic perceptual skills have yet to be thoroughly examined. Two studies have found that birds reared without exposure to adult song show auditory frequency discrimination deficits and poor song recognition skills (Njegovan and Weisman 1997; Sturdy et al. 2001). In summary, songbirds appear to begin using experience of adult song to guide development after hatching, develop behavioral song preferences that are partially due to experience, and form early memories of adult song that direct vocal development and adult song perception, with important consequences for vocal communication.

8.5 Early Experience and Auditory Coding

If early experience of song guides development of song perception and production, then it also guides development of central auditory processing. One approach to identifying the neurons and neural circuits underlying vocal perception is to examine where in the brain the functioning of neurons and circuits covaries with experience. Vocalization coding in the songbird brain is typically examined at the level of single neurons and small groups of neurons using electrophysiology. Understanding how single neurons code song information is crucial to achieving a mechanistic explanation of how the auditory system subserves vocal perception. This experimental approach provides a record of exactly when and how much neurons fire in response to sounds, allowing the quantitative analysis of what information in songs is coded by neurons and therefore what information is important for perception. For example, single-unit studies allow the measurement of response strength, neural selectivity, and spectrotemporal tuning during vocalization processing.

Other studies use the expression of immediate early genes (IEGs) to assay functional changes in neurons that are correlated with hearing vocalizations. *Zenk* is the IEG that is most commonly studied in the songbird auditory system (Mello and Jarvis 2008). Although *zenk* expression is not a direct marker of neural firing (Stripling et al. 1997; London and Clayton 2008), it is thought to be important for synaptic structural plasticity, potentially related to learning processes (Tischmeyer and Grimm 1999). Its expression following vocal processing can be examined in whole brain regions, allowing the detection of functional changes in many more neurons than most electrophysiological studies. *Zenk* expression has been used to identify auditory brain regions that are involved in vocal processing, measure response selectivity, and test the effects of experience manipulations on auditory function. The findings of studies using single and multiunit electrophysiology and immediate early gene expression to examine how developmental experience shapes the auditory processing of vocalizations in midbrain, primary cortex, and secondary cortex are described in Sects. 8.5.1.2, 8.5.2.3, and 8.5.2.5.

8.5.1 Auditory Midbrain

As in mammals, the songbird auditory midbrain is a focus for understanding how the spectral and temporal acoustic features of song are represented by the firing patterns of single neurons and neuronal populations (Fig. 8.2) (Woolley and Portfors 2013). Most midbrain neurons respond robustly to a wide variety of sounds including tones (Woolley and Casseday 2004; Schneider and Woolley 2011), synthetic sounds that mimic specific features of song such as frequency modulated sweeps (Woolley and Casseday 2005), and song (Schneider and Woolley 2010, 2013). The responses of single midbrain neurons are more robust, precise, and reliable than the average response of a cortical neuron; most midbrain neurons produce very similar responses to the same stimulus over multiple presentations and fire at higher rates than do cortex neurons (Schneider and Woolley 2011, 2013). The dense and reliable responses of midbrain neurons to many different sounds allow the direct comparison of responses to song and other sounds in a single neuron. These studies have provided insights into the impact of spectral and temporal correlations in songs and other natural sounds on auditory responses, as well as the contributions of tuning properties of single neurons to perceptual functions such as the recognition of individual songs.

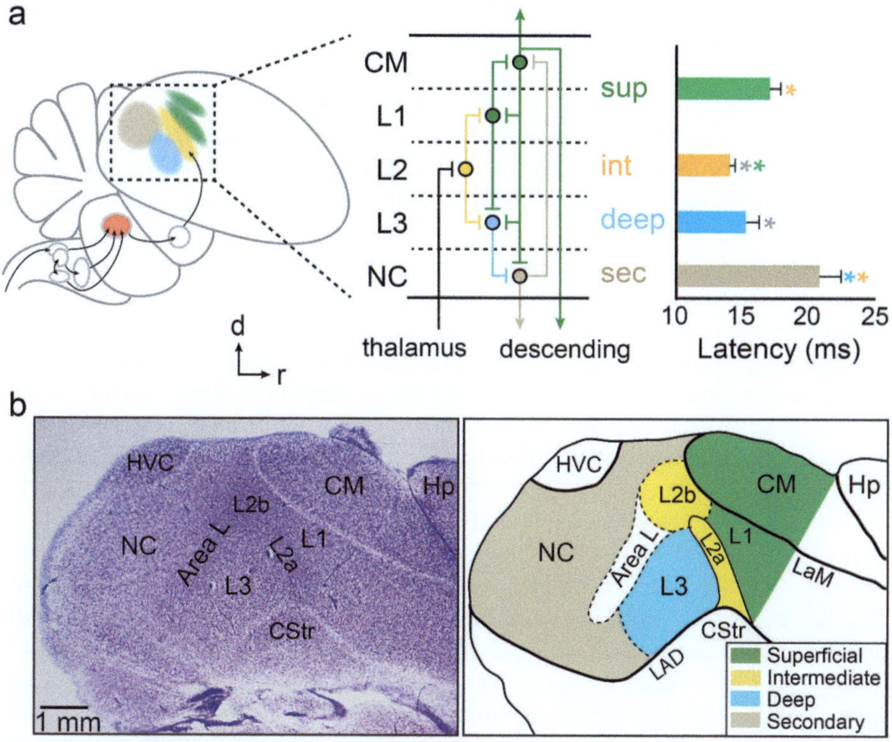

Fig. 8.2 Organization of the songbird auditory system. (**a**) *Left*: Drawing of the songbird brain showing auditory midbrain (orange) and cortex regions, labeled by color. *Center*: Circuit diagram of the songbird auditory cortex showing the major projections and connections among regions. Thalamic input terminates largely in field L2 (intermediate region) and sparsely in field L3 (deep region). Dorsal projections from L2 terminate in L1 and L1 projects to caudal mesopallium (CM; superficial regions). Ventral projections from L2 terminate in L3. Field L3 projects to caudal nidopallium (NC; secondary region), thalamus, and brainstem. CM and NC are reciprocally connected. *Right*: Mean first spike latencies for single neurons recorded in superficial, intermediate, deep, and secondary regions. Latencies are shortest in the intermediate region and significantly longer in superficial and secondary regions. (**b**) *Left*: Nissl stained parasagittal section showing cell bodies and laminae delineating regions. *Right*: Drawing of the same section with regions colored and labeled. Dashed lines indicate boundaries that are defined by transitions in cytoarchitecture. Area L has cytoarchitecture similar to L2b but is not known to receive thalamic input. Hp = hippocampus; LaM = lamina mesopallialis; LAD = lamina arcopallialis dorsalis; CStr = caudal striatum; HVC (proper name, a song system premotor nucleus). [Adapted from Calabrese and Woolley (2015)]

8.5.1.1 Auditory Midbrain Function

Each midbrain neuron has a unique complement of spectral, temporal, and intensity tuning properties, which determines the simple sounds to which it responds (Woolley and Casseday 2004, 2005) and the features of complex sounds such as songs that it encodes (Schneider and Woolley 2010, 2011). The differences in the temporal patterns of the responses that are produced by a neuron provide a basis for measuring the tuning properties of individual neurons; response patterns that differ in response to different songs are due to the unique acoustics of each song and the spectrotemporal tuning of the responding neuron. Accordingly, the responses of midbrain neurons are better characterized by linear–nonlinear models of tuning than are most cortical neurons (Woolley et al. 2006; Schneider and Woolley 2011). This observation suggests that midbrain neurons faithfully encode specific spectrotemporal sounds. When a large number of midbrain neurons, each with distinct spectrotemporal tuning, are considered together, their responses can be used to reconstruct the songs that evoked the responses. Spike trains that are similar across multiple responses to the same song (reliable) but that differ between responses to different songs provide a basis for the neural discrimination of songs sung by different males, potentially contributing to individual recognition during social interactions. The spike trains evoked by presentation of a set of songs can be used to accurately discriminate among individual songs in the set of songs (Schneider and Woolley 2010, 2013).

Comparison of the spectrotemporal tuning of single neurons and the acoustic properties of natural sounds has shown that zebra finch auditory midbrain neurons are tuned for the spectrotemporal modulations that are found in vocalizations (Woolley et al. 2005, 2006) and that the spectrotemporal tuning properties of individual neurons cluster into functional groups that encode specific acoustic features of songs that may relate to perceptual sound features such as pitch, rhythm, and timbre (Woolley et al. 2009). In summary, the responses of auditory midbrain neurons to songs and other sounds show that, while many more sounds than just songs are coded in this region, tuning mechanisms exist at this early level in the auditory processing stream to encode the acoustic properties of songs, forming accurate neural representations that can be used by downstream neurons to decode information such as song "quality," species identity, and individual identity.

8.5.1.2 Early Experience and Auditory Midbrain Function

Early auditory experience influences the development of information coding capacity and firing rate in auditory midbrain neurons. The term "information" refers to a response's capacity to encode information about a stimulus and is a function of firing rate, how much firing rate changes over the time course of a response, how quickly firing rate changes, and response reliability (Theunissen et al. 2004a). Woolley et al. (2010) cross-fostered juvenile zebra finches into the nests of Bengalese finches so that they heard only Bengalese finch vocalizations and learned to sing Bengalese finch song (Fig. 8.3a). Responses of midbrain neurons to zebra and Bengalese finch songs were then recorded from adult cross-fostered zebra finches, normal adult male zebra finches, and normal adult Bengalese finches. Responses of neurons in the zebra finches reared by Bengalese finches were significantly weaker

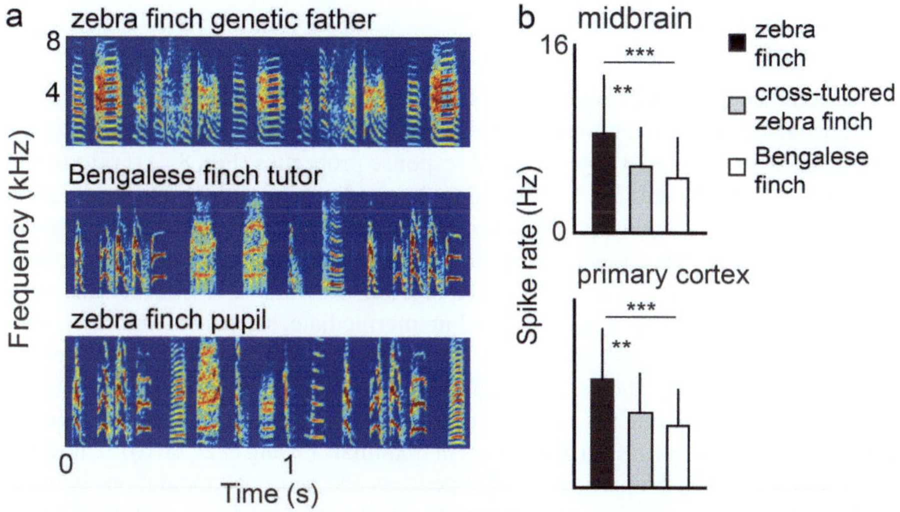

Fig. 8.3 Behavioral and auditory coding effects of atypical song experience. (**a**) Spectrograms of songs show that juvenile zebra finches learn Bengalese finch song when tutored by adult Bengalese finches. *Top*: Song of the genetic father of the zebra finch pupil. *Middle*: Song of the Bengalese finch tutor. *Bottom*: Adult song of the zebra finch pupil. The song of the cross-fostered zebra finch is similar to the Bengalese finch tutor's song and dissimilar from the genetic father's song. (**b**) Spike rates of single midbrain (*top*) and primary cortex (*bottom*) neurons in response to presentation of zebra finch songs. Spike rates are higher in neurons from zebra finches tutored by zebra finches than in neurons from Bengalese finches and from zebra finches tutored by Bengalese finches. Error bars are standard deviations. $^{**}p < 0.01$; $^{***}p < 0.001$. [**b** adapted from Woolley et al. (2010)]

in response to both zebra finch and Bengalese finch songs than were the responses of neurons in zebra finches reared by zebra finches (Woolley et al. 2010) (Fig. 8.3b). Responses of Bengalese finch neurons to the same songs were also lower than those in normal zebra finches. These results suggest that neural responsivity of adult midbrain and primary cortex neurons is experience dependent, and that the species of song a juvenile experiences can affect development of auditory function. It is important to note that the cross-fostering approach in Woolley et al. (2010) manipulated a juvenile's acoustic environment, song experience, and song learning. It is unclear which of these manipulations were causal for the effects of experience that were found. For example, Bengalese finches sing less often and more quietly than do zebra finches (Immelmann 1969), potentially leading to a large difference in the level of acoustic exposure between birds raised by zebra finches and those raised by Bengalese finches. Experiments in which juveniles are reared and tutored by conspecific and heterospecific adults in the same acoustic environment are needed to distinguish among these potential sources of experience-dependent differences in auditory coding.

8.5.2 Auditory Cortex

Avian auditory pallium is composed of six major regions organized into contiguous, interconnected fields of neurons that are distinguished by projections, cell morphology, gene expression, and physiological response properties (Fig. 8.2) (Fortune and Margoliash 1992; Calabrese and Woolley 2015). Despite its lack of organization into thin cell layers, studies of circuitry, gene expression, and electrophysiological responses have revealed functional principles that are common to avian and mammalian pallium (Butler et al. 2011; Calabrese and Woolley 2015). First, the same genetically identified cell types are found in intermediate, superficial, and deep neuronal populations (Dugas-Ford et al. 2012). Second, high-resolution anatomical tracing of connectivity shows that avian auditory cortex is organized into columns with projections spanning the deep, intermediate, and superficial processing regions, akin to connections across cortical layers in mammals (Wang et al. 2010). Third, the same sensory coding properties of single neurons and neuronal populations in intermediate, superficial, and deep regions are observed in birds and mammals (Calabrese and Woolley 2015). For these reasons, avian auditory pallium is now referred to as auditory cortex.

8.5.2.1 Auditory Cortex Organization

The auditory thalamus projects to the intermediate regions, fields L2a and b (L2). Projections from L2 terminate in superficial (L1) and deep (L3) primary cortex. Neurons in L1 project to the dorsal superficial region, caudal mesopallium (CM). The caudal nidopallium (NC) is considered secondary or higher auditory cortex and receives input from L3 and CM, and sends projections out of auditory cortex. Figure 8.2 shows the spatial location, major connections, and anatomical organization of auditory cortex. Region borders are identified by laminae or differences in cytoarchitecture that are visible in Nissl-stained sections at high resolution.

The organization of information flow among cortical region, response properties of neurons in each region, and response properties of major neuron classes are fundamental to understanding how early experience shapes song coding in auditory cortex. Auditory information flows from thalamus to the intermediate region, then to superficial and deep regions, and then to the secondary region, as in mammalian neocortex (Harris and Mrsic-Flogel 2013). Receptive fields are simple in the intermediate region and more complex in superficial and deep regions (Atencio and Schreiner 2010; Calabrese and Woolley 2015). Responses of single neurons and neuronal populations are sparser and more selective in the superficial and deep regions than in the intermediate layer.

Songbird auditory cortex neurons sort into broad-spiking (BS) and narrow-spiking (NS) classes and show the same functional differences in sensory coding as BS and NS neurons in mammalian neocortex (Atencio and Schreiner 2008; Calabrese and Woolley 2015). Broad-spiking neurons are putative excitatory

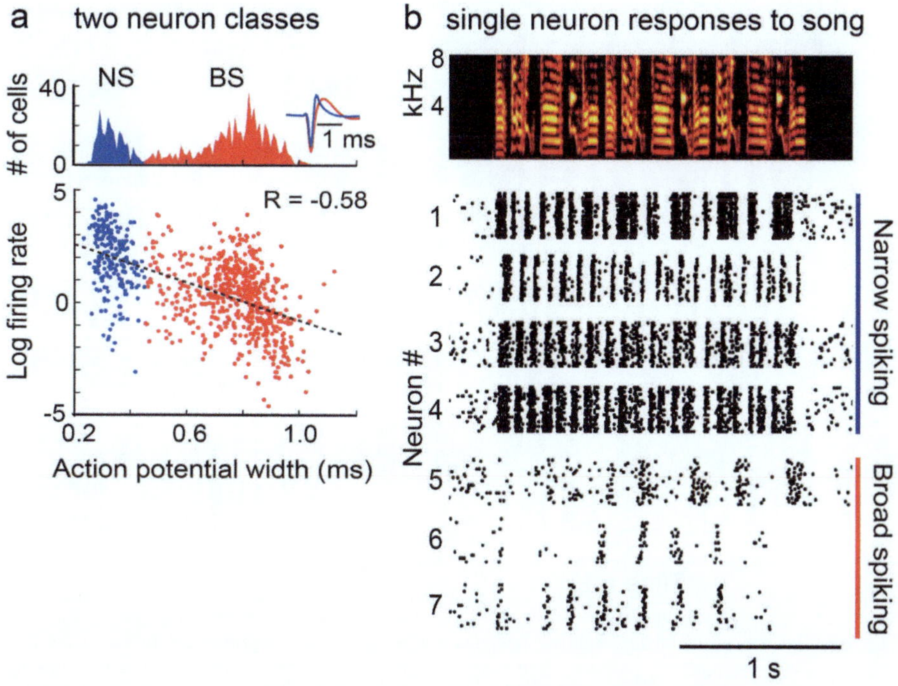

Fig. 8.4 Auditory cortex neurons cluster into two major types based on action potential shape and response properties. (**a**) Narrow spiking neurons (blue) produce shorter duration action potentials and fire at higher rates than do broad spiking neurons (red). (**b**) Raster plots show single neuron spike trains evoked by presentation of the song depicted in the spectrogram above the raster plots. Broad spiking (BS) neurons respond more selectively to acoustic features in song than do narrow spiking (NS) neurons. [Adapted from Calabrese and Woolley (2015)]

neurons that fire wide action potentials at relatively low rates and produce sparse and selective auditory responses (Fig. 8.4). Narrow-spiking neurons are putative inhibitory neurons that fire at higher rates and respond less selectively than do BS cells (Fig. 8.4a). Auditory receptive fields are more complex in BS than in NS neurons (Atencio and Schreiner 2008; Calabrese and Woolley 2015). Figure 8.4b shows example spike trains of BS and NS neurons evoked by the song shown in the spectrogram above the rasters. As in mammalian cortex, BS neurons have more heterogeneous responses and receptive field structures than do NS neurons, in and across regions. Responses of BS neurons are lower in firing rate, higher in stimulus selectivity, and higher in population response sparseness than are those of NS neurons. Neural representations of song and the influences of early experience on song responses differ by cortical region and neuron type in the songbird (Sects. 8.5.2.3 and 8.5.2.5).

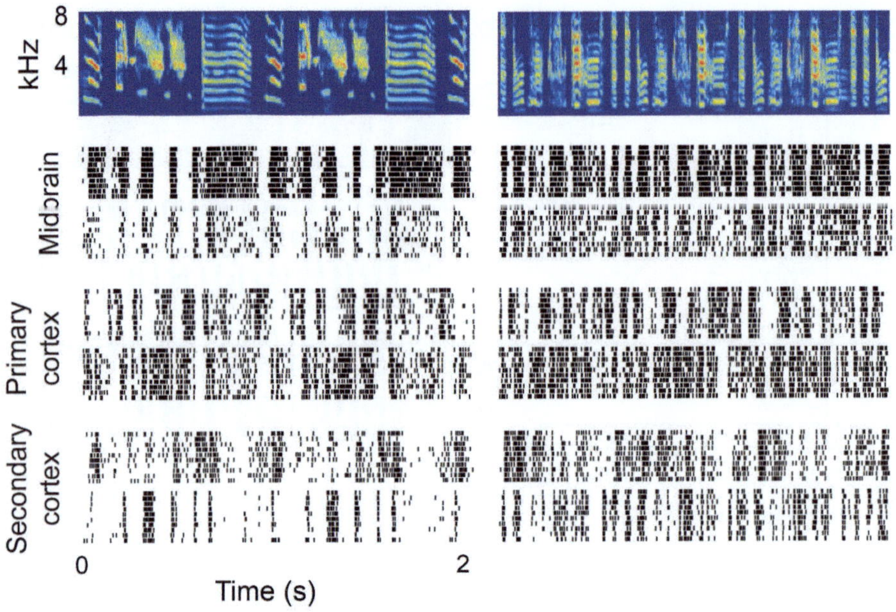

Fig. 8.5 Responses of narrow-spiking neurons are reliable and progressively more sparse and selective along the auditory pathway. Spectrograms of two zebra finch songs and rasters showing spike times of two midbrain (dorsal lateral mesencephalon; MLd), two primary cortex (L2a) and two secondary cortex (NC) neurons' responses to the songs. Responses to 10 presentations of the song are shown for each neuron. Neurons' responses to the same song differ in firing rate and pattern; each neuron reliably responds to different acoustic features of the song. Across regions, firing rates decrease and selectivity increases from midbrain to primary cortex to secondary cortex

8.5.2.2 Primary Auditory Cortex

Primary cortex neurons in field L respond, on average, less robustly and more selectively to songs and other sounds than do midbrain neurons (Fig. 8.5; Schneider and Woolley 2013). The basic spectral and temporal tuning properties of field L neurons overlap with those of midbrain neurons (Woolley et al. 2009). However, unlike midbrain neurons, field L neurons respond, on average, more robustly to song than to sounds such as noise and tone complexes with average power and frequency power spectra matched to song (Theunissen et al. 2004b). Selectivity for song over other sounds is weakly present in juveniles that are just beginning to produce subsong, suggesting that the emergence of song selectivity does not require singing (Amin et al. 2007). The hypothesis that song selectivity in primary cortex develops from auditory experience rather than vocal motor learning is supported by the finding that field L neurons in adult females that do not sing are, on average, selective for song over other sounds (Hauber et al. 2007a). Selectivity for song over other sounds and general responsivity in juvenile field L neurons are lower than in adult field L neurons, suggesting that the coding properties of primary cortex neurons mature over the same period of as song development (Amin et al. 2007).

Responses to song and other sounds differ in strength and selectivity among the distinct regions of primary auditory cortex. Sound-evoked firing rates are highest in the intermediate region and significantly lower in superficial and deep regions (Calabrese and Woolley 2015). Selectivity, measured as the fraction of sounds that do not evoke a response from a neuron, is lowest in the intermediate region and significantly higher at successive processing stages (Meliza and Margoliash 2012; Calabrese and Woolley 2015). Differences in sound-evoked firing rates and selectivity across regions are correlated with the receptive field complexity; receptive fields are simplest in the intermediate region and most complex in the deep region (Nagel and Doupe 2008; Calabrese and Woolley 2015). Together, these region-specific differences indicate that spectrotemporal tuning complexity and response selectivity progressively increase along the cortical processing pathway.

8.5.2.3 Early Experience and Primary Auditory Cortex Function

Juvenile experience of adult song appears to regulate development of cortical function in terms of how robustly neurons respond to songs and how selective neurons' responses are for specific songs. Like midbrain neurons, field L neurons in male zebra finches reared and tutored by Bengalese finches are less responsive to song in general than are field L neurons in normal male zebra finches (Fig. 8.3b) (Woolley et al. 2010). Primary cortex neurons in European starlings and zebra finches reared without adult song experience lack the response selectivity observed in cortical neurons recorded from normal birds (Fig. 8.6). Cousillas et al. (2004) found that raising starlings without adult song experience resulted in abnormally low selectivity; many more multiunit electrophysiological recording sites responded to a variety of song features in isolation-reared birds than in wild caught birds that presumably developed hearing adult song (Fig. 8.6a, b) (Cousillas et al. 2004). Similar studies show that the degree to which responses lacked the selectivity that typifies field L neurons in wild birds correlated with the length of time the young birds were isolated from adult song (George et al. 2004, 2010). For example, the response selectivity of neuronal sites recorded from wild-caught starlings was stronger than in starlings that were isolated from song for their first two years, and the selectivity of recording sites in those birds was stronger than the selectivity of sites in birds that had no exposure to adult song (Fig. 8.6c). Maul et al. (2010) tested the effects of isolation from early song experience on cortical activity in adult male and female zebra finches using functional magnetic resonance imaging (fMRI; Fig. 8.6d). They found that isolation-reared males had highly abnormal cortical responses to songs and syllables, with significant blood oxygen level difference (BOLD) responses in much larger volumes of forebrain than in normal males raised hearing adult song and other vocalizations in a zebra finch colony (Maul et al. 2010). The responses in isolation-reared males were also highly variable and lacked the selectivity for conspecific song over other sounds such as tones that characterized responses in

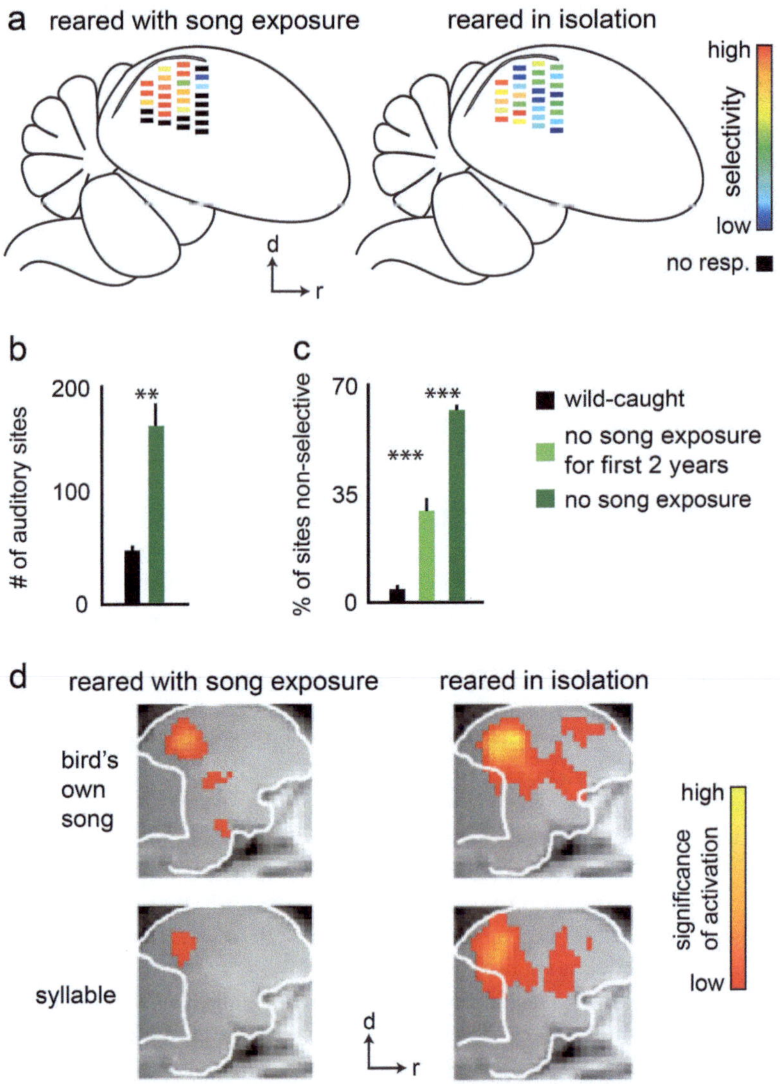

Fig. 8.6 Early experience of song influences response magnitude and selectivity in auditory cortex neurons. (**a**) Multiunit responses in European starlings are more selective and spatially restricted compared to those in European starlings reared in isolation from adult song. (**b**) The number of electrophysiologically mapped sites that respond to auditory stimuli is higher in birds reared without song experience. [Adapted from Cousillas et al. (2004)]. (**c**) Response selectivity of auditory sites to different song features is lowest for birds reared with no adult song experience, intermediate for birds exposed to adult song as young adults, and highest for wild-caught birds. [Adapted from George et al. (2004)]. (**d**) Blood oxygenation level difference (BOLD) to presentation of a song and a song syllable differ in the auditory forebrains of adult male zebra finches reared with song experience and those reared in isolation from song. [Adapted from Maul et al. (2010)]. Error bars are 1 SE. $*p < 0.05$; $**p < 0.01$; $***p < 0.001$

colony-raised males. Just 30 seconds a day of song exposure from audio recordings reduced the abnormal BOLD responses, but responses were still broad in anatomical spread compared to colony-raised males. By contrast, isolation-reared females showed brain activation patterns that were similar to those of colony-raised males. These results suggest that early song exposure affects male auditory function more than female auditory function, possibly because males learn to sing and females do not. Hauber et al. 2007b found no correlation between early song experience and primary cortex response selectivity at the level of the single neuron in nonsinging females (Hauber et al. 2007b). These studies suggest that that the influence of developmental experience of adult song on field L response selectivity may be observed only in vocal learners.

Because song is a social behavior, the importance of developmental experience other than sound has begun to be explored. Isolation-rearing studies often conflate three developmental factors that may influence song coding and perception. First, birds are raised without hearing adult songs, resulting in deficient experience of vocalizations specifically. Second, birds are raised in environments that are acoustically impoverished compared with colony-raised or wild caught birds, resulting in abnormal auditory experience in general. Third, birds are raised with limited social experience compared to birds raised in families, usually in colonies that include several families and both sexes. These factors may all be important for perceptual development, and are difficult to separate experimentally.

The idea that social experience specifically influences auditory development is intriguing, and plausible in the case of song processing because normal song development relies on social interactions. Cousillas et al. (2006) tested the hypothesis that early social interactions shape the processing of vocalizations in auditory cortex. European starlings were raised in three different social settings with varying exposure to other juveniles and adults but with the same auditory experience. The electrophysiological response selectivity of groups of cortical neurons to song stimuli was then compared among birds raised under the different social conditions. Although the sample sizes in experimental groups were low, the neuronal recordings from socially deprived birds showed lower response selectivity than did the recordings from birds raised in a highly social setting, suggesting that early social experience does influence auditory coding in primary cortex neurons (Cousillas et al. 2006). This finding is supported by another study showing that primary cortex neurons in female European starlings raised only by adult males that do not normally serve as tutors for females [females do sing and typically learn from other females (Poirier et al. 2009)] show similar deficits in response selectivity (Cousillas et al. 2008). These studies suggest that social and auditory experience influences the development of auditory coding in songbirds that naturally live in social groups. In summary, the major effects of early experience on auditory coding in primary cortex appear to be the development of neuronal response robustness and selectivity.

8.5.2.4 Secondary Cortex

The medial portion of secondary auditory cortex, NCM, contains neurons with the most complex response properties in the cortex (Mello and Jarvis 2008; Schneider and Woolley 2013). The responses of both NS and BS neurons are selective for specific acoustic features, as demonstrated by reliable responses to songs that are unique to each song (Figs. 8.4 and 8.5) (Schneider and Woolley 2013). But NCM responses are also modulated by recent stimulus history (Schneider and Woolley 2013; Yang and Vicario 2015), the acoustic environment (Terleph et al. 2008; Yang and Vicario 2015), and the behavioral salience of songs (Gobes et al. 2010). For example, NCM expression of *zenk* habituates to repeated presentation of the same song (Mello et al. 1995) and electrophysiological responses to repeated presentation of the same song weaken over presentations (Phan et al. 2006). The complex and context-dependent responses to vocalizations make NCM a particularly interesting region for testing the effects of developmental experience on auditory processing.

8.5.2.5 Early Experience and Secondary Cortex Function

The effects of early experience on song coding in NCM have been examined electrophysiologically by measuring neural activity in birds with differing song experience and by assaying *zenk* expression (Fig. 8.7). Yanagihara and Yazaki-Sugiyama (2016) tested the hypothesis that juvenile NCM neurons develop response selectivity to a tutor's song during song learning (Yanagihara and Yazaki-Sugiyama 2016). Responses of single NCM neurons in juvenile male zebra finches that were being tutored and those kept in isolation were recorded during presentation of fathers' songs, the birds' own songs, other songs, calls, and noise. Whereas the responses of NS neurons were unselective, 13% of BS neurons in the tutored birds were selective for the tutors' songs. This selectivity was not found in the BS neurons of isolated, untutored birds (Fig. 8.7a). Recordings also showed that 12% of BS neurons were selective for the birds' own songs in tutored birds while similar selectivity was found in only 5% of neurons in untutored birds. This study suggests that early experience of adult song promotes the development of response selectivity in NCM BS neurons, with some neurons becoming selective for songs that are instructive for song learning.

This hypothesis is supported by an earlier study showing that NCM neurons in European starlings that have been deprived of adult song experience are less selective for song features than are NCM neurons in wild-caught starlings (Fig. 8.7b) (George et al. 2010). In addition, in zebra finches, cortical *zenk* expression becomes sensitive to song playback at the age that birds begin to memorize tutor songs but not before (Jin and Clayton 1997; Stripling et al. 2001). NCM *zenk* expression following song presentation is lower in juvenile males and females that have not experienced adult song than in juveniles that have heard song, suggesting that *zenk* expression in general is influenced by early auditory experience (Jin and Clayton 1997; Tomaszycki et al. 2006). *Zenk* expression following presentation of the tutor

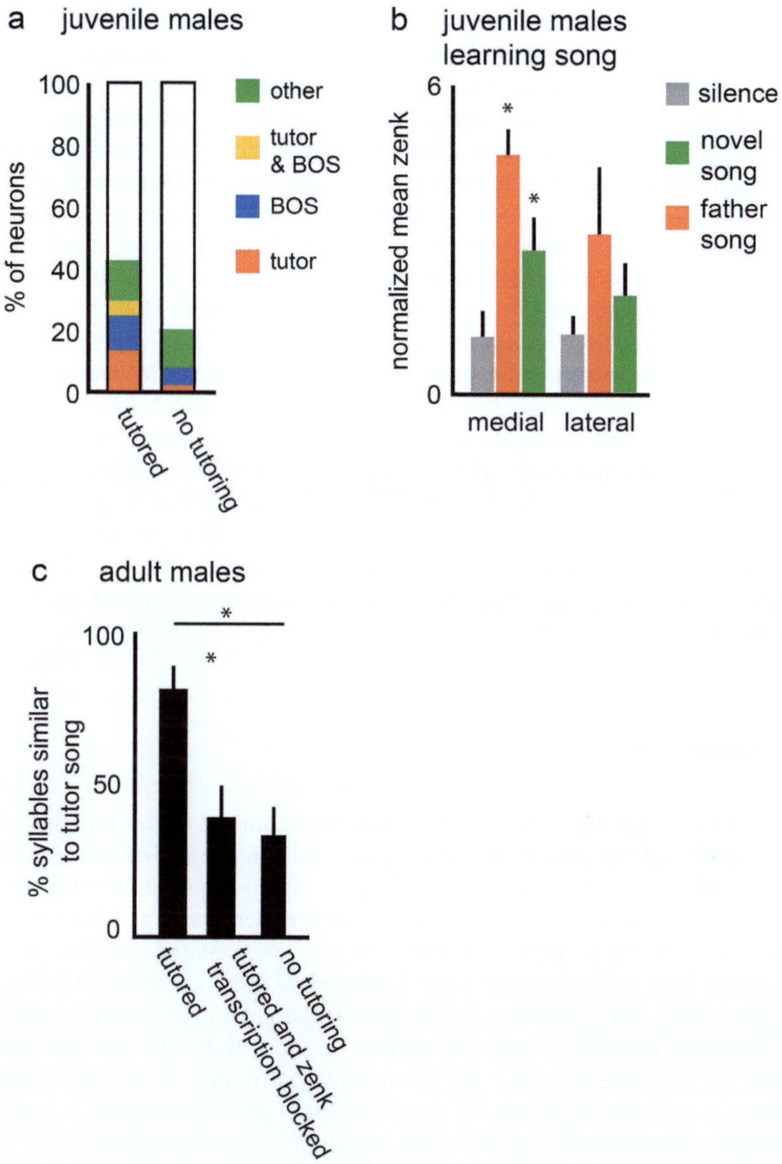

Fig. 8.7 Interaction between song tutoring and secondary cortex responses. (**a**) A higher percentage of medial caudal nidopallium (NCM) neurons respond selectively to specific vocalizations in tutored juveniles compared to isolated juveniles. [Adapted from Yanagihara and Yazaki-Sugiyama (2016)]. (**b**) Father song presentation results in more *zenk* expression in the juvenile male NCM than does presentation of an unfamiliar song. [Adapted from Gobes et al. (2010).] (**c**) Comparison of the percentage of syllables that were similar to tutors' syllables when birds received tutoring and no brain manipulation, tutoring with forebrain *zenk* transcription blocked, and no tutoring. The *zenk* blockade resulted in impaired song learning. [Adapted from London and Clayton (2008).] Error bars are 1 standard error. $*p < 0.05$

song and other songs does not differ in the adult NCM (Terpstra et al. 2004) but it is higher in response to tutor song playback in NCM of juvenile males that are in the process of learning to produce song (Fig. 8.7b) (Gobes et al. 2010). Higher forebrain *zenk* expression levels are also correlated with the accuracy of song learning (Bolhuis et al. 2000; Bolhuis et al. 2001), and song learning is impaired by disruptions of *zenk* expression. London and Clayton (2008) tested the effects of pairing song tutoring sessions and blockade of the signal-regulating kinase that controls *zenk* induction in the auditory forebrain on song production learning in zebra finches (London and Clayton 2008). The adult songs of birds that were tutored while *zenk* expression was blocked were poor copies of the tutor's song (Fig. 8.7c). Tutored birds that did not receive the kinase inhibitor copied significantly more of the same tutor's song. These results and a study in which lesions of NCM disrupted behavioral responses to tutors' songs, but not auditory discrimination learning (Gobes and Bolhuis 2007), provide strong evidence in favor of the idea that NCM is involved in the storage of information about the tutor's song. Electrophysiological data examining the habituation rates of NCM responses to tutor songs also suggest that NCM neurons store information about tutor songs (Phan et al. 2006). In summary, neural selectivity in secondary auditory cortex develops under the influence of early song experience and is important for the tutor song memorization that guides song development.

8.6 Summary

The songbird is proving to be a highly informative animal model for studying the neural coding and perception of learned communication vocalizations. A rich history of behavioral and neurobiological studies on song production learning has paved the way for using songbirds to ask how experience shapes the development of auditory function and perception in the context of vocal communication. The studies reviewed here have identified some fundamental effects of vocal experience on vocal processing and perception, but there is still a great deal to be learned. There are at least four major questions that need to be addressed: (1) How does developmental song exposure influence the development of tuning properties of auditory neurons that encode song? (2) Is the development of perceptual skills such as learning to discriminate among and recognize individual songs dependent on early song experience? Songbird studies linking developmental experience with the maturation of the auditory system and perceptual skills are needed. (3) How does experience facilitate the emergence of neural selectivity that may be important for song perception and learning? (4) How are auditory memories of individual birds' songs represented in neural circuits? The addition of behavioral measures that test the role of early song exposure in perceptual development will be necessary to determine the behavioral relevance of the neural coding effects described here. Approaches that include careful experience manipulations, perceptual tests, and measures of auditory coding at multiple levels of analysis are needed to answer these questions. In particular, the combination of measures at the single neuron, neuronal population, and global neural response (e.g., fMRI) levels will be helpful for understanding how auditory processing leads to the perception of meaningful social messages. Lastly, understanding the mechanisms by which early experience of song shapes auditory development may lead to insights into how early speech exposure shapes auditory processing and speech perception in humans. A better understanding of the mechanisms whereby experience influences auditory function at specific ages may unearth basic principles of critical periods and experience-dependent brain development.

Acknowledgments The author thanks Edwin W Rubel for his scientific findings, ideas, and mentorship that contributed to this work. The author's work was supported by NIH grant R01-DC-009810.

Compliance with Ethics Requirements Sarah M. N. Woolley declares that she has no conflict of interest.

References

Amin, N., Doupe, A., & Theunissen, F. E. (2007). Development of selectivity for natural sounds in the songbird auditory forebrain. *Journal of Neurophysiology*, 97(5), 3517–3531.

Atencio, C. A., & Schreiner, C. E. (2008). Spectrotemporal processing differences between auditory cortical fast-spiking and regular-spiking neurons. *The Journal of Neuroscience*, 28(15), 3897–3910.

Atencio, C. A., & Schreiner, C. E. (2010). Laminar diversity of dynamic sound processing in cat primary auditory cortex. *Journal of Neurophysiology*, 103(1), 192–205.

Beecher, M. D., & Brenowitz, E. A. (2005). Functional aspects of song learning in songbirds. *Trends in Ecology & Evolution*, 20(3), 143–149.

Bennur, S., Tsunada, J., Cohen, Y. E., & Liu, R. C. (2013). Understanding the neurophysiological basis of auditory abilities for social communication: A perspective on the value of ethological paradigms. *Hearing Research*, 305, 3–9.

Bolhuis, J. J., Zijlstra, G. G., den Boer-Visser, A. M., & Van Der Zee, E. A. (2000). Localized neuronal activation in the zebra finch brain is related to the strength of song learning. *Proceedings of the National Academy of Sciences of the USA*, 97(5), 2282–2285.

Bolhuis, J. J., Hetebrij, E., Den Boer-Visser, A. M., De Groot, J. H., & Zijlstra, G. G. (2001). Localized immediate early gene expression related to the strength of song learning in socially reared zebra finches. *European Journal of Neuroscience*, 13(11), 2165–2170.

Braaten, R. F., & Reynolds, K. (1999). Auditory preference for conspecific song in isolation-reared zebra finches. *Animal Behaviour*, 58(1), 105–111.

Braaten, R. F., Petzoldt, M., & Colbath, A. (2006). Song perception during the sensitive period of song learning in zebra finches (*Taeniopygia guttata*). *Journal of Comparative Psychology*, 120(2), 79–88.

Brenowitz, E. A., & Beecher, M. D. (2005). Song learning in birds: Diversity and plasticity, opportunities and challenges. *Trends in Neuroscience*, 28(3), 127–132.

Butler, A. B., Reiner, A., & Karten, H. J. (2011). Evolution of the amniote pallium and the origins of mammalian neocortex. *Annals of the New York Academy of Sciences*, 1225, 14–27.

Calabrese, A., & Woolley, S. M. (2015). Coding principles of the canonical cortical microcircuit in the avian brain. *Proceedings of the National Academy of Sciences of the USA*, 112(11), 3517–3522.

Catchpole, C. K., & Slater, P. J. B. (2008). *Bird song: Biological themes and variations*, 2nd ed. Cambridge: Cambridge University Press.
Cate, C. T., & Mug, G. (1984). The development of mate choice in zebra finch females. *Behaviour*, 90(1–3), 125–150.
Clayton, N. S. (1988). Song learning and mate choice in estrildid finches raised by 2 species. *Animal Behaviour*, 36, 1589–1600.
Cousillas, H., Richard, J. P., Mathelier, M., Henry, L., et al. (2004). Experience-dependent neuronal specialization and functional organization in the central auditory area of a songbird. *European Journal of Neuroscience*, 19(12), 3343–3352.
Cousillas, H., George, I., Mathelier, M., Richard, J. P., et al. (2006). Social experience influences the development of a central auditory area. *Naturwissenschaften*, 93(12), 588–596.
Cousillas, H., George, I., Henry, L., Richard, J. P., & Hausberger, M. (2008). Linking social and vocal brains: Could social segregation prevent a proper development of a central auditory area in a female songbird? *PLoS ONE*, 3(5), e2194.
Dooling, R. J. (1982). Ontogeny of song recognition in birds. *American Zoologist*, 22, 571–580.
Doupe, A. J., & Kuhl, P. K. (1999). Birdsong and human speech: Common themes and mechanisms. *Annual Review of Neuroscience*, 22, 567–631.
Dugas-Ford, J., Rowell, J. J., & Ragsdale, C. W. (2012). Cell-type homologies and the origins of the neocortex. *Proceedings of the National Academy of Sciences of the USA*, 109(42), 16974–16979.
Eales, L. A. (1985). Song learning in zebra finches: Some effects of song model availability on what is learnt and when. *Animal Behaviour*, 33, 1293–1300.
Fortune, E. S., & Margoliash, D. (1992). Cytoarchitectonic organization and morphology of cells of the field-L complex in male zebra finches (*Taenopygia guttata*). *The Journal of Comparative Neurology*, 325(3), 388–404.
George, I., Cousillas, H., Vernier, B., Richard, J. P., et al. (2004). Sound processing in the auditory-cortex homologue of songbirds: Functional organization and developmental issues. *Journal of Physiology Paris*, 98(4–6), 385–394.
George, I., Alcaix, S., Henry, L., Richard, J. P., et al. (2010). Neural correlates of experience-induced deficits in learned vocal communication. *PLoS ONE*, 5(12), e14347.
Gleich, O., & Langemann, U. (2011). Auditory capabilities of birds in relation to the structural diversity of the basilar papilla. *Hearing Research*, 273(1–2), 80–88.
Gobes, S. M., & Bolhuis, J. J. (2007). Birdsong memory: A neural dissociation between song recognition and production. *Current Biology*, 17(9), 789–793.
Gobes, S. M., Zandbergen, M. A., & Bolhuis, J. J. (2010). Memory in the making: Localized brain activation related to song learning in young songbirds. *Proceedings Biological Sciences*, 277(1698), 3343–3351.
Harris, K. D., & Mrsic-Flogel, T. D. (2013). Cortical connectivity and sensory coding. *Nature*, 503(7474), 51–58.
Hauber, M. E., Cassey, P., Woolley, S. M., & Theunissen, F. E. (2007a). Neurophysiological response selectivity for conspecific songs over synthetic sounds in the auditory forebrain of non-singing female songbirds. *Journal of Comparative Physiology A: Neuroethology, Sensory, Neural, and Behavioral Physiology*, 193(7), 765–774.
Hauber, M. E., Woolley, S. M., & Theunissen, F. E. (2007b). Experience-dependence of neural responses to social vs. isolate conspecific songs in the forebrain of female zebra finches. *Journal of Ornithology*, 148.2, 231–239.
Hauber, M. E., Campbell, D. L. M., & Woolley, S. M. (2010). Functional role and female perception of male song in zebra finches. *Emu – Austral Ornithology*, 110, 209–218.
Heffner, H. E., & Heffner, R. S. (2008). Audition. In S. F. Davis (Ed.), *Handbook of research methods in experimental psychology* (pp. 413–440). Hoboken, NJ: Wiley-Blackwell.
Heffner, H. E., & Heffner, R. S. (2007). Hearing ranges of laboratory animals. *Journal of the American Association of Laboratory Animal Science*, 46(1), 20–22.

Immelmann, K. (1969). Song development in the zebra finch and other estrildid finches. In R. A. Hinde (Ed.), *Bird vocalizations* (pp. 61–77). Cambridge: Cambridge University Press.
Jin, H., & Clayton, D. F. (1997). Localized changes in immediate-early gene regulation during sensory and motor learning in zebra finches. *Neuron*, 19(5), 1049–1059.
Karten, H. J. (2013). Neocortical evolution: Neuronal circuits arise independently of lamination. *Current Biology*, 23(1), R12–15.
Konishi, M. (1964). Effects of deafening on song development in two species of juncos. *Condor*, 66, 85–102.
Konishi, M. (2004). The role of auditory feedback in birdsong. *Annals of the New York Academy of Sciences*, 1016, 463–475.
Köppl, C. (2011). Birds—same thing, but different? Convergent evolution in the avian and mammalian auditory systems provides informative comparative models. *Hearing Research*, 273(1–2), 65–71.
Kuhl, P. K. (2007). Is speech learning 'gated' by the social brain? *Developmental Science*, 10(1), 110–120.
Kuhl, P. K. (2010). Brain mechanisms in early language acquisition. *Neuron*, 67(5), 713–727.
Lauay, C., Gerlach, N. M., Adkins-Regan, E., & Devoogd, T. J. (2004). Female zebra finches require early song exposure to prefer high-quality song as adults. *Animal Behaviour*, 68(6), 1249–1255.
London, S. E., & Clayton, D. F. (2008). Functional identification of sensory mechanisms required for developmental song learning. *Nature Neuroscience*, 11(5), 579–586.
Marler, P. (2004). Innateness and the instinct to learn. *Anais da Academia Brasileira de Ciências*, 76(2), 189–200.
Marler, P., & Waser, M. S. (1977). Role of auditory feedback in canary song development. *Journal of Comparative and Physiology Psychology*, 91(1), 8–16.
Maul, K. K., Voss, H. U., Parra, L. C., Salgado-Commissariat, D., et al. (2010). The development of stimulus-specific auditory responses requires song exposure in male but not female zebra finches. *Developmental Neurobiology*, 70(1), 28–40.
Meliza, C. D., & Margoliash, D. (2012). Emergence of selectivity and tolerance in the avian auditory cortex. *The Journal of Neuroscience*, 32(43), 15158–15168.
Mello, C., Nottebohm, F., & Clayton, D. (1995). Repeated exposure to one song leads to a rapid and persistent decline in an immediate early gene's response to that song in zebra finch telencephalon. *The Journal of Neuroscience*, 15(10), 6919–6925.
Mello, C. V., & Jarvis, E. D. (2008). Behavior-dependent expression of inducible genes in vocal learning birds. In H. P. Zeigler & P. Marler (Eds.), *Neuroscience of birdsong*. Cambridge: Cambridge University Press.
Miller, D. B. (1979). Long-term recognition of father's song by female zebra finches. *Nature*, 280, 389–391.
Nagel, K. I., & Doupe, A. J. (2008). Organizing principles of spectro-temporal encoding in the avian primary auditory area field L. *Neuron*, 58(6), 938–955.
Njegovan, M., & Weisman, R. (1997). Pitch discrimination in field- and isolation-reared black capped chickadees (*Parus atricapillus*). *Journal of Comparative Psychology*, 111, 294–301.
Okanoya, K., & Dooling, R. J. (1987). Hearing in passerine and psittacine birds: A comparative study of absolute and masked auditory thresholds. *Journal of Comparative Psychology*, 101(1), 7–15.
Phan, M. L., Pytte, C. L., & Vicario, D. S. (2006). Early auditory experience generates long-lasting memories that may subserve vocal learning in songbirds. *Proceedings of the National Academy of Sciences of the USA*, 103(4), 1088–1093.
Poirier, C., Boumans, T., Verhoye, M., Balthazart, J., & Van der Linden, A. (2009). Own-song recognition in the songbird auditory pathway: Selectivity and lateralization. *The Journal of Neuroscience*, 29(7), 2252–2258.
Poremba, A., Bigelow, J., & Rossi, B. (2013). Processing of communication sounds: Contributions of learning, memory, and experience. *Hearing Research*, 305, 31–44.

Richner, H. (2016). Interval singing links to phenotypic quality in a songbird. *Proceedings of the National Academy of Sciences of the USA*, 113(45), 12763–12767.
Riebel, K. (2009). Song and female mate choices in zebra finches—a review. *Advances in the Study of Behavior*, 40, 197–238.
Saffran, J. R., Werker, J. F., & Werner, L. A. (2006). The infant's auditory world: Hearing, speech and the beginnings of language. In R. Seigler & D. Kuhn (Eds.), *Handbook of child development* (pp. 58–108). Hoboken, NJ: John Wiley & Sons.
Sanes, D. H., & Woolley, S. M. (2011). A behavioral framework to guide research on central auditory development and plasticity. *Neuron*, 72(6), 912–929.
Schneider, D. M., & Woolley, S. M. (2010). Discrimination of communication vocalizations by single neurons and groups of neurons in the auditory midbrain. *Journal of Neurophysiology*, 103(6), 3248–3265.
Schneider, D. M., & Woolley, S. M. (2011). Extra-classical tuning predicts stimulus-dependent receptive fields in auditory neurons. *The Journal of Neuroscience*, 31(33), 11867–11878.
Schneider, D. M., & Woolley, S. M. (2013). Sparse and background-invariant coding of vocalizations in auditory scenes. *Neuron*, 79(1), 141–152.
Stripling, R., Volman, S. F., & Clayton, D. F. (1997). Response modulation in the zebra finch neostriatum: Relationship to nuclear gene regulation. *The Journal of Neuroscience*, 17(10), 3883–3893.
Stripling, R., Kruse, A. A., & Clayton, D. F. (2001). Development of song responses in the zebra finch caudomedial neostriatum: Role of genomic and electrophysiological activities. *Journal of Neurobiology*, 48(3), 163–180.
Sturdy, C. B., Phillmore, L. S., Sartor, J. J., & Weisman, R. G. (2001). Reduced social contact causes auditory perceptual deficits in zebra finches, (*Taeniopygia guttata*). *Animal Behaviour*, 62, 1207–1218.
Terleph, T. A., Lu, K., & Vicario, D. S. (2008). Response properties of the auditory telencephalon in songbirds change with recent experience and season. *PLoS ONE*, 3(8), e2854.
Terpstra, N. J., Bolhuis, J. J., & den Boer-Visser, A. M. (2004). An analysis of the neural representation of birdsong memory. *The Journal of Neuroscience*, 24(21), 4971–4977.
Theunissen, F. E., Woolley, S. M., Hsu, A., & Fremouw, T. (2004a). Methods for the analysis of auditory processing in the brain. *Annals of the New York Academy of Sciences*, 1016, 187–207.
Theunissen, F. E., Amin, N., Shaevitz, S. S., Woolley, S. M., et al. (2004b). Song selectivity in the song system and in the auditory forebrain. *Annals of the New York Academy of Sciences*, 1016, 222–245.
Tischmeyer, W., & Grimm, R. (1999). Activation of immediate early genes and memory formation. *Cellular and Molecular Life Sciences*, 55, 564–574.
Tomaszycki, M. L., Sluzas, E. M., Sundberg, K. A., Newman, S. W., & DeVoogd, T. J. (2006). Immediate early gene (ZENK) responses to song in juvenile female and male zebra finches: Effects of rearing environment. *Journal of Neurobiology*, 66(11), 1175–1182.
Wang, Y., Brzozowska-Prechtl, A., & Karten, H. J. (2010). Laminar and columnar auditory cortex in avian brain. *Proceedings of the National Academy of Sciences of the USA*, 107(28), 12676–12681.
Werker, J. F., & Tees, R. C. (1999). Influences on infant speech processing: Toward a new synthesis. *Annual Review of Psychology*, 50, 509–535.
Woolley, S. M. (2008). Auditory feedback and singing in adult birds. In H. P. Zeigler & P. Marler (Eds.), *Neuroscience of birdsong* (pp. 228–239). Cambridge: Cambridge University Press.
Woolley, S. M., & Casseday, J. H. (2004). Response properties of single neurons in the zebra finch auditory midbrain: Response patterns, frequency coding, intensity coding, and spike latencies. *Journal of Neurophysiology*, 91(1), 136–151.
Woolley, S. M., & Casseday, J. H. (2005). Processing of modulated sounds in the zebra finch auditory midbrain: Responses to noise, frequency sweeps, and sinusoidal amplitude modulations. *Journal of Neurophysiology*, 94(2), 1143–1157.

Woolley, S. M., & Moore, J. M. (2011). Coevolution in communication senders and receivers: Vocal behavior and auditory processing in multiple songbird species. *New Perspectives on Neurobehavioral Evolution*, 1225, 155–165.

Woolley, S. M., & Portfors, C. V. (2013). Conserved mechanisms of vocalization coding in mammalian and songbird auditory midbrain. *Hearing Research*, 305, 45–56.

Woolley, S. M., & Rubel, E. W. (1999). High-frequency auditory feedback is not required for adult song maintenance in Bengalese finches. *The Journal of Neuroscience*, 19(1), 358–371.

Woolley, S. M., Wissman, A. M., & Rubel, E. W. (2001). Hair cell regeneration and recovery of auditory thresholds following aminoglycoside ototoxicity in Bengalese finches. *Hearing Research*, 153(1–2), 181–195.

Woolley, S. M., Fremouw, T. E., Hsu, A., & Theunissen, F. E. (2005). Tuning for spectro-temporal modulations as a mechanism for auditory discrimination of natural sounds. *Nature Neuroscience*, 8(10), 1371–1379.

Woolley, S. M., Gill, P. R., & Theunissen, F. E. (2006). Stimulus-dependent auditory tuning results in synchronous population coding of vocalizations in the songbird midbrain. *The Journal of Neuroscience*, 26(9), 2499–2512.

Woolley, S. M., Gill, P. R., Fremouw, T., & Theunissen, F. E. (2009). Functional groups in the avian auditory system. *The Journal of Neuroscience*, 29(9), 2780–2793.

Woolley, S. M., Hauber, M. E., & Theunissen, F. E. (2010). Developmental experience alters information coding in auditory midbrain and forebrain neurons. *Developmental Neurobiology*, 70(4), 235–252.

Yanagihara, S., & Yazaki-Sugiyama, Y. (2016). Auditory experience-dependent cortical circuit shaping for memory formation in bird song learning. *Nature Communications*, 7, 11946.

Yang, L. M., & Vicario, D. S. (2015). Exposure to a novel stimulus environment alters patterns of lateralization in avian auditory cortex. *Neuroscience*, 285, 107–118.

Zann, R. A. (1996). *The zebra finch: A synthesis of field and laboratory studies*. Oxford: Oxford University Press.

MIX
Papier aus verantwortungsvollen Quellen
Paper from responsible sources
FSC® C105338

If you have any concerns about our products,
you can contact us on
ProductSafety@springernature.com

In case Publisher is established outside the EU,
the EU authorized representative is:
Springer Nature Customer Service Center GmbH
Europaplatz 3, 69115 Heidelberg, Germany

Printed by Libri Plureos GmbH
in Hamburg, Germany